Stem Cells Between Regeneration and Tumorigenesis

Edited By:

Monica Neagu
Immunobiology Laboratory
"Victor Babeş" National Institute of Pathology
Bucharest, Romania

&

Cristiana Tanase
"Victor Babes" National Institute of Pathology
Biochemistry-Proteomics Department
Bucharest, Romania

advertisements or ideas contained in the Work.

Limitation of Liability:

In no event will Bentham Science Publishers, its staff, editors and/or authors, be liable for any damages, including, without limitation, special, incidental and/or consequential damages and/or damages for lost data and/or profits arising out of (whether directly or indirectly) the use or inability to use the Work. The entire liability of Bentham Science Publishers shall be limited to the amount actually paid by you for the Work.

General:

1. Any dispute or claim arising out of or in connection with this License Agreement or the Work (including non-contractual disputes or claims) will be governed by and construed in accordance with the laws of the U.A.E. as applied in the Emirate of Dubai. Each party agrees that the courts of the Emirate of Dubai shall have exclusive jurisdiction to settle any dispute or claim arising out of or in connection with this License Agreement or the Work (including non-contractual disputes or claims).
2. Your rights under this License Agreement will automatically terminate without notice and without the need for a court order if at any point you breach any terms of this License Agreement. In no event will any delay or failure by Bentham Science Publishers in enforcing your compliance with this License Agreement constitute a waiver of any of its rights.
3. You acknowledge that you have read this License Agreement, and agree to be bound by its terms and conditions. To the extent that any other terms and conditions presented on any website of Bentham Science Publishers conflict with, or are inconsistent with, the terms and conditions set out in this License Agreement, you acknowledge that the terms and conditions set out in this License Agreement shall prevail.

Bentham Science Publishers Ltd.
Executive Suite Y - 2
PO Box 7917, Saif Zone
Sharjah, U.A.E.
Email: subscriptions@benthamscience.org

BENTHAM SCIENCE

CONTENTS

FOREWORD

Crucial findings in research and clinical application results have proved stem cell to be endowed with dual characteristics, on one hand exploited for beneficial regeneration processes and source of uncontrolled tumoral proliferation, on the other.

Statistics predicts that there will be over 20 million new cancer cases worldwide by the end of 2014, because the incidence of major cancers and population growth are steadily increasing. Hence we will be facing 56% more cancer cases diagnosed in 2030 in comparison to 2012 [1, 2].

In contrast to cancer world-wide statistics, where there are clear and accurate statistics, the figures for tissue regeneration domain are more on the economic side. Therefore the global tissue engineering and regeneration market is estimated to $20.8 billion in 2014 and will triple in 2019 [3].

In this global image of human diseases, stem cells are *walking a thin thread*, balancing two competing ideas: one is stem cell as main tumorigenesis player and the other is stem cells as the crucial regeneration trigger. Between these two sides of the *barricade* stands the book that I am kindly inviting you to decipher.

Almost all of the chapters are elaborating on issues that stem cells are endowed with. Preserving the genetic information and retaining the proliferative potential of the tissue in which these cells reside are sustained by their capacity to self-renew and the ability to differentiate into different cell types. Stem cells have a long life span, slow cycling, but their quiescence may turn into high proliferation ability when exposed to certain stimuli. Various factors, including the ones resident in their niche can influence the fate of stem cells.

You will discover the latest up-dates in hematopoietic stem cells, cancer stem cells in oral pathology, stem cells that are involved in skin's regeneration and tumorigenesis, in neurodegeneration and brain tumours, in endocrine pathology, how the immune system controls stem cells, the proteomics behind these cells and the latest technologies involved in stem cells identification.

The book raises its voice in the chorus of publications that would enhance the information regarding stem cells development as therapy approaches in precision medicine.

Giovanni Pellacani
Department of Dermatology,
University of Modena
Modena MO, Italy

[1] Ferlay J, Soerjomataram I, Ervik M, *et al.* GLOBOCAN 2012 v1.0, Cancer Incidence and Mortality Worldwide: IARC CancerBase No. 11 [Internet]. Lyon, France: International Agency for Research on Cancer; 2013. Available from: http://globocan.iarc.fr, accessed December 2013.

[2] Bray F, Jemal A, Grey N, *et al.* Global cancer transitions according to the Human Development Index (2008-2030): a population-based study. Lancet Oncol 2012; 13:790-801

[3] Available at: http://www.reportlinker.com/p02313354-summary/Tissue-Engineering-and-Regeneration-Technologies-and-Global-Markets.html

PREFACE

The proposed book will enrich the previous publications regarding stem cell, by promoting information in two research/clinical renewable areas, such as regeneration and tumorigenesis. Recent information gathered both from research and clinical application proved stem cell to have a "Janus face" characteristics, on one hand exploited for beneficial regeneration processes and source of uncontrolled tumoral proliferation, on the other. Therefore, these cells while used in experimental regeneration they can induce unwanted tumorigenesis pathways.

In this respect, signal transduction, stem cell markers, immune-related processes, "omics" technologies as up-dated identification, as well as pharmacological trends will emphasize the relationship between research and clinical behaviour of this controversy.

This book is emerging from the need to create a clear image regarding the complex mechanisms governing the involvement of stem cell in both regeneration and tumorigenesis.

The giants of cellular biology, have shown that the reawakening of pluripotency inherent in all cells have challenged forever our notions of cellular identity. The implications of the new reprogramming paradigm in biomedicine is steadily enhancing our understanding of cell differentiation and prospects for cellular therapies and *in vivo* regeneration.

The multi-authored book that we propose focuses specifically on various approaches in terms of organ and specific pathways that trigger the two opposite ways that a stem cell can follow. By definition, a stem cell has both self-renewal and multi-potentiality abilities. For this unique dual capacity, stem cells interpret signalling pathways in specialized ways. Adult stem cell for the treatment of damaged or diseased tissues relies on the ability of stem cells to produce paracrine factors that have a trophic effect on existing tissue cells, improving their functional capacity and develops the complex process of regeneration. On the other side of the *barricade*, the notion of cancer stem cells gained prominence in recent years but whether they are only the initiators and/or perpetuator of neoplasia, is still a matter of intense debate. Local factors from the microenvironment (niche) can sustain the self-renewal potential and possibly guide towards multiple stem cell populations.

The chapters will follow the balance between regeneration and tumorigenesis focusing several tissue and systems types: neuro-, haemato-, oro-, dermato-, digestive, endocrine domain. Chapters that focus on the immune processes that regulate and control the stem cell duality and state of the art identification *omics* technologies and pharmacological approaches. This viewpoint will have as central pillar the stem cell and its capacity to be main target for solving system medicine issues.

This e-book begins with the chapter elaborated by *Chen et al.* showing that the discovery of hematopoietic stem cells (HSCs) ushered in a new era in stem cell and life science research. The therapeutic benefits of HSCs have long been recognized; as bone marrow transplants have saved many lives, and with an increasing understanding of HSC biology and its translation to the clinic, studies of HSCs will continue benefit the health of humans and our curiosity of life in general.

The chapter elaborated by *Slootweg and Zurac* focuses on the epithelial stem cells in oral mucosa. Normal epithelial stem cells are characterized, their location, methods of identification and stemness markers are presented. The concept of cancer stem cells in oral cancer, their origin and markers are discussed along with cancer stem cells niche and the interference with treatment. The chapter discuss the role of oral stem cells in oral mucosa wound repair.

The chapter dedicated to skin stem cells elaborated by *Caruntu et al.* highlights the main characteristics of processes like regeneration and tumorigenesis in skin. The chapter describes the skin regeneration pathways and it elaborates on an interesting link regarding regeneration triggering tumorigenesis in cutaneous tissue. Stem cells that can trigger non-melanoma and melanoma skin cancers are described in separate sections.

The chapter focusing on adult pituitary stem cells elaborated by *Gheorghişan-Galateanu* shows the recent reports of potential populations of stem cells in the pituitary. The nature of pituitary stem cells remains a matter of debate. The variety of markers and approaches used to identify pituitary progenitors and stem cells makes it difficult to compare results and integrate the findings.

There are two chapters that focus on stem cells in brain. One chapter elaborated by *Enciu* shows the neuroregeneration of mammalian brain, where a functional stem cell niche and proper molecular cues is needed. Neural stem cells have been initially considered, as a putative source for lost neurons, but in terms of cognitive rescue, the results have been disappointing. The chapter discusses the regenerative potential of stem cells therapy in the modified cellular and molecular context of the aged brain.

The other chapter focusing on brain, developed by *Cruceru and Popa* presents a quick view of the entangled signaling pathways involved in the cancer stem cells in brain tumors with aggressive behavior such as glioblastoma, presenting possible targets for future personalized therapies with improved outcome.

The chapter that brings data regarding the main issues triggered by the immune response in stem cell approaches is elaborated by *Neagu and Constantin*. The chapter characterizes the immunogenicity of stem cells, where major histocompatibility expression is the *immune mould* that can drive toward regeneration or tumorigenesis. The chapter shows the processes that are involved in stem cells modulating the immune system elements. Immune cells are important players for stem cell differentiation in both regeneration and tumorigenesis processes.

New proteomic insights in this domain are elaborated by *Tanase et al.* showing the studies on stem cells and protein interactions using proteomics approaches. Development of stem cell approaches has evolved in the post-genomic era and the implementation of proteomic applications represent the great challenge. Current proteomics studies of stem cell signaling pathway can lead to the discovery of molecular mechanisms that govern cell-cell interaction and/or with stem cell niche.

The last chapter elaborated by *Larisa-Emilia Cheran et al.* gives an overview of the new micro- and nano-technologies designed to monitor stem cell differentiation in the context of their potential applications in disease modeling, tissue engineering, regenerative medicine, as well as drug screening and toxicology.

The reader will gain a good insight on the complexity and controversy surrounding the "stem cell paradigms", namely the dual stem cell behaviour, regeneration *versus* tumorigenesis. Intimate processes like stem cells promoting the maintenance of other stem cells, background for optimization of tailored therapy intending the close collaboration between bench and bedside will be the crucial target in personalized medicine.

Monica Neagu
Faculty of Biology
University of Bucharest
Bucharest
Romania
&
Cristiana Tanase
Faculty of Medicine
"Titu Maiorescu" University
Bucharest
Romania

List of Contributors

Adrian Claudiu Popa	Department of Cellular and Molecular Medicine, Carol Davila University of Medicine and Pharmacy, Bucharest, Romania Army Centre for Medical Research, Bucharest, Romania
Alin Cheran	Ross University School of Medicine, New Jersey, USA
Ana Caruntu	"Dan Theodorescu" Oral and Maxillofacial Surgery Hospital, Bucharest, Romania
Ana-Maria Enciu	Department of Cellular and Molecular Medicine, Carol Davila University of Medicine and Pharmacy, Bucharest, Romania V.Babes National Institute of Pathology, Bucharest, Romania
Ancuta Augustina Gheorghisan-Galateanu	Department of Cellular and Molecular Biology and Histology, "Carol Davila" University of Medicine and Pharmacy, Bucharest, Romania "C.I.Parhon" National Institute of Endocrinology, Bucharest, Romania
Andreea-Roxana Lupu	Immunobiology Laboratory, "Victor Babeş" National Institute of Pathology, Bucharest, Romania
Carolina Constantin	Immunobiology Laboratory, "Victor Babeş" National Institute of Pathology, Bucharest, Romania
Caterina Longo	Dermatology and Skin Cancer Unit, Arcispedale S Maria Nuova, Reggio Emilia, Italy
Constantin Caruntu	Immunology Department, "Victor Babes" National Institute of Pathology, Bucharest, Romania Department of Physiology, "Carol Davila" University of Medicine and Pharmacy, Bucharest, Romania
Cristiana Pistol Tanase	"Victor Babes" National Institute of Pathology, Biochemistry-Proteomics Department, Bucharest, Romania Faculty of Medicine, "Titu Maiorescu" University, Bucharest, Romania
Dana Cucu	"Dan Setlacec" Center of General Surgery and Liver Transplantation, Fundeni Clinical Institute, Bucharest, Romania The Department of Anatomy, Physiology and Biophysics, Faculty of Biology, University of Bucharest, Bucharest, Romania
Daniel Boda	Dermatology Research Laboratory, "Carol Davila" University of Medicine and Pharmacy, Bucharest, Romania
Elena Codrici	"Victor Babes" National Institute of Pathology, Biochemistry-Proteomics Department, Bucharest, Romania
Ionela Daniela Popescu	"Victor Babes" National Institute of Pathology, Biochemistry-Proteomics Department, Bucharest, Romania

Irinel Popescu — "Dan Setlacec" Center of General Surgery and Liver Transplantation, Fundeni Clinical Institute, Bucharest, Romania

Larisa-Emilia Cheran — Department of Chemistry, University of Toronto, Toronto, Canada

Laura Necula — "Victor Babes" National Institute of Pathology, Biochemistry-Proteomics Department, Bucharest, Romania
"Stefan S. Nicolau", Institute of Virology, Cellular and Molecular Pathology, Bucharest, Romania

Maria Linda Cruceru — Department of Cellular and Molecular Medicine, Carol Davila University of Medicine and Pharmacy, Bucharest, Romania

Monica Neagu — Immunobiology Laboratory, "Victor Babeş" National Institute of Pathology, Bucharest, Romania
Faculty of Biology, University of Bucharest, Bucharest, Romania

Nicolae Bacalbasa — "Carol Davila" University of Medicine and Pharmacy, Bucharest, Romania

Pieter J Slootweg — Department of Pathology, Radboud University Medical Centre Nijmegen, Netherlands

Qiang Huang — School of Pharmaceutical Sciences, Sun Yat-Sen University, Guangzhou, China
Centre for Cellular & Structural biology, Sun Yat-Sen University, Guangzhou, China

Qing Li — School of Pharmaceutical Sciences, Sun Yat-Sen University, Guangzhou, China
Centre for Cellular & Structural biology, Sun Yat-Sen University, Guangzhou, China

Radu Albulescu — "Victor Babes" National Institute of Pathology, Biochemistry-Proteomics Department, Bucharest, Romania
Faculty of Medicine, "Titu Maiorescu" University, Bucharest, Romania
National Institute for Chemical Pharmaceutical R&D, Bucharest, Romania

Sabina Zurac — Department of Pathology, University of Medicine and Pharmacy Carol Davila, Colentina University Hospital, Bucharest, Romania

Simona Mihai — "Victor Babes" National Institute of Pathology, Biochemistry-Proteomics Department, Bucharest, Romania

Simona Olimpia Dima — "Dan Setlacec" Center of General Surgery and Liver Transplantation, Fundeni Clinical Institute, Bucharest, Romania

Siqi Chen — School of Pharmaceutical Sciences, Sun Yat-Sen University, Guangzhou, China
Centre for Cellular & Structural biology, Sun Yat-Sen University, Guangzhou, China

Traian Popescu	National Institute of Materials Physics, Magurele, Romania
Valeria Tica	"Dan Setlacec" Center of General Surgery and Liver Transplantation, Fundeni Clinical Institute, Bucharest, Romania
Yawei Liu	Health Division of Guard Bureau, General Staff Department of PLA, Beijing, China
Zhong Wang	School of Pharmaceutical Sciences, Sun Yat-Sen University, Guangzhou, China Centre for Cellular & Structural biology, Sun Yat-Sen University, Guangzhou, China

Stem Cells Between Regeneration and Tumorigenesis

CHAPTER 1

Stem Cells in Hematopoietic Processes and Therapy Tools

Siqi Chen[1,2], Qiang Huang[1,2], Qing Li[1,2], Yawei Liu[3] and **Zhong Wang[1,2,*]**

[1] *School of Pharmaceutical Sciences, Sun Yat-Sen University, Guangzhou, China*

[2] *Centre for Cellular & Structural biology, Sun Yat-Sen University, Guangzhou, China*

[3] *Health Division of Guard Bureau, General Staff Department of PLA, Beijing, China*

Abstract: The discovery of hematopoietic stem cells (HSCs) ushered in a new era in stem cell and life science research. Much of the technology and knowledge that has been gained through HSC studies is now applied in many fields of biology and has fundamentally changed our understanding of stem cells. For example, cell identification and purification using cell surface markers, which were developed and fine-tuned in HSC studies, are now routinely used to investigate the developmental stages of different cell populations not only in hematopoiesis but also in the development of other organs and in tumor biology. The therapeutic benefits of HSCs have long been recognized as bone marrow transplants have saved many lives, and with an increasing understanding of HSC biology and its translation to the clinic, studies of HSCs will continue to benefit the health of humans and our curiosity of life in general. In this chapter, we will briefly describe the discovery, regulation, and therapeutic application of HSCs. We apologize that many studies are not included here due to the nature of this review.

Keywords: Gene therapy, Hematopoietic niche, Hematopoietic stem cells, Transplant.

[*] **Corresponding author Zhong Wang:** Centre for Cellular & Structural biology, Sun Yat-Sen University, Guangzhou, China Health Division of Guard Bureau, General Staff Department of PLA, Beijing, China; Tel/Fax: 86-020-39943426; E-mails: 18101116733@139.com, wangzh357@mail.sysu.edu.cn.

Cristiana Tanase & Monica Neagu (Eds.)

HEMATOPOIETIC STEM CELLS (HSCS)

Hematopoiesis

Blood accounts for ~7% of the weight of the human body; the peripheral blood of an adult human averages five liters and is composed of different types of blood cells suspended in plasma. Erythrocytes (red blood cells) comprise the vast majority proportion of blood cells, representing 45% of whole blood by volume, whereas leukocytes (white blood cells) account for approximately 0.7%; and plasma accounts for approximately 50% of the blood volume. Different types of blood cells vary significantly in terms of life span, for example, a few hours for certain granulocytes, 120 days for red blood cells, and many years for certain lymphocytes. As a tissue, an adult human's blood contains $3x10^{13}$ cells, with approximately 10^{12} cells being replenished every day, making blood one of the most highly regenerative tissues. The process of regenerating new blood cells, or hematopoiesis, is one of the most actively studied topics and has led to knowledge of the detailed regulation of hematopoiesis and the discovery of hematopoietic stem cells.

The Discovery of Hematopoietic Stem Cells (HSCs)

It is currently accepted that the regeneration of blood cells begins with the self-replication and differentiation of hematopoietic stem cells (HSCs). However, the actual discovery of HSCs did not occur until more than a hundred years after the concept of HSCs was introduced in the 19th century, when anatomists noticed a wide variety of cellular morphologies while examining human bone marrow. In the early 20th century, the Russian scientists A. Maximow *et al.* postulated that hematopoiesis in humans could be characterized as a cellular hierarchy that was derived from one precursor cell, the hematopoietic stem cell [1]. However, no direct evidence or substantial experiments supported the existence of the HSC until much later. Extensive research on the hematopoietic system did not begin until the end of World War II.

The atomic bombing on Hiroshima and Nagasaki caused numerous deaths due to the severe failure of hematopoietic function resulting from the lethal dosage of ionizing radiation. Scientists began studying the damage caused by ionizing radiation, especially to the hematopoietic system, and how to repair such damage. Jacoboson *et al.* discovered that certain cells from the bone marrow (BM) and spleen of mice could completely reconstitute the hematopoietic system of recipient animals that had suffered a lethal dosage of ionizing radiation [2, 3], indicating the existence of cells with self-renewal capacity that were transplantable. Further evidence of hematopoietic cell self-renewal came in 1951 when Lorenz *et al.* discovered that after the injection of spleen cells or bone marrow cells from radiation-free donors into recipients who had received ionizing radiation, the patients' hematopoietic function improved, and none died [4]. In the same year, Brecher *et al.* confirmed that the transplantation of bone morrow from rats that had not received radiation could fully repair the severe bone morrow failure of rats that had suffered lethal radiation [5]. The existence of hematopoietic stem cells was demonstrated by these series of studies, which also confirmed that the failure of the hematopoietic system could be reestablished *via* cell transplantation, indicating the existence of hematopoietic stem cells and their transplant ability.

In the 1960s, the existence of a multi-lineage hematopoietic stem cell pool was confirmed by a series of *in vivo* clonal repopulation tests, which represented the regenerative ability of hematopoietic stem cells [6, 7]. The visible colony-forming unit-spleen (CFU-S) assay was also developed to identify the multipotent cells of different blood lineages [7]. Nonetheless, due to the poor understanding of the hematopoietic system, results obtained with the CFU-S were incorrectly interpreted as confirmation of the existence of hematopoietic stem cells in mice. Furthermore, it was not clear whether HSCs consisted of one type of cells that can differentiate into all other types of mature cells or two or more different cell types that have different functions. Nevertheless, the CFU-S and transplant assays were the foundation for later functional studies of HSCs and progenitor cells.

The Identification of HSCs

In the late 20th century, the rapid advancement of molecular biology, immunology, flow cytometry, and other technologies aided in breakthrough discoveries in the areas of hematopoietic system and hematopoietic stem cells. Indeed, the general understanding of the development and function of hematopoietic system and its relationship with hematopoietic stem cells progressed very rapidly. It has now been firmly established that the hematopoietic system is a developmental hierarchy that consists of multi-potent hematopoietic stem cells at the peak, progenitor cells in the middle, and all types of terminally differentiated mature cells at the bottom (Fig. **1**).

Hematopoietic stem cells were first purified from mice based on their cell surface markers using flow cytometry. Thy-1^{low}Lin-Sca^{+} cells, which represent 0.05% of the cells in mouse bone marrow, are able to reconstitute the hematopoietic system completely in mice that have been exposed to lethal radiation, and produce all the defined myelolymphoid lineages [8]. This study pioneered the use of cell surface markers combined with flow cytometry technology to isolate distinct hematopoietic stem cell populations and study their functions. Following this work, further studies revealed that even a single hematopoietic stem cell can give rise to long-term hematopoiesis and produce a steady-state pool of between 20,000 and 100,000 hematopoietic stem cells and over 10^9 blood cells daily in mice [9, 10]. Numerous studies since then have focused on using cell markers combined with flow cytometry, transplantation, and functional analyses to examine different cell populations of the hematopoietic system. The development of hematopoietic cells starting from HSCs to mature cells can now be mapped based on their cell surface markers (Fig. **1**) [11 - 15].

Human [16] and murine [8] hematopoietic stem cells are extremely rare, constituting less than 0.1% of bone marrow cells. Murine HSCs were identified by the surface markers c-Kit^{+}, Thy-1.1^{low}, lineage $marker^{-/low}$, Sca-1^{+}, $Slamf1^{+}$, $Flk2^{-}$, and $CD34^{-}$ [8, 14, 15, 17 - 19]. The profile of human HSCs consists of $CD34^{+}$ and Thy-1^{+}, but lacks $CD38^{-}$, $CD45RA^{-}$, and mature lineage markers [16].These cells are capable of giving rise long-term hematopoiesis after transplantation in mice or human xenogenic mouse models [20 - 22]. Clinical trials have also confirmed that blood formation can be rescued and sustainable hematopoiesis can be maintained in recipients upon autologous HSC transplantation [23].

Compared with other precursor cells and mature effector cells, the most prominent features of hematopoietic cells are their self-renewal ability and differentiation potential. Through self-renewal, the total number of HSCs remains at a stable level to maintain normal hematopoiesis; through differentiation, HSCs can give rise to all types of mature cells, each of which has a unique physiological function.

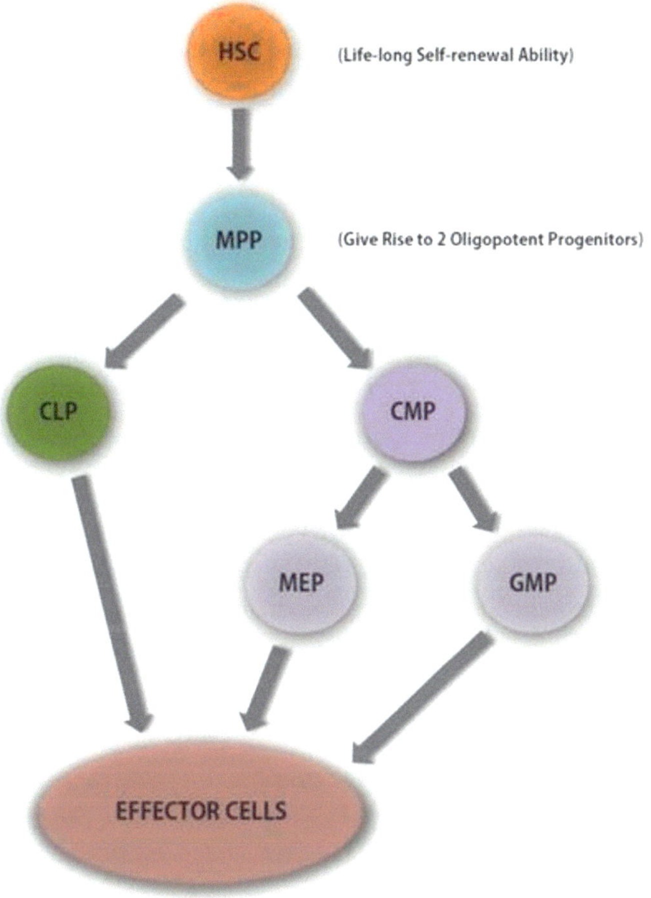

Fig. (1). The diagram of hematopoietic development. Hematopoietic stem cells (HSCs) are stem cells that have self-renewal capability and give rise to all types of blood cells. Long-term HSCs will give rise to MPP (multi-potential progenitor) cells and then develop into either CMP (common myeloid progenitor) or CLP (common lymphoid progenitor) cells. CMPs will further develop into MEP (megakaryocyteerythrocyte progenitor) or GMP (granulocyte-macrophage progenitor) cells. The developmental process will eventually produce all the effector cells or mature blood cells, including red blood cells (erythrocytes), megakaryocytes, neutrophiles, Stem Cells between Regeneration and Tumorigenesis, Stem Cells in Hematopoietic Process and Therapy Tools 9 basophils, eosinophils, monocytes, macrophages, dendritic cells, T cells, B cells, and natural-killer cells.

BIOLOGICAL REGULATION OF HEMATOPOIETIC STEM CELLS

The self-renewal and differentiation potential of HSCs are regulated by many signal transduction pathways. The ability to isolate HSCs and other specific cell populations has enabled detailed studies of the function and regulation of these cells. During the past few decades, with the advance of many other technologies, such as gene transfer and mouse genetic models, many labs have been able to study in detail the regulation of HSCs and other progenitor cells. Here, we briefly summarize some of the studies.

HSC Niche

The self-renewal, differentiation, apoptosis, and even migration of HSCs depend on the unique microenvironment, or the hematopoietic stem cell niche, which is the anatomical location in which HSCs reside and self-renew in the bone marrow. The fundamental role of the HSC niche is to maintain HSCs in a quiescent state to control self-renewal and differentiation into progenitors. The precise controls over the fate of HSCs are ensured by specific interactions between the niche and HSCs. When migrating out of the niche, HSCs cease to self-renew and start to differentiate into progenitors, ultimately differentiating to mature blood cells. Although the concept of the HSC niche was proposed more than 30 years ago [23], the HSC niche was observed only when Askenasy *et al.* applied fluorescent labeling to bone marrow cells of recipient mice [23]. The HSC niche was defined as a three-dimensional functional unit composed of stromal tissue, extracellular matrix, and bone surface. The stromal components are a mixture of tissues and various types of cells, including blood vessels, endothelial cells, adipocytes, and other stromal cells. Specific membrane-bound and secreted paracrine factors are also present in the niche. Specialized osteoblasts are the principle cellular components of the HSC niche and produce certain factors: Angiopoietin 1 and thrombopoietin are produced to maintain HSCs in a quiescent state; Wnt and interferon γ to induce proliferation; CXCL12 to control migration and localization; SCF, thrombopoietin, and interleukins to maintain vitality. High concentrations of extracellular calcium have also been reported to participate in the HSC-niche interaction [24].

Molecular Pathways in HSC Regulation

In both adult mice and humans, most HSCs are maintained in a quiescent state. The balance between proliferation and self-renewal is regulated by many signaling pathways, some of which have been extensively studied, including the PI3K, Wnt, and Hedgehog (Hh) pathways. Here, we briefly describe the impact of some of these pathways on HSCs and hematopoiesis.

PI3K Signaling Pathway

In response to a number of extracellular signals, PI3K is activated and generates 3' phosphoinositide lipids, which elicit cellular responses by acting upon downstream targets, such as the mTOR signaling pathway and the Akt/Foxo pathways. The tumor suppressor protein PTEN is the major lipid phosphatase and can weaken PI3K signaling by dephosphorylating 3' phosphoinositide lipids. PTEN deficiency or activated PI3K will promote tumorigenesis.

In the hematopoietic system, PTEN plays an important role in maintaining HSCs and leukemogenesis. When PTEN is deleted using the Mx1-Cre transgenic model, the PI3K signaling pathway is activated; the mice present a transient increase in the proportion of HSCs that are rapidly cycling, which will cause the depletion of normal HSCs [25 - 28]. The depletion of PTEN also results in the generation of leukemic stem cells [28]. In transplantation assays, PTEN-deficient HSCs only possess the capacity of short-term multi-lineage reconstitution of the hematopoietic system while losing the ability of long-term engraftment. The loss of PTEN will increase significantly with the depletion of the bone marrow HSC pool and break the quiescent state of HSCs. The loss of PTEN also results in the accumulation of HSCs in extramedullary hematopoietic sites that cannot sustain HSC self-renewal [28].

mTOR

mTOR, a member of the PI3K-related kinase family, plays an important role in sensing and responding to environmental changes. Both mTOR complexes, mTOR complex 1 (mTORC1) and mTOR complex 2 (mTORC2), are associated with the phosphorylation of multiple substrates [29, 30]. Recent research has revealed that the fine-tuning activity of the two mTORCs is responsible for the maintenance of HSCs as well as the suppression of leukemogenesis [31].

Both the mTOR and PI3K-Akt pathways are down-regulated in the bone marrow niche [32]. mTORC1 suppression is essential for the quiescence of HSCs. The loss of Tscl, an essential component of mTORC1 complex, will induce HSCs to enter into the cell cycle and cause a large depletion of HSCs [33, 34]. Tscl deletion can also increase the concentrations of reactive oxygen species (ROS), which can then active p53. Some studies have suggested that increased mTORC1 activity is to some extent responsible for the decline in HSC function with age [35].

Rapamycin, an immune suppression drug used in organ transplantation, is an inhibitor of mTOR and is widely used for studying mTOR functions. Recent studies show that rapamycin treatment combined with the activation of canonical Wnt-β-catenin signaling can sustain HSCs in both humans and mice and activate the generation of long-term HSCs [36]. Rapamycin can even rescue the defects of PTEN-deficient HSCs [28]. Interestingly, knocking out Rictor, one essential component of the mTORC2 complex, leads to a decrease in AKT phosphorylation and also restores the function of PTEN-deficient HSCs [37].

Protein Tyrosine Phosphatase Shp2

Shp2, a protein tyrosine phosphatase, participates in many signaling cascades, including the RAS-mitogen-activated protein (MAP) kinase pathways, the PI3K-Akt pathway, and the JAK-STAT pathway, which explains its multiple roles in cell differentiation, division, and migration [38]. Unlike other protein tyrosine phosphatases, which normally inactivate kinases and serve as negative regulators of cell functions, Shp2 up-regulates positive signaling pathways and down-regulates negative signaling pathways in cell proliferation [39].

Shp2 is important in the initial steps of embryonic stem cell differentiation [40]. In hematopoiesis, when Shp2 activity is lost, HSC function decreases mainly due to reduced HSC dormancy and increased apoptosis. Shp2-deficient HSCs exhibit less quiescence, with fewer cells in the G_0 and G_1 phases of the cell cycle. Furthermore, p53-independent apoptosis in HSCs and progenitors is increased, resulting in cytopenia in both the peripheral blood and the bone marrow [41, 42]. The number of LSK cells, consisting of long-term hematopoietic stem cells (LT-HSCs), short-term HSCs (ST-HSCs), and multipotent progenitors (MPPs), along with CMPs, GMPs, and MEPs, is remarkably decreased in Shp2-deficient animal models, making Shp2-deficient animals more sensitive to 5-FU, a cell cycle-specific myeloablative drug [42]. Shp2-deficient HSCs also lose the capacity of reconstituting lethally irradiated recipients, generating very few blood cells with compromised homing and initial engraftment [42].

Ubiquitin Proteasome System

The ubiquitin proteasome system (UPS) is important for many cellular functions. The UPS regulates HSCs through multiple mechanisms, such as mediating transcription, the cell cycle, and protein quality control [43]. Abnormal UPS function leads to many severe diseases, including hematopoietic malignancies.

The UPS functions through an enzymatic cascade that includes a ubiquitin-activating enzyme, E1, the ubiquitin-conjugating enzyme, E2, and the ubiquitin-protein ligase—E3 [44]. Several proto-oncogenes, c-Cbl, Itch, and Fbw-7, which function as E3 ubiquitin ligases, regulate HSCs.

The E3-ligase c-Cbl inhibits the activity of Notch1, c-Kit, and STAT5, which are responsible for the maintenance of HSCs [45]. In Cbl knock-out mice models, the number of HSCs increases without an increase in mature cells [46]. In bone marrow transplantation assays and BrdU incorporation assays, $Cbl^{-/-}$ HSCs exhibit increased reconstitution and proliferation abilities.

Itch, which belongs to the HECT family of E3 ligases, regulates more than 20 cellular targets, including the Notch signaling pathway. Similar to c-Cbl, $Itch^{-/-}$ mice exhibit enhanced HSC self-renewal and abnormal proliferation in bone marrow transplantation experiments and BrdU assays without increases in mature cells [46]. $Itch^{-/-}$ HSCs exhibit an enhanced hematopoietic recovery capacity under myeloablation by 5-FU, and $Itch^{-/-}$ HSCs also show enhanced progenitor ability in colony-forming assays [47].

Fbw-7, another known RING finger E3 ligase, regulates over 20 proteins, including Notch and c-myc. A conditional Fbw-7$^{-/-}$ knockout in the hematopoietic system led to an increased number of actively cycling LSKs [48, 49], eventually leading to diminished HSCs [48]. C-Myc, one of the targets of Fbw-7, is suggested to be responsible for the decreased frequency and eventual loss of HSCs. Similar to c-Cbl$^{-/-}$ and Itch$^{-/-}$ mouse models, Fbw-7$^{-/-}$ mice also show the increased cycling of progenitor cells.

VHL, another RING finger E3 ligase, induces the degradation of HIF-1α, a transcription factor that plays an important role under conditions of hypoxia. The HSC niche is maintained at a low-oxygen state for HSC quiescence, and HSCs express a high level HIF-1α to promote survival in this less than favorable environment. Defects in VHL will result in the stabilization of HIF-1α and induce HSC quiescence [50]. The Vhl$^{-/-}$ LSK population is less capable of transplantation, showing defects in homing and accelerated apoptosis compared with the wild-type LSK population [47].

Transcription Regulators

The unique gene expression program determines the fate of HSCs, and its ability to self-renew or differentiate [51]. The changes of gene expression in the differentiation progress of HSCs are generally accompanied by epigenetic changes in gene regulatory regions to drive the cells into a dormant or active state [52]. Transcription regulators such as HoxB4, Bmi1, HLF, HES1, and others, play pivotal roles in determining the fate of HSCs and their differentiation progress.

Although the molecular mechanism is still not well understood, HoxB4, a homeobox protein, is a known transcriptional factor involved in stem cell biology. After retroviral transduction into mouse hematopoietic stem cells, HoxB4 can increase the total number of HSCs by approximately 1000-fold by activating symmetric self-renewal divisions [53]. However, when human HoxB4 was transduced into CD34^{+} cells, only a limited expansion of stem cell activity was observed [54]. Furthermore, HoxB4 can force mouse embryonic stem cells, but not human embryonic stem cells, to differentiate into a hematopoietic fate [55, 56].

Bmi1 is another transcription regulator that participates in stem cell function.

Bmi1 knock-down in mice resulted in the gradual loss of proliferative ability in both hematopoietic stem cells and progenitors and thus lead to anemia [57]. Similarly, the knock-down of Bmi1 in human $CD34^+$ cells decreased $CD34^+$ cell clonal potential [57, 58]. Bmi1 over-expression plays an important role in the tumor-initiating cells of a wide range of solid tumors and myeloid leukemia [59]. These observations suggested that the function of Bmi1 is conserved across species in both normal and malignant stem cells. Thus, Bmi1 may be a valid target for therapeutic intervention in various types of tumors [59].

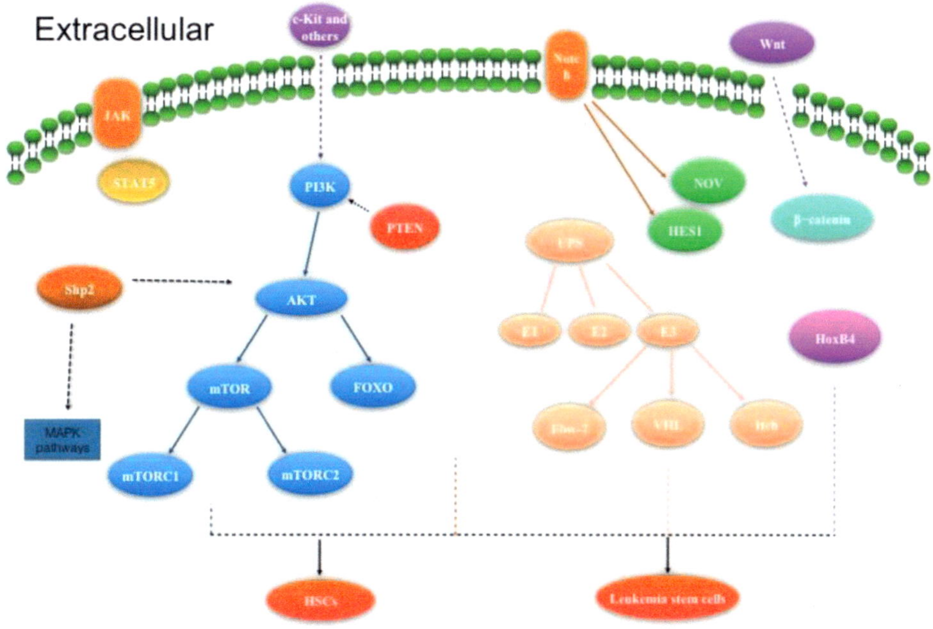

Fig. (2). The summary of pathways regulating hematopoietic stem cells. Hematopoietic stem cells (HSCs) are regulated and influenced by many signaling pathways and molecules, from extracellular signaling molecules, such as Wnt, c-kit, to downstream signaling molecules, including PI3K, MAPK, Wnt, Notch, and others. Due to space limitation, only molecules described in this chapter are listed here. For many of these molecules and other molecules involved in HSC regulation, mutations or deregulation of these molecules can generate leukemia stem cells and eventually lead to leukemia.

The Notch pathway is involved in cell proliferation and has been suggested to play an important role in stem cell biology. The over-expression of HES1, an HLF transcription factor in the Notch pathway, can enhance the repopulation potential of hematopoietic stem cells[60]. When CCN3 (NOV), one of the extracellular modulators of the Notch pathway, binds to the extracellular region of Notch, Hes1 expression is significantly increased[61]. A reduction of NOV is often accompanied by the decreased expression of HES1. Although Notch does not appear to have an obligatory function in adult hematopoietic stem cells, the constitutive activation of the Notch pathway nonetheless promotes the proliferation and inhibits the differential potential of hematopoietic stem cells [62].

The Wnt pathway regulates many cellular functions, including HSCs. Wnt recombinant proteins can increase the expansion of HSCs *in vivo* [36]. When the exons encoding the negative regulatory sequences of β–catenin are deleted using the Mx1-Cre model, Wnt signaling is acutely activated. This event leads to a transient increase in the number of LSK cells as well as the number of LSK cells in the cell cycle [63, 64]. Furthermore, HSCs lose the ability to sustain hematopoiesis, which may result in bone marrow failure within a few weeks due to the depletion of HSC function and the developmental block of multiple progenitors [64, 65].

THERAPEUTIC APPLICATION OF HSCS

As described above, the discovery of HSCs is closely linked to their therapeutic application. Bone marrow transplants were attempted even before the identification and isolation of HSCs. As the first cell therapy or organ transplantation, bone marrow transplantation has saved numerous lives since its introduction. Furthermore, our understanding of HSCs and stem cell biology has improved tremendously. Although bone marrow transplants and the most current HSC therapies do not utilize highly purified HSCs, the long-term benefits of transplantation are the result of HSC engraftment. More and more studies are focusing on applying the highly purified HSCs in clinic. The interplay between basic research of HSCs and their application in clinic has and will continue to benefit patients.

Hematopoietic Cell Transplantation (HCT) (or Bone Marrow Transplantation)

Replacement of a patient's malfunctioning HSCs with normal HSCs from a healthy donor is a logical therapeutic approach for hematopoietic diseases, such as immunodeficiency, autoimmunity, hemoglobinopathies, and hematologic malignancies, which are largely the result of defective HSCs. The first successful hematopoietic cell transplantation (HCT) was performed in 1956 when Donnal and colleagues intravenously infused bone marrow from healthy donors into patients exposed to radiation and chemotherapy [66].

Two types of HCTs, allogeneic and autogeneic, are currently used in the clinic. Allogeneic bone marrow transplantation is an alternative treatment option for hematological disorders that are incurable with standard therapies. This procedure involves the replacement of a patient's HSCs with normal donor HSCs, provided that sufficient hematopoietic chimerism is achieved, HCT can effectively reverse non-malignant genetic hematologic disorders such as sickle cell anemia, (thalassemia, and primary immune deficiencies [67].

HCT and Immuno-Diseases

One major function of hematopoiesis is to generate a large amount of immune cells as a defense mechanism. However, over- or self-active T cells and/or B cells can lead to autoimmune diseases, such as diabetes mellitus type 1 (DM1), multiple sclerosis (MS), and systemic lupus erythematosus (SLE). Although these diseases are multi-factorial, often involving environmental components, they are frequently associated with genetic deficiencies in which HSCs differentiate to generate self-reactive effector T-cells and/or B-cells that recognize and attack host tissues [68]. Theoretically, such diseases can be cured by elimination of the self-reactive T/B-cells in hosts, followed by transplantation and generation of a new non-self-reactive T/B cell compartment from a disease resistant donor(s).

In the current transplantation procedures, host immune cells are generally eliminated first, along with other cells. Thus, autoimmune disease will be suppressed initially either autologous or allogeneic HCT. However, for autologous HCT, the transplanted HSCs can again differentiate into self-reactive immune cells in the recipient, leading to relapse. Indeed, this has been observed in mouse models. In the NOD mouse model of spontaneous autoimmune diabetes mellitus, syngeneic transplantation of purified HSC offered no long-term survival benefit [69]. Clinically, the outcome of autologous transplantation for autoimmune diseases is mixed. Depending on environmental factors, the long-term remission of disease was observed in some patients, whereas symptom relapse was observed in others [70]. Allogeneic HSC transplantation is likely to be the more effective therapy, at least according to a pilot study. When the NOD mouse model with spontaneous autoimmune diabetes mellitus was transplanted with allogeneic HSCs, non-self-reactive immune cells were generated, all HSC chimeras showed decreased or no sign of hyperglycemia with the attenuation of islet lesions, demonstrating that allogeneic HSC transplants can help allo- and autoimmunity [69]. HCT and HSC transplantations have also effectively blocked the disease pathogenesis of DM1 [71], MS [72] and SLE [73, 74].

Drawbacks of HCT Transplantation

Although allogeneic HCT is likely the most effective therapy for many intrinsic blood or immune diseases, the procedure is rarely performed in the clinic due to its high mortality rate, approximately 10-20%, which is far too high for routine clinical operation [75]. Thus, allogeneic HCT is restricted to life-threatening disorders, including hematologic malignancies and bone marrow failures, when HCT is the only therapeutic option.

Graft *vs.* host disease (GvHD) is one of the most frequent and severe complications after allogeneic HCT [76]. It is a complex inflammatory response in which the immune cells in the graft recognize the host as "foreign" and attack the host's body cells, especially the immune system, skin, liver, and gut [77, 78]. Although the possibility and severity of GvHD can be decreased by choosing a transplantation donor with increased matching histocompatibility [79], the risks and outcome of GvHD still remain beyond acceptable, dramatically limiting the application of hematopoietic cell transplantation.

Recent studies suggest that GvHD can be alleviated by the transplantation of mesenchymal stem cells (MSCs) [80], which are also present in the bone marrow.

MSCs are composed of adherent, fibroblast-like cells involved in the maintenance of mesenchymal tissues. Clinical trials have demonstrated that using MSCs for severe, steroid-resistant GvHD treatment achieved positive results [81, 82]. The intravenous infusion of MSCs in leukemia patients promoted HSC engraftment and platelet recovery speed and lowered the incidence of GvHD[83, 84]. MSCs possibly modulate immune cell responses through the secretion of transforming growth factor-beta (TGF-β), prostaglandin E2, and indoleamine 2,3-dioxygenase, resulting in the suppression of T-cell proliferation and activation [85, 86].

Another potential solution to GvHD is to purify HSCs prior to transplantation. Although the term hematopoietic stem cell transplantation (HSCT) and hematopoietic cell transplantation (HCT) are often used interchangeably in the literature, the two actually differ greatly with regard to clinical methodology. HCT refers to the transplantation of bone marrow, composed of mixed types of hematopoietic cells, including stem, progenitor, and mature blood cells. HSCT refers to the transplantation of highly purified HSCs [87]. Both methods rely on HSCs to reconstitute hematopoiesis in the recipient, yet it is the cell populations other than HSCs that make HCT different from HSCT.

The purification of HSCs is the only effective path to purge contaminating hematopoietic populations, mainly host-reactive mature T-cells, from transplantable HSC populations [88]. HSC purification prior to transplantation eliminates GvHD and leads to shortened immune reconstitution, further increasing transplantation safety [89]. The efficacy of this approach was verified clinically in both autologous and allogeneic transplantation. In clinical cases of autologous purified HSC transplantation, a long-term hematolymphoid repopulating capacity was observed while the recipient remained free of metastasized cancer cells [90]. In allogeneic transplantation, as T-cells are removed from the transplantable purified HSCs, the complete elimination of GvHD was observed [91].

Although GvHD can be avoided by implementing HSC transplantation, this approach has its own drawbacks. Prior to HSCT, recipients must undergo a myeloablative conditioning process such as irradiation and high-dose chemotherapy. Such conditioning prepares the recipient for transplantation and eradicates tumor cells given a malignant situation, yet side effects are reported, such as infertility, secondary malignancies, endocrine dysfunction and organ damage [92]. Additionally, HSC purification also means the elimination of other bone marrow cell populations, without which engraftment can be difficult to achieve. For example, T-cell depletion prior of HSCT has resulted in increased graft failure [93].

Organ Transplantation

In addition to hematological diseases, HCT transplantation has also shown promise in the transplantation of other organs. The use of HCT may dramatically abrogate the complications of solid organ transplantation, extend organ longevity, decrease infection probability, and eliminate the life-long administration of immune-suppressive drugs.

Although immune-suppressive drugs have improved the success rate of organ transplantation significantly, immune rejection is still a very challenging problem that must be addressed in organ transplantation. Experiments with rodents have shown that immune tolerance was induced and hematopoietic chimeras were created in the process of bone marrow transplantation, leading to the acceptance of skin grafts from donors [94]. These studies revealed the possibility of applying HCT as a therapeutic modality for organ transplantation, which currently requires lifelong administration of immune-suppressive drugs after organ transplantation [95 - 97].

The permanent transplant tolerance resulting from HCT is attributed to the elimination of donor-reactive T cells. Generation of new immune cells in a chimeric microenvironment from donor HSCs leads to the deletion of reactive immune effector cells against both host- and donor-reactive antigen receptors [77, 98]. In recent trials of kidney/bone marrow transplant patients, long-term immune tolerance to donor tissues by the host was observed [96, 99]. Recent studies of allo-transplantation of purified HSCs either prior to or concurrent with the transplantation of matched donor heart tissue proved to be effective for the prevention of injury and subsequent rejection of donor organs [100].

Viral Infection

HSC transplantation may also be an effective therapeutic strategy for viral infections that are not curable by standard therapy, such as HIV. For example, Timothy Ray Brown, an HIV-infected patient, was treated with HCT using donors with a CCR5 delta32 mutation, which renders T-cells impenetrable to HIV, and was reported to be cured of HIV in 2009 [101]. Such a practice illustrates the potential therapeutic use of HCT in such "incurable" diseases, and its long-term outcome is promising. However, as donor(s) who is both resistant to viral infection and matched for transplantation is extremely difficult to find, its application is very limited [102].

HSC Gene Therapy

Another application of HSCs is gene therapy for various genetic hematolymphoid diseases. Indeed, curing human disease by gene introduction into cells, thereby eliminating the cause of the disease, has long been a goal for scientists and physicians, and HSCs are among the most promising targets in gene therapy due to their unique capacity for self-renewal and ability to reconstitute the entire blood system. This special capability ensures that even with a small portion of HSCs transfected with the corrected target gene would be able to provide normal blood function for the whole body. The application of HSC gene therapy began in the 1990s, though the results were disappointing, mainly due to the low efficiency of gene transfection. Only after improvements in gene transduction efficiency and vector design since 2000 did HSC gene therapy demonstrate success in clinical trials. To date, HSC gene therapy has been employed with success in a small number of patients suffering from X-linked severe combined immunodeficiency (SCID-X1) [103, 104], adenosine deaminase deficiency (ADA-SCID) [105, 106], adrenoleukodystrophy (ALD) [107], thalassemia, chronic granulomatous disease (CGD) [108, 109] and Wiskott-Aldrich syndrome (WAS) [110]. We describe here just a few examples of HSC gene therapy in immuno-deficient disorders.

HSC Gene Therapy for SCID-X1

SCID-X1 is an X-linked inherited disease caused by a mutation in the gene encoding the γ common subunit of interleukin-2, -4, -7, -9, and -15 receptors

[111]. Patients with X-SCID lack mature T and natural killer (NK) cells and subsequently have defective B cell function, leading to severe immunodeficiency. These patients suffer from infection very early in life and usually die in early childhood [112, 113]. The first report of successful HSC gene therapy was reported in 2000 for SCID-X1, offering tremendous optimism to the field [103]. In this work, autologous bone marrow CD34+ cells were transduced with a complementary DNA of the wild-type gene in a replication-defective Moloney retrovirus-derived vector, and re-infused into the patients without any myeloablation [103]. During a 10-month follow-up period, T and NK cells with γC transgene expression were detected. In addition, the T, B, and NK cell counts and functions, including antigen-specific responses, were similar to those of the same age healthy children [103]. During a long-term 9-year follow-up, the immunodeficiency associated with SCID-X1 in 9 patients receiving HSC gene therapy was completely corrected [104].

However, 3 years after receiving HSC gene therapy, clonal vector-associated leukemias occurred in several patients in the pioneering SCID-X1 HSC gene therapy trial [114, 115]. In the 2 youngest patients, aberrant LMO2 gene expression activated by proviral insertions was discovered in leukemic cells [114]. Up to 2008, 5 of 20 patients, have unfortunately developed T cell leukemias between 23 and 68 months after receiving the gene therapy with γ-retroviral vectors since 1999 [116, 117]. Among the 5 leukemia cases, 1 death occurred, and 4 were treated with chemotherapy and/or allogeneic transplantation [118, 119]. The insertion of the proviral enhancer-activated proto-oncogene LMO2 and CCND2 was documented in 4 and 1 patient, respectively [118, 119].

HSC Gene Therapy for ADA-SCID

ADA-SCID is another immuno-deficiency disease caused by ADA mutations. Although several trials of HSC gene therapy were performed for ADA-SCID in the 1990s, because of the low HSC transduction efficiency and/or transgene expression, no significant immunologic improvements or clinical benefits were achieved [120, 121] until 2002 [106]. In the 2002 trial, 2 patients, for whom enzyme replacement therapy was unavailable, received an infusion of $CD34^+$ HSCs transfected with a normal human ADA gene in γ-retroviral Moloney murine leukemia virus (MLV) vector [106]. Sustained engraftment of the modified HSCs was observed, as was the differentiation of multiple cell populations [106]. According to another long-term follow-up study reported in 2009, 9 of 10 patients in this clinical trial had immune reconstitution and normal T cell function [105]. However, different from the SCID-X1 trial, no patients had developed clonal hematopoietic malignancies, with a median period of 3 years and the longest period of 9 years [116, 122], even though the insertion sites were identified as proximal to proto-oncogenes or genes involved in cell proliferation or self-renewal, including LMO2 [123]. These data suggested that insertions in potentially dangerous genomic sites *per se* are not sufficient to induce proviral-insertion-associated leukemia. The properties of the transduced progenitor, the underlying disease, and the transgene characteristics could all contribute to the likelihood of clonal leukemia progression [124].

HSC Gene Therapy for β-Thalassemia

β-thalassemias are inherited anemias, caused by mutations in the β-globin-coding region or its promoter/enhancer region. The disease is characterized by the lack of or reduced β-globin protein production. For the severely ill patients, allogeneic hematopoietic stem cell transplantation offered the only potential for a lifelong cure, however, with the risk of rejection or graft-*versus*-host disease [125]. The use of autologous HSCs with HSC gene therapy offers another potential alternative for these disorders.

In 2007, a β(E)/β(o)-thalassemia patient received autologous CD34+ cells transduced with HIV-based lentiviral vector to express wild-type β-globin [126]. Twelve months after the treatment, the patient became transfusion independent for the following 21 months.

Most of the therapeutic benefits resulted from a dominant myeloid/erythroid clone containing an insertion in HMGA2. Further analysis revealed that the integrated vector caused the transcriptional activation of HMGA2 in erythroid cells. However, no evidence of further expansion or a malignant or pre-malignant state was observed [126]. It remains unclear whether the observed clonal dominance was coincidental or would occur in additional patients or whether these safety-modified lentiviruses will still lead to unacceptable toxicity.

One drawback that prevents the wide use of gene therapy in clinical settings is that it is highly individual specific. Gene therapy is also limited by the lack of appropriate methods to achieve reliable gene transduction without insertional mutagenesis [127]. Intensive research efforts are being conducted to solve these issues, including the use of iPS cells.

iPSC-based Hematologic Disease Treatment

Bone marrow transplantation is the most successful and widely used approach for hematopoietic and lymphatic system regeneration, which relies on the self-renewal and multi-lineage differentiation ability of hematopoietic stem cells. However, the scarcity of HSCs and the lack of efficient methods for HSC *ex vivo* expansion significantly limit the broad application of HSCs in both clinics and basic research. However, induced pluripotent stem cell (iPSC) technology has provided a promising alternative for HSC-based therapy.

Human embryonic stem cells (ESCs) differ from HSCs in that they can be maintained and expanded in culture while maintaining normal karyotypes and pluripotency. However, ESCs are not patient specific. Furthermore, as ESCs are derived from the blastocyst stage of early embryos, ethical issues are raised concerning the use of human ESCs. iPSCs overcome these limitations of ESCs.

Through the retro-viral delivery of four genes into fibroblast cells, Yamanaka's group generated the induced pluripotent stem cells (iPSCs) [128, 129]. Later, different groups confirmed that iPSCs can be generated from many other types of mature cells [128, 129]. iPSCs have the same properties as ESCs, but the source cells can be easily obtained theoretically from any individual. Although the retro-viral delivery method is still one of the most reliable and widely applied approaches for iPSC generation, due to safety concerns, several virus-free methods have been developed, such as the direct delivery of synthetic modified mRNA [130], administration of drugs [131], or introduction of reprogramming proteins [132].

Pluripotent iPSCs provide a powerful alternative for establishing models for organ development, function and disease [133]. iPSCs are patient-specific and disease-specific, retaining all the genetic identify of the parental cells. iPSCs are clonally derived and expandable and can theoretically differentiate into all cell types. Compared to most adult stem cells, iPSCs are more susceptible to genetic modifications [133]. Although only developed recently, iPSCs have already demonstrated their great therapeutic potential.

iPSCs in Blood Diseases

Patient-specific iPSCs is perhaps the best option for blood cell therapy. Autologous iPSC-derived cells are unlikely to cause immune rejection or GvHD when transplanted. Diseases with monogenic mutations are ideal targets for iPSC-based cell therapy, as iPSCs can be easily modified genetically. Several iPSC-based blood diseases have been studied as potential therapy approach in human [134 - 139].

Sickle cell anemia mouse model is the first demonstrated proof-of-principle study [140]. In this study, iPSCs were derived from mouse skin fibroblasts using transcription factors. The human sickle hemoglobin allele was then knocked into the iPSCs, which then differentiated into hematopoietic progenitors *in vitro*. The humanized sickle cell anemia mice were able to be rescued by the transplantation of these iPSC-derived hematopoietic progenitor cells [140].

X-linked chronic granulomatous disease (X-CGD) is an inherited immuno-deficient disease caused by mutation of gp91^{phox}, resulting in a lack of reactive oxygen species (ROS) production in mature neutrophils. iPSCs were generated from an X-CGD patient[137], and transduced with gp91^{phox}. Stable expression of gp91^{phox} was achieved, and neutrophil ROS production was restored [137].

In addition to inherited disease modeling, iPSCs also show potential in acquired disease modeling, such as myeloproliferative neoplasms (MPNs), aplastic anemia, myelodysplastic syndrome (MDS), and many types of leukemias. As most disease-associated genetic mutations of such diseases occurred in hematopoietic lineages, iPSCs generated from fibroblasts would not recapitulate the disease. Polycythemia vera (PV), caused by an acquired JAK2-V617F somatic mutation, was studied using iPSC model[134]. iPSCs were generated from peripheral blood

CD34+ cells of PV patients. The iPSCs with JAK2-V617F mutations showed a normal phenotype, karyotype, and pluripotency with an enhanced erythropoiesis, recapitulating the major clinical feature in PV patients [134]. Thus, further genetic manipulation of iPSCs may be able to cure this disease.

Potential of iPSCs in Drug Development

The potential application of iPSCs in drug development is also limitless. Traditional drug development relies on cell line-based and animal-based pre-clinical testing. Although many developments in drug screening technologies, such as genetic animal models, have helped to hasten drug development, direct relevance to human disease is still lacking in these systems, resulting in high failure rates for drug development [141]. Disease-specific iPSCs, which can provide a large quantity of disease specific cells *in vitro*, can be a vital source or at least a supplement to the current technologies in drug development. This property is especially useful when access to primary patient samples is limited. Indeed, the application of iPSCs in drug screening has been reported [142]. Although iPSC technology is only in an early stage, the future of its application is bright and promising.

Summary

Studies of HSCs have enriched our understanding of hematopoiesis, which begins from HSCs at the top of the developmental hierarchy. The balance between the self-renewal and differentiation of HSCs is regulated by many different signaling pathways and the interactions between HSCs and their niche. Improvements in technology for isolating and studying HSCs are a prerequisite for investigating HSCs and are closely intertwined with our understanding of these cells. Recently, many extremely powerful technologies, such as single-cell sequencing and single-cell mass cytometry, have been developed or are in development to enable scientists to study HSCs in a more detailed manner. These studies will help us understand the finely tuned regulation of HSCs.

Along with the progress of understanding HSC biology, the clinical application of HSCs is expanding. Therapeutic approaches using stem cells are no longer limited to traditional bone marrow transplantation. Indeed, clinicians have found many ways to harness the power of HSCs in an effort to help patients. However, the translation of much of the basic research is still lagging. With the overwhelming amount of information generated from basic research, bench scientists and clinicians need to interact more closely and work together to translate their studies from bench to the bedside. Such a partnership will help translate knowledge into improving the health of patients and all human beings.

CONFLICT OF INTEREST

The authors confirm that they have no conflict of interest to declare for this publication.

ACKNOWLEDGEMENTS

Declared none.

REFERENCES

[1] Maximow A. Der Lymphozyt als gemeinsame Stammzelle der verschiedenen Blutelemente in der embryonalen Entwicklung und im postfetalen Leben der Saugetiere. Folia Haematol (Frankf) 1909; 8: 125-34.

[2] Jacobson LO, Simmons EL, Marks EK, Eldredge JH. Recovery from radiation injury. Science 1951; 113(2940): 510-1.
[http://dx.doi.org/10.1126/science.113.2940.510] [PMID: 14828383]

[3] Jacobson LO, Zirkle RE. Effect of spleen protection on mortality following x-irradiation. J Lab Clin Med 1949; 34: 1538-43.

[4] Lorenz E, Uphoff D, Reid TR, Shelton E. Modification of irradiation injury in mice and guinea pigs by bone marrow injections. J Natl Cancer Inst 1951; 12(1): 197-201.
[PMID: 14874130]

[5] Brecher G, Cronkite EP. Post-radiation parabiosis and survival in rats. Proc Soc Exp Biol Med 1951; 77(2): 292-4.
[http://dx.doi.org/10.3181/00379727-77-18754] [PMID: 14854024]

[6] Becker AJ, McCULLOCH EA, Till JE. Cytological demonstration of the clonal nature of spleen colonies derived from transplanted mouse marrow cells. Nature 1963; 197: 452-4.
[http://dx.doi.org/10.1038/197452a0] [PMID: 13970094]

[7] Till JE, McCULLOCH EA. A direct measurement of the radiation sensitivity of normal mouse bone marrow cells. Radiat Res 1961; 14: 213-22.
[http://dx.doi.org/10.2307/3570892] [PMID: 13776896]

[8] Spangrude GJ, Heimfeld S, Weissman IL. Purification and characterization of mouse hematopoietic stem cells. Science 1988; 241(4861): 58-62.

[http://dx.doi.org/10.1126/science.2898810] [PMID: 2898810]

[9] Smith LG, Weissman IL, Heimfeld S. Clonal analysis of hematopoietic stem-cell differentiation *in vivo.* Proc Natl Acad Sci USA 1991; 88(7): 2788-92. [http://dx.doi.org/10.1073/pnas.88.7.2788] [PMID: 1672767]

[10] Osawa M, Hanada K, Hamada H, Nakauchi H. Long-term lymphohematopoietic reconstitution by a single CD34-low/negative hematopoietic stem cell. Science 1996; 273(5272): 242-5. [http://dx.doi.org/10.1126/science.273.5272.242] [PMID: 8662508]

[11] Morrison SJ, Weissman IL. The long-term repopulating subset of hematopoietic stem cells is deterministic and isolatable by phenotype. Immunity 1994; 1(8): 661-73. [http://dx.doi.org/10.1016/1074-7613(94)90037-X] [PMID: 7541305]

[12] Kondo M, Weissman IL, Akashi K. Identification of clonogenic common lymphoid progenitors in mouse bone marrow. Cell 1997; 91(5): 661-72. [http://dx.doi.org/10.1016/S0092-8674(00)80453-5] [PMID: 9393859]

[13] Akashi K, Traver D, Miyamoto T, Weissman IL. A clonogenic common myeloid progenitor that gives rise to all myeloid lineages. Nature 2000; 404(6774): 193-7. [http://dx.doi.org/10.1038/35004599] [PMID: 10724173]

[14] Christensen JL, Weissman IL. Flk-2 is a marker in hematopoietic stem cell differentiation: a simple method to isolate long-term stem cells. Proc Natl Acad Sci USA 2001; 98(25): 14541-6. [http://dx.doi.org/10.1073/pnas.261562798] [PMID: 11724967]

[15] Kiel MJ, Yilmaz OH, Iwashita T, Yilmaz OH, Terhorst C, Morrison SJ. SLAM family receptors distinguish hematopoietic stem and progenitor cells and reveal endothelial niches for stem cells. Cell 2005; 121(7): 1109-21. [http://dx.doi.org/10.1016/j.cell.2005.05.026] [PMID: 15989959]

[16] Baum CM, Weissman IL, Tsukamoto AS, Buckle AM, Peault B. Isolation of a candidate human hematopoietic stem-cell population. Proc Natl Acad Sci USA 1992; 89(7): 2804-8. [http://dx.doi.org/10.1073/pnas.89.7.2804] [PMID: 1372992]

[17] Whitlock CA, Tidmarsh GF, Muller-Sieburg C, Weissman IL. Bone marrow stromal cell lines with lymphopoietic activity express high levels of a pre-B neoplasia-associated molecule. Cell 1987; 48(6): 1009-21. [http://dx.doi.org/10.1016/0092-8674(87)90709-4] [PMID: 3493849]

[18] Matsuoka S, Ebihara Y, Xu M, *et al.* CD34 expression on long-term repopulating hematopoietic stem cells changes during developmental stages. Blood 2001; 97(2): 419-25. [http://dx.doi.org/10.1182/blood.V97.2.419] [PMID: 11154218]

[19] Adolfsson J, Borge OJ, Bryder D, *et al.* Upregulation of Flt3 expression within the bone marrow Lin(-)Sca1(+)c-kit(+) stem cell compartment is accompanied by loss of self-renewal capacity. Immunity 2001; 15(4): 659-69. [http://dx.doi.org/10.1016/S1074-7613(01)00220-5] [PMID: 11672547]

[20] Majeti R, Park CY, Weissman IL. Identification of a hierarchy of multipotent hematopoietic progenitors in human cord blood. Cell Stem Cell 2007; 1(6): 635-45. [http://dx.doi.org/10.1016/j.stem.2007.10.001] [PMID: 18371405]

[21] Matsuzaki Y, Kinjo K, Mulligan RC, Okano H. Unexpectedly efficient homing capacity of purified murine hematopoietic stem cells. Immunity 2004; 20(1): 87-93. [http://dx.doi.org/10.1016/S1074-7613(03)00354-6] [PMID: 14738767]

[22] Osawa M, Hanada K, Hamada H, Nakauchi H. Long-term lymphohematopoietic reconstitution by a single CD34-low/negative hematopoietic stem cell. Science 1996; 273(5272): 242-5. [http://dx.doi.org/10.1126/science.273.5272.242] [PMID: 8662508]

[23] Michallet M, Philip T, Philip I, *et al.* Transplantation with selected autologous peripheral blood CD34+Thy1+ hematopoietic stem cells (HSCs) in multiple myeloma: impact of HSC dose on engraftment, safety, and immune reconstitution. Exp Hematol 2000; 28(7): 858-70. [http://dx.doi.org/10.1016/S0301-472X(00)00169-7] [PMID: 10907648]

[24] Adams GB, Chabner KT, Alley IR, *et al.* Stem cell engraftment at the endosteal niche is specified by the calcium-sensing receptor. Nature 2006; 439(7076): 599-603. [http://dx.doi.org/10.1038/nature04247] [PMID: 16382241]

[25] Tothova ZK, Huntly BJ, Lee BH, Castrillon DH, Cullen DE, McDowell EP. Lazo-, Kallanian S WI, Sears C, Armstrong SA, Passegue E, DePinho RA, Gilliland DG. FoxOs are critical mediators of hematopoietic stem cell resistance to physiologic oxidativestress. Cell 2007; 2007(128): 325-39. [http://dx.doi.org/10.1016/j.cell.2007.01.003] [PMID: 17254970]

[26] Zhang J, Grindley JC, Yin T, *et al.* PTEN maintains haematopoietic stem cells and acts in lineage choice and leukaemia prevention. Nature 2006; 441(7092): 518-22. [http://dx.doi.org/10.1038/nature04747] [PMID: 16633340]

[27] Miyamoto K, Araki KY, Naka K, *et al.* Foxo3a is essential for maintenance of the hematopoietic stem cell pool. Cell Stem Cell 2007; 1(1): 101-12. [http://dx.doi.org/10.1016/j.stem.2007.02.001] [PMID: 18371339]

[28] Yilmaz ÒH, Valdez R, Theisen BK, *et al.* Pten dependence distinguishes haematopoietic stem cells from leukaemia-initiating cells. Nature 2006; 441(7092): 475-82. [http://dx.doi.org/10.1038/nature04703] [PMID: 16598206]

[29] Ma XM, Blenis J. Molecular mechanisms of mTOR-mediated translational control. Nat Rev Mol Cell Biol 2009; 10(5): 307-18. [http://dx.doi.org/10.1038/nrm2672] [PMID: 19339977]

[30] Laplante M, Sabatini DM. mTOR signaling in growth control and disease. Cell 2012; 149(2): 274-93. [http://dx.doi.org/10.1016/j.cell.2012.03.017] [PMID: 22500797]

[31] Hirao A, Hoshii T. Mechanistic / mammalian target protein of rapamycin signaling in hematopoietic stem cells and leukemia. Cancer Sci 2013; 104(8): 977-82. [http://dx.doi.org/10.1111/cas.12189] [PMID: 23648144]

[32] Yamazaki S, Iwama A, Takayanagi S, *et al.* Cytokine signals modulated *via* lipid rafts mimic niche signals and induce hibernation in hematopoietic stem cells. EMBO J 2006; 25(15): 3515-23. [http://dx.doi.org/10.1038/sj.emboj.7601236] [PMID: 16858398]

[33] Chen C, Liu Y, Liu R, *et al.* TSC-mTOR maintains quiescence and function of hematopoietic stem cells by repressing mitochondrial biogenesis and reactive oxygen species. J Exp Med 2008; 205(10): 2397-408.

[http://dx.doi.org/10.1084/jem.20081297] [PMID: 18809716]

[34] Gan B, Sahin E, Jiang S, *et al.* mTORC1-dependent and -independent regulation of stem cell renewal, differentiation, and mobilization. Proc Natl Acad Sci USA 2008; 105(49): 19384-9. [http://dx.doi.org/10.1073/pnas.0810584105] [PMID: 19052232]

[35] Chen C, Liu Y, Liu Y, Zheng P. mTOR regulation and therapeutic rejuvenation of aging hematopoietic stem cells. Sci Signal 2009; 2(98): ra75. [http://dx.doi.org/10.1126/scisignal.2000559] [PMID: 19934433]

[36] Huang J, Nguyen-McCarty M, Hexner EO, Danet-Desnoyers G, Klein PS. Maintenance of hematopoietic stem cells through regulation of Wnt and mTOR pathways. Nat Med 2012; 18(12): 1778-85. [http://dx.doi.org/10.1038/nm.2984] [PMID: 23142822]

[37] Magee JA, Ikenoue T, Nakada D, Lee JY, Guan K-L, Morrison SJ. Temporal changes in PTEN and mTORC2 regulation of hematopoietic stem cell self-renewal and leukemia suppression. Cell Stem Cell 2012; 11(3): 415-28. [http://dx.doi.org/10.1016/j.stem.2012.05.026] [PMID: 22958933]

[38] Liu X, Qu C-K. Protein Tyrosine Phosphatase SHP-2 (PTPN11) in Hematopoiesis and Leukemogenesis. J Signal Transduct 2011.

[39] Xu D, Qu C-K. Protein tyrosine phosphatases in the JAK/STAT pathway. Front Biosci 2008; 13: 4925-32. [http://dx.doi.org/10.2741/3051] [PMID: 18508557]

[40] Chan RJ, Johnson SA, Li Y, Yoder MC, Feng G-S. A definitive role of Shp-2 tyrosine phosphatase in mediating embryonic stem cell differentiation and hematopoiesis. Blood 2003; 102(6): 2074-80. [http://dx.doi.org/10.1182/blood-2003-04-1171] [PMID: 12791646]

[41] Chan G, Cheung LS, Yang W, *et al.* Essential role for Ptpn11 in survival of hematopoietic stem and progenitor cells. Blood 2011; 117(16): 4253-61. [http://dx.doi.org/10.1182/blood-2010-11-319517] [PMID: 21398220]

[42] Zhu HH, Ji K, Alderson N, *et al.* Kit-Shp2-Kit signaling acts to maintain a functional hematopoietic stem and progenitor cell pool. Blood 2011; 117(20): 5350-61. [http://dx.doi.org/10.1182/blood-2011-01-333476] [PMID: 21450902]

[43] Bhat K, Greer S. Proteolytic and non-proteolytic roles of ubiquitin and the ubiquitin proteasome system in transcriptional regulation. Biochim Biophys Acta 2011; 1809(2): 150-5.

[44] Pickart C, Eddins M. Ubiquitin: structures, functions, mechanisms. Biochim Biophys Acta 2004; 1695(1): 55-72.

[45] Zeng S, Xu Z, Lipkowitz S, Longley JB. Regulation of stem cell factor receptor signaling by Cbl family proteins (Cbl-b/c-Cbl). Blood 2005; 105(1): 226-32. [http://dx.doi.org/10.1182/blood-2004-05-1768] [PMID: 15315962]

[46] Rathinam C, Thien CB, Langdon WY, Gu H, Flavell RA. The E3 ubiquitin ligase c-Cbl restricts development and functions of hematopoietic stem cells. Genes Dev 2008; 22(8): 992-7. [http://dx.doi.org/10.1101/gad.1651408] [PMID: 18413713]

[47] Moran-Crusio K, Reavie LB, Aifantis I. Regulation of hematopoietic stem cell fate by the ubiquitin proteasome system. Trends Immunol 2012; 33(7): 357-63. [http://dx.doi.org/10.1016/j.it.2012.01.009] [PMID: 22349458]

[48] Matsuoka S, Oike Y, Onoyama I, *et al.* Fbxw7 acts as a critical fail-safe against premature loss of hematopoietic stem cells and development of T-ALL. Genes Dev 2008; 22(8): 986-91. [http://dx.doi.org/10.1101/gad.1621808] [PMID: 18367647]

[49] Thompson BJ, Jankovic V, Gao J, *et al.* Control of hematopoietic stem cell quiescence by the E3 ubiquitin ligase Fbw7. J Exp Med 2008; 205(6): 1395-408. [http://dx.doi.org/10.1084/jem.20080277] [PMID: 18474632]

[50] Takubo K, Goda N, Yamada W, *et al.* Regulation of the HIF-1alpha level is essential for hematopoietic stem cells. Cell Stem Cell 2010; 7(3): 391-402. [http://dx.doi.org/10.1016/j.stem.2010.06.020] [PMID: 20804974]

[51] Neumüller RA, Knoblich JA. Dividing cellular asymmetry: asymmetric cell division and its implications for stem cells and cancer. Genes Dev 2009; 23(23): 2675-99. [http://dx.doi.org/10.1101/gad.1850809] [PMID: 19952104]

[52] Bernstein BE, Mikkelsen TS, Xie X, *et al.* A bivalent chromatin structure marks key developmental genes in embryonic stem cells. Cell 2006; 125(2): 315-26. [http://dx.doi.org/10.1016/j.cell.2006.02.041] [PMID: 16630819]

[53] Antonchuk J, Sauvageau G, Humphries RK. HOXB4-induced expansion of adult hematopoietic stem cells *ex vivo.* Cell 2002; 109(1): 39-45. [http://dx.doi.org/10.1016/S0092-8674(02)00697-9] [PMID: 11955445]

[54] Amsellem S, Pflumio F, Bardinet D, *et al. Ex vivo* expansion of human hematopoietic stem cells by direct delivery of the HOXB4 homeoprotein. Nat Med 2003; 9(11): 1423-7. [http://dx.doi.org/10.1038/nm953] [PMID: 14578882]

[55] Kyba M, Perlingeiro RC, Daley GQ. HoxB4 confers definitive lymphoid-myeloid engraftment potential on embryonic stem cell and yolk sac hematopoietic progenitors. Cell 2002; 109(1): 29-37. [http://dx.doi.org/10.1016/S0092-8674(02)00680-3] [PMID: 11955444]

[56] Wang L, Menendez P, Shojaei F, *et al.* Generation of hematopoietic repopulating cells from human embryonic stem cells independent of ectopic HOXB4 expression. J Exp Med 2005; 201(10): 1603-14. [http://dx.doi.org/10.1084/jem.20041888] [PMID: 15883170]

[57] Lessard J, Sauvageau G. Bmi-1 determines the proliferative capacity of normal and leukaemic stem cells. Nature 2003; 423(6937): 255-60. [http://dx.doi.org/10.1038/nature01572] [PMID: 12714970]

[58] Rizo A, Olthof S, Han L, Vellenga E, de Haan G, Schuringa JJ. Repression of BMI1 in normal and leukemic human CD34(+) cells impairs self-renewal and induces apoptosis. Blood 2009; 114(8): 1498-505. [http://dx.doi.org/10.1182/blood-2009-03-209734] [PMID: 19556423]

[59] Schuringa JJ, Vellenga E. Role of the polycomb group gene BMI1 in normal and leukemic hematopoietic stem and progenitor cells. Curr Opin Hematol 2010; 17(4): 294-9. [http://dx.doi.org/10.1097/MOH.0b013e328338c439] [PMID: 20308890]

[60] Shojaei F, Trowbridge J, Gallacher L, *et al.* Hierarchical and ontogenic positions serve to define the molecular basis of human hematopoietic stem cell behavior. Dev Cell 2005; 8(5): 651-63. [http://dx.doi.org/10.1016/j.devcel.2005.03.004] [PMID: 15866157]

[61] Sakamoto K, Yamaguchi S, Ando R, *et al.* The nephroblastoma overexpressed gene (NOV/ccn3) protein associates with Notch1 extracellular domain and inhibits myoblast differentiation *via* Notch signaling pathway. J Biol Chem 2002; 277(33): 29399-405. [http://dx.doi.org/10.1074/jbc.M203727200] [PMID: 12050162]

[62] Varnum-Finney B, Xu L, Brashem-Stein C, *et al.* Pluripotent, cytokine-dependent, hematopoietic stem cells are immortalized by constitutive Notch1 signaling. Nat Med 2000; 6(11): 1278-81. [http://dx.doi.org/10.1038/81390] [PMID: 11062542]

[63] Kühn R, Schwenk F, Aguet M, Rajewsky K. Inducible gene targeting in mice. Science 1995; 269(5229): 1427-9. [http://dx.doi.org/10.1126/science.7660125] [PMID: 7660125]

[64] Scheller M, Huelsken J, Rosenbauer F, *et al.* Hematopoietic stem cell and multilineage defects generated by constitutive beta-catenin activation. Nat Immunol 2006; 7(10): 1037-47. [http://dx.doi.org/10.1038/ni1387] [PMID: 16951686]

[65] Kirstetter P, Anderson K, Porse BT, Jacobsen SE, Nerlov C. Activation of the canonical Wnt pathway leads to loss of hematopoietic stem cell repopulation and multilineage differentiation block. Nat Immunol 2006; 7(10): 1048-56. [http://dx.doi.org/10.1038/ni1381] [PMID: 16951689]

[66] Thomas ED, Lochte HL Jr, Lu WC, Ferrebee JW. Intravenous infusion of bone marrow in patients receiving radiation and chemotherapy. N Engl J Med 1957; 257(11): 491-6. [http://dx.doi.org/10.1056/NEJM195709122571102] [PMID: 13464965]

[67] Barth E, Malorgio C, Tamaro P. Allogeneic bone marrow transplantation in hematologic disorders of childhood: new trends and controversies. Haematologica 2000; 85(11) (Suppl.): 2-8. [PMID: 11268319]

[68] Todd JA, Bell JI, McDevitt HO. HLA-DQ beta gene contributes to susceptibility and resistance to insulin-dependent diabetes mellitus. Nature 1987; 329(6140): 599-604. [http://dx.doi.org/10.1038/329599a0] [PMID: 3309680]

[69] Beilhack GF, Scheffold YC, Weissman IL, *et al.* Purified allogeneic hematopoietic stem cell transplantation blocks diabetes pathogenesis in NOD mice. Diabetes 2003; 52(1): 59-68. [http://dx.doi.org/10.2337/diabetes.52.1.59] [PMID: 12502494]

[70] Krauss AC, Kamani NR. Hematopoietic stem cell transplantation for pediatric autoimmune disease: where we stand and where we need to go. Bone Marrow Transplant 2009; 44(3): 137-43. [http://dx.doi.org/10.1038/bmt.2009.147] [PMID: 19597421]

[71] Nikolic B, Takeuchi Y, Leykin I, Fudaba Y, Smith RN, Sykes M. Mixed hematopoietic chimerism allows cure of autoimmune diabetes through allogeneic tolerance and reversal of autoimmunity. Diabetes 2004; 53(2): 376-83. [http://dx.doi.org/10.2337/diabetes.53.2.376] [PMID: 14747288]

[72] van Gelder M, Kinwel-Bohré EP, van Bekkum DW. Treatment of experimental allergic

encephalomyelitis in rats with total body irradiation and syngeneic BMT. Bone Marrow Transplant 1993; 11(3): 233-41.
[PMID: 8467289]

[73] Smith-Berdan S, Gille D, Weissman IL, Christensen JL. Reversal of autoimmune disease in lupus-prone New Zealand black/New Zealand white mice by nonmyeloablastive transplantation of purified allogeneic hematopoietic stem cells. Blood 2007; 110(4): 1370-8.
[http://dx.doi.org/10.1182/blood-2007-03-081497] [PMID: 17435112]

[74] Traynor AE, Schroeder J, Rosa RM, *et al.* Treatment of severe systemic lupus erythematosus with high-dose chemotherapy and haemopoietic stem-cell transplantation: a phase I study. Lancet 2000; 356(9231): 701-7.
[http://dx.doi.org/10.1016/S0140-6736(00)02627-1] [PMID: 11085688]

[75] Michlitsch JG, Walters MC. Recent advances in bone marrow transplantation in hemoglobinopathies. Curr Mol Med 2008; 8(7): 675-89.
[http://dx.doi.org/10.2174/156652408786241393] [PMID: 18991653]

[76] Barnes DW, Loutit JF, Micklem HS. "Secondary disease" of radiation chimeras: a syndrome due to lymphoid aplasia. Ann N Y Acad Sci 1962; 99: 374-85.
[http://dx.doi.org/10.1111/j.1749-6632.1962.tb45321.x] [PMID: 13969359]

[77] Keever CA, Flomenberg N, Brochstein J, *et al.* Tolerance of engrafted donor T cells following bone marrow transplantation for severe combined immunodeficiency. Clin Immunol Immunopathol 1988; 48(3): 261-76.
[http://dx.doi.org/10.1016/0090-1229(88)90020-7] [PMID: 3042210]

[78] Ferrara JL, Deeg HJ. Graft-*versus*-host disease. N Engl J Med 1991; 324(10): 667-74.
[http://dx.doi.org/10.1056/NEJM199103073241005] [PMID: 1994250]

[79] Schierman LW, Nordskog AW. Influence of the B bloodgroup-histocompatibility locus in chickens on a graft-*versus*-host reaction. Nature 1963; 197: 511-2.
[http://dx.doi.org/10.1038/197511a0] [PMID: 13991668]

[80] Yañez R, Lamana ML, García-Castro J, Colmenero I, Ramírez M, Bueren JA. Adipose tissue-derived mesenchymal stem cells have *in vivo* immunosuppressive properties applicable for the control of the graft-*versus*-host disease. Stem Cells 2006; 24(11): 2582-91.
[http://dx.doi.org/10.1634/stemcells.2006-0228] [PMID: 16873762]

[81] Le Blanc K, Rasmusson I, Sundberg B, *et al.* Treatment of severe acute graft-*versus*-host disease with third party haploidentical mesenchymal stem cells. Lancet 2004; 363(9419): 1439-41.
[http://dx.doi.org/10.1016/S0140-6736(04)16104-7] [PMID: 15121408]

[82] Ringdén O, Uzunel M, Rasmusson I, *et al.* Mesenchymal stem cells for treatment of therapy-resistant graft-*versus*-host disease. Transplantation 2006; 81(10): 1390-7.
[http://dx.doi.org/10.1097/01.tp.0000214462.63943.14] [PMID: 16732175]

[83] Lazarus HM, Koc ON, Devine SM, *et al.* Cotransplantation of HLA-identical sibling culture-expanded mesenchymal stem cells and hematopoietic stem cells in hematologic malignancy patients. Biol Blood Marrow Transplant 2005; 11(5): 389-98.
[http://dx.doi.org/10.1016/j.bbmt.2005.02.001] [PMID: 15846293]

[84] Lee ST, Jang JH, Cheong J-W, *et al.* Treatment of high-risk acute myelogenous leukaemia by myeloablative chemoradiotherapy followed by co-infusion of T cell-depleted haematopoietic stem cells and culture-expanded marrow mesenchymal stem cells from a related donor with one fully mismatched human leucocyte antigen haplotype. Br J Haematol 2002; 118(4): 1128-31. [http://dx.doi.org/10.1046/j.1365-2141.2002.03767.x] [PMID: 12199796]

[85] Aggarwal S, Pittenger MF. Human mesenchymal stem cells modulate allogeneic immune cell responses. Blood 2005; 105(4): 1815-22. [http://dx.doi.org/10.1182/blood-2004-04-1559] [PMID: 15494428]

[86] Nauta AJ, Fibbe WE. Immunomodulatory properties of mesenchymal stromal cells. Blood 2007; 110(10): 3499-506. [http://dx.doi.org/10.1182/blood-2007-02-069716] [PMID: 17664353]

[87] Czechowicz A, Weissman IL. Purified hematopoietic stem cell transplantation: the next generation of blood and immune replacement. Immunol Allergy Clin North Am 2010; 30(2): 159-71. [http://dx.doi.org/10.1016/j.iac.2010.03.003] [PMID: 20493393]

[88] Sprent J, Miller JF. Fate of H2-activated T lymphocytes in syngeneic hosts. II. Residence in recirculating lymphocyte pool and capacity to migrate to allografts. Cell Immunol 1976; 21(2): 303-13. [http://dx.doi.org/10.1016/0008-8749(76)90058-7] [PMID: 769995]

[89] Tsao GJ, Allen JA, Logronio KA, Lazzeroni LC, Shizuru JA. Purified hematopoietic stem cell allografts reconstitute immunity superior to bone marrow. Proc Natl Acad Sci USA 2009; 106(9): 3288-93. [http://dx.doi.org/10.1073/pnas.0813335106] [PMID: 19223585]

[90] Negrin RS, Atkinson K, Leemhuis T, *et al.* Transplantation of highly purified CD34+Thy-1+ hematopoietic stem cells in patients with metastatic breast cancer. Biol Blood Marrow Transplant 2000; 6(3): 262-71. [http://dx.doi.org/10.1016/S1083-8791(00)70008-5] [PMID: 10871151]

[91] Shizuru JA, Jerabek L, Edwards CT, Weissman IL. Transplantation of purified hematopoietic stem cells: requirements for overcoming the barriers of allogeneic engraftment. Biol Blood Marrow Transplant 1996; 2(1): 3-14. [PMID: 9078349]

[92] Ferry C, Socié G. Busulfan-cyclophosphamide *versus* total body irradiation-cyclophosphamide as preparative regimen before allogeneic hematopoietic stem cell transplantation for acute myeloid leukemia: what have we learned? Exp Hematol 2003; 31(12): 1182-6. [http://dx.doi.org/10.1016/j.exphem.2003.09.008] [PMID: 14662323]

[93] Kernan NA, Bordignon C, Heller G, *et al.* Graft failure after T-cell-depleted human leukocyte antigen identical marrow transplants for leukemia: I. Analysis of risk factors and results of secondary transplants. Blood 1989; 74(6): 2227-36. [PMID: 2804361]

[94] Main JM, Prehn RT. Successful skin homografts after the administration of high dosage X radiation and homologous bone marrow. J Natl Cancer Inst 1955; 15(4): 1023-9. [PMID: 13233946]

[95] Starzl TE, Demetris AJ, Murase N, Ildstad S, Ricordi C, Trucco M. Cell migration, chimerism, and graft acceptance. Lancet 1992; 339(8809): 1579-82. [http://dx.doi.org/10.1016/0140-6736(92)91840-5] [PMID: 1351558]

[96] Scandling JD, Busque S, Dejbakhsh-Jones S, *et al.* Tolerance and chimerism after renal and hematopoietic-cell transplantation. N Engl J Med 2008; 358(4): 362-8. [http://dx.doi.org/10.1056/NEJMoa074191] [PMID: 18216356]

[97] Millan MT, Shizuru JA, Hoffmann P, *et al.* Mixed chimerism and immunosuppressive drug withdrawal after HLA-mismatched kidney and hematopoietic progenitor transplantation. Transplantation 2002; 73(9): 1386-91. [http://dx.doi.org/10.1097/00007890-200205150-00005] [PMID: 12023614]

[98] Shizuru JA, Weissman IL, Kernoff R, Masek M, Scheffold YC. Purified hematopoietic stem cell grafts induce tolerance to alloantigens and can mediate positive and negative T cell selection. Proc Natl Acad Sci USA 2000; 97(17): 9555-60. [http://dx.doi.org/10.1073/pnas.170279297] [PMID: 10920206]

[99] Kawai T, Cosimi AB, Spitzer TR, *et al.* HLA-mismatched renal transplantation without maintenance immunosuppression. N Engl J Med 2008; 358(4): 353-61. [http://dx.doi.org/10.1056/NEJMoa071074] [PMID: 18216355]

[100] Gandy KL, Weissman IL. Tolerance of allogeneic heart grafts in mice simultaneously reconstituted with purified allogeneic hematopoietic stem cells. Transplantation 1998; 65(3): 295-304. [http://dx.doi.org/10.1097/00007890-199802150-00001] [PMID: 9484743]

[101] Hütter G, Nowak D, Mossner M, *et al.* Long-term control of HIV by CCR5 Delta32/Delta32 stem-cell transplantation. N Engl J Med 2009; 360(7): 692-8. [http://dx.doi.org/10.1056/NEJMoa0802905] [PMID: 19213682]

[102] van Griensven J, De Clercq E, Debyser Z. Hematopoietic stem cell-based gene therapy against HIV infection: promises and caveats. AIDS Rev 2005; 7(1): 44-55. [PMID: 15875660]

[103] Cavazzana-Calvo M, Hacein-Bey S, de Saint Basile G, *et al.* Gene therapy of human severe combined immunodeficiency (SCID)-X1 disease. Science 2000; 288(5466): 669-72. [http://dx.doi.org/10.1126/science.288.5466.669] [PMID: 10784449]

[104] Hacein-Bey-Abina S, Hauer J, Lim A, *et al.* Efficacy of gene therapy for X-linked severe combined immunodeficiency. N Engl J Med 2010; 363(4): 355-64. [http://dx.doi.org/10.1056/NEJMoa1000164] [PMID: 20660403]

[105] Aiuti A, Cattaneo F, Galimberti S, *et al.* Gene therapy for immunodeficiency due to adenosine deaminase deficiency. N Engl J Med 2009; 360(5): 447-58. [http://dx.doi.org/10.1056/NEJMoa0805817] [PMID: 19179314]

[106] Aiuti A, Slavin S, Aker M, *et al.* Correction of ADA-SCID by stem cell gene therapy combined with nonmyeloablative conditioning. Science 2002; 296(5577): 2410-3. [http://dx.doi.org/10.1126/science.1070104] [PMID: 12089448]

[107] Cartier N, Hacein-Bey-Abina S, Bartholomae CC, *et al.* Hematopoietic stem cell gene therapy with a lentiviral vector in X-linked adrenoleukodystrophy. Science 2009; 326(5954): 818-23.

[http://dx.doi.org/10.1126/science.1171242] [PMID: 19892975]

[108] Ott MG, Schmidt M, Schwarzwaelder K, *et al.* Correction of X-linked chronic granulomatous disease by gene therapy, augmented by insertional activation of MDS1-EVI1, PRDM16 or SETBP1. Nat Med 2006; 12(4): 401-9.
[http://dx.doi.org/10.1038/nm1393] [PMID: 16582916]

[109] Kang EM, Choi U, Theobald N, *et al.* Retrovirus gene therapy for X-linked chronic granulomatous disease can achieve stable long-term correction of oxidase activity in peripheral blood neutrophils. Blood 2010; 115(4): 783-91.
[http://dx.doi.org/10.1182/blood-2009-05-222760] [PMID: 19965657]

[110] Boztug K, Schmidt M, Schwarzer A, *et al.* Stem-cell gene therapy for the Wiskott-Aldrich syndrome. N Engl J Med 2010; 363(20): 1918-27.
[http://dx.doi.org/10.1056/NEJMoa1003548] [PMID: 21067383]

[111] Buckley RH. Molecular defects in human severe combined immunodeficiency and approaches to immune reconstitution. Annu Rev Immunol 2004; 22: 625-55.
[http://dx.doi.org/10.1146/annurev.immunol.22.012703.104614] [PMID: 15032591]

[112] Leonard WJ. The molecular basis of X-linked severe combined immunodeficiency: defective cytokine receptor signaling. Annu Rev Med 1996; 47: 229-39.
[http://dx.doi.org/10.1146/annurev.med.47.1.229] [PMID: 8712778]

[113] Sugamura K, Asao H, Kondo M, *et al.* The interleukin-2 receptor gamma chain: its role in the multiple cytokine receptor complexes and T cell development in XSCID. Annu Rev Immunol 1996; 14: 179-205.
[http://dx.doi.org/10.1146/annurev.immunol.14.1.179] [PMID: 8717512]

[114] Hacein-Bey-Abina S, Von Kalle C, Schmidt M, *et al.* LMO2-associated clonal T cell proliferation in two patients after gene therapy for SCID-X1. Science 2003; 302(5644): 415-9.
[http://dx.doi.org/10.1126/science.1088547] [PMID: 14564000]

[115] Stein S, Ott MG, Schultze-Strasser S, *et al.* Genomic instability and myelodysplasia with monosomy 7 consequent to EVI1 activation after gene therapy for chronic granulomatous disease. Nat Med 2010; 16(2): 198-204.
[http://dx.doi.org/10.1038/nm.2088] [PMID: 20098431]

[116] Fischer A, Hacein-Bey-Abina S, Cavazzana-Calvo M. 20 years of gene therapy for SCID. Nat Immunol 2010; 11(6): 457-60.
[http://dx.doi.org/10.1038/ni0610-457] [PMID: 20485269]

[117] Gaspar HB, Parsley KL, Howe S, *et al.* Gene therapy of X-linked severe combined immunodeficiency by use of a pseudotyped gammaretroviral vector. Lancet 2004; 364(9452): 2181-7.
[http://dx.doi.org/10.1016/S0140-6736(04)17590-9] [PMID: 15610804]

[118] Hacein-Bey-Abina S, Garrigue A, Wang GP, *et al.* Insertional oncogenesis in 4 patients after retrovirus-mediated gene therapy of SCID-X1. J Clin Invest 2008; 118(9): 3132-42.
[http://dx.doi.org/10.1172/JCI35700] [PMID: 18688285]

[119] Howe SJ, Mansour MR, Schwarzwaelder K, *et al.* Insertional mutagenesis combined with acquired somatic mutations causes leukemogenesis following gene therapy of SCID-X1 patients. J Clin Invest

2008; 118(9): 3143-50.
[http://dx.doi.org/10.1172/JCI35798] [PMID: 18688286]

[120] Kohn DB, Hershfield MS, Carbonaro D, *et al.* T lymphocytes with a normal ADA gene accumulate after transplantation of transduced autologous umbilical cord blood CD34+ cells in ADA-deficient SCID neonates. Nat Med 1998; 4(7): 775-80.
[http://dx.doi.org/10.1038/nm0798-775] [PMID: 9662367]

[121] Blaese RM, Culver KW, Miller AD, *et al.* T lymphocyte-directed gene therapy for ADA- SCID: initial trial results after 4 years. Science 1995; 270(5235): 475-80.
[http://dx.doi.org/10.1126/science.270.5235.475] [PMID: 7570001]

[122] Cassani B, Montini E, Maruggi G, *et al.* Integration of retroviral vectors induces minor changes in the transcriptional activity of T cells from ADA-SCID patients treated with gene therapy. Blood 2009; 114(17): 3546-56.
[http://dx.doi.org/10.1182/blood-2009-02-202085] [PMID: 19652199]

[123] Aiuti A, Cassani B, Andolfi G, *et al.* Multilineage hematopoietic reconstitution without clonal selection in ADA-SCID patients treated with stem cell gene therapy. J Clin Invest 2007; 117(8): 2233-40.
[http://dx.doi.org/10.1172/JCI31666] [PMID: 17671653]

[124] Wu C, Dunbar CE. Stem cell gene therapy: the risks of insertional mutagenesis and approaches to minimize genotoxicity. Front Med 2011; 5(4): 356-71.
[http://dx.doi.org/10.1007/s11684-011-0159-1] [PMID: 22198747]

[125] Yannaki E, Emery DW, Stamatoyannopoulos G. Gene therapy for β-thalassaemia: the continuing challenge. Expert Rev Mol Med 2010; 12: e31.
[http://dx.doi.org/10.1017/S1462399410001626] [PMID: 20883576]

[126] Cavazzana-Calvo M, Payen E, Negre O, *et al.* Transfusion independence and HMGA2 activation after gene therapy of human β-thalassaemia. Nature 2010; 467(7313): 318-22.
[http://dx.doi.org/10.1038/nature09328] [PMID: 20844535]

[127] McCormack MP, Rabbitts TH. Activation of the T-cell oncogene LMO2 after gene therapy for X-linked severe combined immunodeficiency. N Engl J Med 2004; 350(9): 913-22.
[http://dx.doi.org/10.1056/NEJMra032207] [PMID: 14985489]

[128] Takahashi K, Tanabe K, Ohnuki M, *et al.* Induction of pluripotent stem cells from adult human fibroblasts by defined factors. Cell 2007; 131(5): 861-72.
[http://dx.doi.org/10.1016/j.cell.2007.11.019] [PMID: 18035408]

[129] Mali P, Ye Z, Hommond HH, *et al.* Improved efficiency and pace of generating induced pluripotent stem cells from human adult and fetal fibroblasts. Stem Cells 2008; 26(8): 1998-2005.
[http://dx.doi.org/10.1634/stemcells.2008-0346] [PMID: 18511599]

[130] Warren L, Manos PD, Ahfeldt T, *et al.* Highly efficient reprogramming to pluripotency and directed differentiation of human cells with synthetic modified mRNA. Cell Stem Cell 2010; 7(5): 618-30.
[http://dx.doi.org/10.1016/j.stem.2010.08.012] [PMID: 20888316]

[131] Lin T, Ambasudhan R, Yuan X, *et al.* A chemical platform for improved induction of human iPSCs. Nat Methods 2009; 6(11): 805-8.

[http://dx.doi.org/10.1038/nmeth.1393] [PMID: 19838168]

[132] Kim D, Kim C-H, Moon J-I, *et al.* Generation of human induced pluripotent stem cells by direct delivery of reprogramming proteins. Cell Stem Cell 2009; 4(6): 472-6. [http://dx.doi.org/10.1016/j.stem.2009.05.005] [PMID: 19481515]

[133] Ye Z, Chou B-K, Cheng L. Promise and challenges of human iPSC-based hematologic disease modeling and treatment. Int J Hematol 2012; 95(6): 601-9. [http://dx.doi.org/10.1007/s12185-012-1095-9] [PMID: 22619021]

[134] Ye Z, Zhan H, Mali P, *et al.* Human-induced pluripotent stem cells from blood cells of healthy donors and patients with acquired blood disorders. Blood 2009; 114(27): 5473-80. [http://dx.doi.org/10.1182/blood-2009-04-217406] [PMID: 19797525]

[135] Raya A, Rodríguez-Pizà I, Guenechea G, *et al.* Disease-corrected haematopoietic progenitors from Fanconi anaemia induced pluripotent stem cells. Nature 2009; 460(7251): 53-9. [http://dx.doi.org/10.1038/nature08129] [PMID: 19483674]

[136] Sebastiano V, Maeder ML, Angstman JF, *et al.* *In situ* genetic correction of the sickle cell anemia mutation in human induced pluripotent stem cells using engineered zinc finger nucleases. Stem Cells 2011; 29(11): 1717-26. [http://dx.doi.org/10.1002/stem.718] [PMID: 21898685]

[137] Zou J, Sweeney CL, Chou B-K, *et al.* Oxidase-deficient neutrophils from X-linked chronic granulomatous disease iPS cells: functional correction by zinc finger nuclease-mediated safe harbor targeting. Blood 2011; 117(21): 5561-72. [http://dx.doi.org/10.1182/blood-2010-12-328161] [PMID: 21411759]

[138] Zou J, Mali P, Huang X, Dowey SN, Cheng L. Site-specific gene correction of a point mutation in human iPS cells derived from an adult patient with sickle cell disease. Blood 2011; 118(17): 4599-608. [http://dx.doi.org/10.1182/blood-2011-02-335554] [PMID: 21881051]

[139] Müller LU, Milsom MD, Harris CE, *et al.* Overcoming reprogramming resistance of Fanconi anemia cells. Blood 2012; 119(23): 5449-57. [http://dx.doi.org/10.1182/blood-2012-02-408674] [PMID: 22371882]

[140] Hanna J, Wernig M, Markoulaki S, *et al.* Treatment of sickle cell anemia mouse model with iPS cells generated from autologous skin. Science 2007; 318(5858): 1920-3. [http://dx.doi.org/10.1126/science.1152092] [PMID: 18063756]

[141] DiMasi JA, Hansen RW, Grabowski HG. The price of innovation: new estimates of drug development costs. J Health Econ 2003; 22(2): 151-85. [http://dx.doi.org/10.1016/S0167-6296(02)00126-1] [PMID: 12606142]

[142] Lee G, Papapetrou EP, Kim H, *et al.* Modelling pathogenesis and treatment of familial dysautonomia using patient-specific iPSCs. Nature 2009; 461(7262): 402-6. [http://dx.doi.org/10.1038/nature08320] [PMID: 19693009]

CHAPTER 2

Epithelial Oral Stem Cells

Pieter J. Slootweg[1] and **Sabina Zurac**[2,*]

[1] *Department of Pathology, Radboud University Medical Centre Nijmegen the Netherlands*

[2] *Department of Pathology, University of Medicine and Pharmacy Carol Davila, Colentina University Hospital, Bucharest, Romania*

Abstract: This chapter focuses on the epithelial stem cells in oral mucosa. It starts with the characterization of the normal epithelial stem cells (location, methods of identification, stemness markers). Next, it discusses the concept of cancer stem cells: origin, cancer stem cell markers, cancer stem cells niche and interference with treatment. Finally, the chapter discusses the role of oral stem cells in oral mucosa wound repair.

Keywords: Oral cancer stem cells, Oral mucosa regeneration, Oral stem cells, Stemness.

INTRODUCTION

Epithelial stem cells within oral mucosa have features similar to other adult stem cells: capacity of self-renewal and the ability to differentiate into mature cell types [1]. Identification of oral stem cells follows the gold standards used for the identification of any adult stem cells: genetic labeling of a putative stem cell (with subsequent transmission of the label to that cell progenitors) and transplantation (with subsequent formation of the tissue in the new location) [2]. Since these techniques are difficult to use, efforts were made in order to identify oral stem cell markers.

* **Corresponding author Sabina Zurac:** Department of Pathology, University of Medicine and Pharmacy Carol Davila Colentina University Hospital, Bucharest, Romania; Tel/Fax: +40213156612; E-mail: sabina_zurac@yahoo.com.

Cristiana Tanase & Monica Neagu (Eds.)

Genetic Labeling

Stem cells are very slow cycling cells; they are capable of retaining a certain genetic label for a long time [1]. Several labeled DNA nucleotides were experimentally tested in order to identify the low turnover rate cells, thus tritiated-thymidine (3H-TdR), 5-bromo-29-deoxyuridine (BrdU) and 5-iodo-2'-deoxyuridine (IdU) were used to label the cells. 3H-TdR retaining cells were identified 240 days after labeling in palate and buccal mucosa of the mice [3] and 69 days after labeling in tongue mucosa of the hamsters [4]. BrdU studies revealed the presence of 3-7% of BrdU retaining cells in oral mucosa in rats after 10 weeks post-labeling [5] and 1.45-4.28% of BrdU retaining cells in oral mucosa in mice after 35 days post-labeling [6]. No studies of genetic label retaining cells in human oral mucosa are available yet but an elegant study of normal human esophagus and stomach identifies IdU-retaining cells (less than 0,1% of epithelial cells) with stem cells features (long living uncommitted cells located in stem cell niches) [7].

Clone Formation

Stem cells have high *in vitro* proliferation potential. Older studies showed the capability of keratinocytes to form clones. The number and morphologic appearance of these clones depends on the size of founding cells: holoclones (cells smaller than 11 microns – most likely stem cells) form large round smooth edges colonies with reduced tendency towards terminal differentiation with small cells in the periphery; meroclones (larger cells with high proliferative potential and more differentiate phenotype - transit amplifying cells) form smaller colonies with wrinkled irregular edges with intermediate growing potential; paraclones (large terminally differentiating cells) form small highly irregular terminal colonies of large cells with short *in vitro* lifespan [8]. Further studies confirmed the presumption that holoclones are indeed stem cells. Magnetically separated oral human keratinocytes with $CD71^-$ phenotype are the smallest and most clonogenic cells, also expressing cytokeratin 19 and p63 keratinocyte stem cell markers but lacking cytokeratin 10 and involucrin (differentiation markers) [9]; Krt5-GFP(hi) cells in a line of transgenic mice also function as holoclones [10].

Stem Cell Markers

Several putative oral stem cell markers were proposed. Nestin, nanog and c-kit (CD117) (otherwise stem cells markers for other tissues) were not reported in putative oral epithelial stem cells [11]. These markers were identified in dental pulp and periodontal ligament stem cells [12 - 14] and also in *lamina propria* of oral mucosa [15] and palate [16], nestin positive stem cells from these locations being able to differentiate into cell-types of cranial neural crest ontology [17]. Most of the oral stem cells markers were first identified in stem cell of the hair follicle and epidermis; in oral mucosa they are also expressed by almost all the basal cells, thus preventing the isolation of pure populations of stem cells [1]. On these issues, there is still no consensus, several studies identifying a plethora of putative stem cells markers in different areas of oral mucosa: masticatory mucosa -α6β4 integrin, β1 integrin, CD44, CD71, melanoma chondroitin sulphate proteoglycan (MCSP), p75, CK15, CK19, p63, Oct3/4; lining mucosa - β1 integrin, collagen IV, p75, CK14, ABCG2; specialized mucosa - β1 integrin, CK5, CK14, sox2. All cells positive for these markers proved either self-renewal capabilities by high colony forming properties and/or differentiation towards oral epithelial equivalents in cell culture [1, 9 - 11, 18 - 22].

Location of Oral Epithelial Stem Cells

Most likely, oral epithelial stem cells are located within the basal layer of the squamous epithelium, representing 1-7% of the cells [5].

There are differences concerning their density between different areas of oral mucosa. BrdU retaining cells (putative stem cells) are more frequent in masticatory mucosa (most frequent in gingiva, followed by palate – thick keratinized mucosa with prominent rete ridges), less frequent in lining mucosa (alveolar and buccal mucosa and inferior surface of tongue – thin mostly non-keratinized mucosa almost devoid of rete ridges) and least frequent in specialized mucosa (*dorsum linguae* – mucosa with lingual papillas) [6]. Similar data concerning location were noted in humans based on *in vitro* studies using both 3H-TdR and BrdU in single or sequential double-labeling technique: buccal mucosa showing higher S phase labeling than mandibular gingiva [23].

In addition, intraepithelial stem cell location varies depending on mucosa type; they are present in the basal layer within bottom of the rete ridges in thick mucosa and randomly distributed within thin mucosa [6]. In specialized mucosa such as dorsum of the tongue, stem cells are located mostly at the boundaries of the papillary and interpapillary epithelium near the anterior and posterior columns of the filiform papillae [1].

Stem Cell Niche

Stem cells reside in a special environment called the stem cell niche. This recently studied structure is complex including several types of cells (epithelial, mesenchymal and immune cells), extracellular matrix components and particularly vascularization. The role of stem cell niche is extremely wide, from regulating proliferation and apoptosis resistance of stem cells, maintaining stemness, to recruiting new stem cells from nearby cells [24 - 26].

Interesting is the fact that the role of stem cell niche depends on the type of stem cells. In case of hematopoietic and neural stem cells, the stem cells can maintain the stemness outside the complex microenvironment of the niche [27, 28]; in case of intestinal stem cell niche, ablation of Paneth cells (cells proved to have important stem cell maintenance function) does not interfere with stemness of the corresponding stem cells [29]. Experiments in the *Drosophila* and mice proved that local soluble factors are able to confer stemness to local partially differentiated cells (*Drosophila* cystocytes, respectively mice spermatogonia dedifferentiate to germline stem cells), thus extending the initial role of the niche from supporting stem cell (either maintaining it in a quiescent state, or, in case of local insult, stimulating self-renewal and/or differentiation) to controlling the actual formation of stem cells [30, 31]. Hair follicle stem cell niche, either complete or depleted of stem cell, is not able to re-trigger stemness to cells that have already engaged the differentiation lineage [32], but interfollicular epidermal stem cells are able to produce their own self-renewing and differentiating signals for their progeny [33].

The base of self-renewal and differentiation characteristics of the stem cells resides in cell division [34]. Several factors from the niche control stem cell division (such as cell-cell interaction – between stem cell and adjacent non-stem cells from the niche or, rarely, other stem cells, paracrine factors (growth factors, cytokines and metabolites), local pH and calcium levels, ground substances and collagen fibers) control the circumstances under which the stem cell will divide but, no matter which factor determines stem cell mitosis, each daughter cell will receive two identical sets of chromosomes [35].

However, stem cells can divide either symmetrically (thus enlarging the stem cell population) or asymmetrically (one stem cell will produce another stem cell – self-renewal – and a daughter cell already engaged in differentiation). Since the niche itself is not sufficient to explain all the processes involved in stem cell division, additional mechanisms are involved, making possible that different information is transmitted from one cell generation to the next one; differential transmission of the cellular memory between consecutive generations during mitosis is possible by intervention of several processes mainly based on non-random segregation of the mitotic apparatus between the daughter cells [34].

One such process is the asymmetric segregation of centrosomes. Each centrosome presents two centrioles. During mitosis the old centrioles (mother centrioles) remain on the old centrosome, the new ones (new centrioles – daughter centrioles) assembling on the site of the old ones. The mother centrosome (with the old set of centrioles) and the daughter centrosome (with the new set of centrioles) have some structural differences; if they segregate non-randomly between the cells, an asymmetric cell division occurs; in some cases (*Drosophila* male germ line stem cells) the mother centrosome always goes to the stem cells [36], in other situations (*Drosophila* larval neuroblasts), quite the opposite, the daughter centrosome being transferred to the stem cells [37].

Another process involved in asymmetric division is the inheritance of the midbody. Midbody is a spindle midzone-derived structure located between two daughter cells with a possible role as barrier. Abscission occurs on the two sides of the midbody but the severing process is sequential, the midbody being still connected to one cell while the separation from the other is complete. Normally, these structures deteriorate after mitosis and are eliminated as residual body in the extracellular space, sometimes being engulfed by other cells [38]. Some cells may retain post-mitotic midbody after mitosis; they can have non-cytokinetic functions, being involved in intercellular communication [39], polarity determination [40] or stemness determination (neural stem cells retain midbodies and retaining a midbody within a differentiated cell reprogram that cell to pluripotency) [41].

Cancer Stem Cells (CSC)

The cancer stem cell (CSC) is defined as "a cell within a tumor that possesses the capacity to self-renew and to cause the heterogeneous lineages of cancer cells that comprise the tumor." Terms such as tumor initiating cell and tumorigenic cell are used to designate potential cancer stem cells [42].

The Origin of Cancer Cells; CSC Provenience

Origin of putative CSC is debatable, three major hypotheses being proposed, each of them with pros and cons. CSC may originate directly from normal stem cells; this hypothesis stipulates that long living stem cells acquire sufficient mutations necessary for malignant transformation; the CSC simply take advantage of the stemness of normal stem cell; however, since the normal stem cells are quite rare, the malignant transformation will address a very small number of cells. The second hypothesis stipulates that CSC arise from progenitor cells, they are more numerous than stem cells and they preserve a certain self-renewal capacity. The third hypothesis states that normal mature differentiated cells (the major cellular compartment) will acquire oncogenic mutations, some of them being also capable to drive dedifferentiation towards CSC; epithelial-mesenchymal transition (EMT) may also be involved in stemness acquiring of tumor cells [43 - 47].

EMT is the process normally involved in embryologic development and wound healing; it allows an epithelial polarized cell to switch to a mobile mesenchymal cell phenotype by a series of progressive functional transformation including loss of cell polarities, break of intercellular tight junctions and degradation of the basement membrane [48]. Furthermore, the cells break cell–matrix connections and migrate away from the epithelium of origin [49]. At the end of the process, the transformed cell has acquired enhanced motility and invasiveness [50]. Drawing attention on EMT hallmarks, loss of E-cadherin and expression of Snail are of utmost importance; E-cadherin is an independent prognostic factor for oral squamous cell carcinoma (OSCC) and Cadherin switching (characterized by up-regulation of Snail and N-cadherin and down-regulation of E-cadherin) correlates with OSCC progression [48, 51 - 53].

CSC Identification

Identification of CSC implies demonstration of both self-renewal and tumor propagation, features that can be optimally evaluated by serial transplantation in animal models as gold standard assays [42]. Various techniques were used for CSC isolation, flow cytometry and anchorage-independent culture or dye exclusion assays being widely used [54]. However, as previously discussed, the methods are laborious and difficult to apply since they imply the destruction of the stem cell niche and transplantation of the putative CSC elsewhere. Moreover, significant bias of the experiments may occur since BrdU labeled putative stem cells can die during transplantation, thus releasing the label which is further taken up by injured local cells [55]. Due to these problems, identification and utilization of stem cell markers offer the possibility of studying CSC in their local microenvironment in relation with the neoplasm and local response.

CD44

The first oral CSC marker ever discussed was CD44. CD44 is a surface glycoprotein acting as a hyaluronic acid receptor; it is involved in cell migration and adhesion. This protein interacts with osteopontin and matrix metalloproteases, and its presence is necessary for MMP9 activity [56 - 59]. The interaction of CD44 with MMP9 activates TGF beta, thus favoring angiogenesis and tumor invasion; tumor cells in the invasive front (area concentrating CSCs) co-express CD44, ALDH1, CK1 and MMP9 [60, 61].

CD44 was previously identified in several malignancies (breast, central nervous system, colon, prostate and pancreas cancers) [62]. In an immunodeficient mouse model, Prince *et al.* identified that a minor population (less than 10% of the cells in OSCC) from the pool of CD44+ cancer cells has both self-renewal and differentiation capacity and that all the tumorigenic potential was contained within the CD44+Lin- phenotype population of the tumor [63]. CD44 expression co-localized with keratin 5/14 expression (basal cell marker), overexpresses Bmi1 (stemness marker) [63], induces resistance to apoptosis-inducing stimuli [64] and correlates with ALDH and phosphorylated STAT3 in high grade OSCC [54, 65]. CD44 positive cells present several up-regulated genes that are related to chemoresistance (such as ABCB1, ABCG2, CYP2C8, and TERT) [66]. Recurring OSCC presents an increased number of CD44+ cells in comparison with non-recurring tumors [67].

Overall, there is a strong correlation between CD44 and stemness property in correlation to OSCC aggressiveness [59]. The main drawback of using CD44 expression as oral CSC marker is that, this surface glycoprotein is also expressed by normal and neoplastic oral epithelium [54].

ALDH

ALDH (aldehyde dehydrogenase) belongs to a family of cytosolic enzymes mostly present in the liver. These enzymes convert retinol into retinoic acid, thus playing a role in the differentiation of stem cells [68], and detoxify toxic intracellular aldehyde metabolites by oxidation into carboxylic acid [69] such as toxic aldehyde metabolites that occur in alcohol metabolism [70] or chemotherapy [69]. ALDH positive CSCs were identified in breast, colon, lung and brain tumors as self-renewing highly tumorigenic cells [71 - 74]. In OSCC, cells expressing ALDH present the highest self-renewal ability, tumorigenicity, and radioresistance [69, 75, 76]. Tumor cells coexpressing CD44+CD24-ALDH+ present CSC features and correspondingly have high co-expression of the Snail protein; both self-renewal ability and tumorigenicity are inhibited by the knockdown of Snail in these cells; moreover, overexpression of Snail in ALDH1 negative OSCC tumor cells increases both malignant features (local invasivity and metastatic capability) and radio-chemoresistant properties, favoring the transformation of ALDH1 negative cells into ALDH1 positive ones [69].

CD133

CD133 (prominin-1) is a complex glycosylated protein consisting of two large extracellular loops and five transmembrane domains, this protein is localized on cell's membrane protrusions and was initially described as a hematopoietic stem cells marker [77]. To date, CD133 was identified in epithelial cells [78] and in adult stem cells from prostate, kidney and nervous system [59]; as CSC marker, CD133 was identified in brain, prostate, breast, lung and colon cancers [79 - 81]. Putative CD133 positive CSCs in human OSCC are highly clonogenic and have higher invasiveness and tumorigenesis than CD133 negative counterpart; moreover, CD133 positive cells from either OSCC or laryngeal cancers present chemoresistant properties against paclitaxel [82, 83]. Co-expression of Oct-4 and Nanog associated with an over-expression of CD133 correlates with a poorer prognosis in OSCC [84]. CD44+ cancer stem-like cells expressed higher CD133 levels than CD44 negative cells in OSCC [66]. Some other authors failed to identify CD133 expression in putative CSCs in colon cancer [85] or in OSCC [86]. These results may be the consequence of false negative results due to incapacity of some antibodies to detect non-glycosylated CD133 or due to the presence of alternatively spliced CD133 isoforms since the tissue distribution of CD133 mRNA is more widespread than the expression of one of the epitopes [86].

Oct-4, Sox2 and Nanog

Oct-4, Sox2 and *Nanog* are human embryonic stem cell genes. Their transcription factors are capable to reprogram somatic cell nuclei to an undifferentiated state. Yu *et al.* showed that *Oct4, Sox2, Nanog* and *Lin28* are sufficient to reprogram adult human somatic cells to pluripotent stem cells (called induced pluripotent stem cells) with all the characteristics of embryonic stem cells (normal karyotypes, telomerase activity, capacity to differentiate into derivatives of all three primary germ layers) [87].

Normally, *Oct-4* and *Nanog* are expressed only in pluripotent embryonic stem and germ cells and they are down-regulated during differentiation. Their functions in embryonic cells are accomplished by targeting genes encoding components of chromatin remodeling and histone-modifying complexes (SMARCAD1, MYST3, and SET) and interacting with several transcription factors and cell signaling molecules (*STAT3*, HesX1, Zic3, TCF3, FGF2, LEFTY2) [66, 88 - 92].

In oral mucosa, stem cells from lamina propria express Oct4, Sox2 and Nanog;

transplantation of these cells (normal non-tumoral stem cells) in immunodeficient mice determines tumor formation of osteoblastic, chondroblastic, adipocytic and neural lineage [93].

Oct-4, Sox2 and Nanog are expressed in cancer cells with stem-like features [54, 59]. They are expressed in poorly differentiated tumors when compared with well differentiated ones in mammary cancers and bladder carcinomas [94]. Oct4 and Sox2 are expressed by cancer stem-like cells isolated from head and neck squamous cell carcinomas as sphere-forming cells; moreover, these spheres are chemoresistant to cisplatin, 5-fluorouracil, paclitaxel and doxetaxel; the tumorigenic potential of these cells is impressive (less than 100 cells are capable to produce tumors in nude mice) [95]. Other authors also confirmed that Oct4 and Nanog expressions in CSC derived from OSCC associate with cisplatin resistance [96].

c-Met

c-Met is proto-oncogene coding a tyrosine kinase receptor for hepatocyte growth factor (HGF) and it is normally expressed by stem cells and progenitor cells only. CSCs isolated from OSCC may express c-Met [54, 97]; when they do, their tumorigenic potential is higher when compared to CD44+ cells; moreover, cells co-expressing c-Met and CD44 are endowed with a significantly higher tumorigenic potential than cells with singular expression of either c-Met or CD44 [98] or the cells co-expressing c-Met and ALDH1 [98]. Tumor expression of c-Met is usually associated with local invasion and metastatic potential, angiogenesis, chemoresistance and decreased survival [98 - 100].

Side Populations (SPs)

Cells able to eliminate the fluorescent DNA binding dye *Hoechst 33342* are called "side populations". The efflux of this chemical compound is facilitated by the presence of high levels of molecules belonging to ABC (ATP-binding cassette) transporter superfamily (MDR1, MRP1, ABCB5, ABCG2, ABCB5) [101, 102] that also facilitate the efflux of other drugs, thus conferring chemoresistance [54, 59, 103, 104]. They were first identified in the bone marrow [105], present stem cell properties (endowed with clonogenicity and highly tumorigenic) and express several stem cell markers (CD44, Oct-4, Bmi-1) [102]. To date, SP cells were identified in both normal tissues (skin, lung, brain and liver) and malignant tumors (cancers of gastrointestinal, hepatic, ovarian, mammary, thyroid or nasopharyngeal origin and also in gliomas and neuroblastomas) [59, 106, 107]. Recently, presence of SP cells was investigated in head and neck squamous cell carcinomas; these cells expressed Oct-4, CD44 and Bmi1 (stem cells markers) and presented high tumorigenic and metastatic potential [108]. Moreover, expression of ABC transporter family members such as ABCG2 and ABCB5 were identified as negative prognostic factors in proliferative oral epithelial lesions. Oral leukoplakia with ABCG2 expression has 3.24-fold increased risk of malignant transformation [109]; high ABCB5 expression was identified as an independent prognostic factor for tumor progression and recurrence [110].

Using SPs-related assays is quite simple from the technical point of view and this technique has the advantage of avoiding the problems related to antibody binding efficiency unavoidable when using stem cells markers [54, 59]. However, SPs assays may offer biased results due to the fact that Hoechst 33342 dye is toxic to live cells, thus isolated cells are merely cells able to escape from effects of the dye; moreover, since the efflux of the dye is a dynamic process, several factors may influence SP features such as dye concentration, cellular concentrations or the time of staining [106].

All these data, sometimes confusing, converge to the conclusion that present methods have significant limitations and more identification and isolation methods are urgently needed in order to properly identify and select CSC.

Cancer Stem Cell Niche

Similarly to normal cell stems, CSCs rely on a stem cell niche that hosts a complex interaction between tumor cells and stromal microenvironment. Similarly to normal stem cell niche, the CSC niche preserves stemness (self-renewal and proliferation with differentiation) and recruits new CSCs from differentiated nontumorigenic tumor cells. Furthermore, CSC niche offers proper protection to CSC from different factors, thus ensuring chemo and/or radiotherapy resistance [24, 111, 112].

Oral CSC niche has perivascular location, more likely within a 100 microns radius of a blood vessel. Endothelial cells are one of the most important cells present in CSC niche. There are several studies supporting the role of endothelial cells: ablation of tumor associated endothelial cells correlates with a significant loss of CSC [113]; survival and angiogenic capacity of endothelial cells are supported by tumor cells through VEGF secretion (VEGF induces VEGFR1 accumulation in endothelial cells, favoring endothelial-cell survival) *via* constitutive ubiquitination inhibition due to Akt/ERK pathway activation and downregulation of VEGFR2 (which controls capillary tube formation) *via* ubiquitination due to JNK/c-Jun pathway activation [114]; conversely, endothelial cells secrete several mediators - interleukin-6 (IL- 6), interleukin-8 (CXCL8), and epidermal growth factor (EGF) - able to activate *via* phosphorylation STAT3, Akt and ERK in OSCC tumor cells, thus enhancing survival and migration of tumor cells [115].

Other cells within the oral CSC niche with a major contribution in providing support to CSCs are the myofibroblasts. They are able to induce nontumorigenic tumor cells to dedifferentiate by activating the Wnt pathway [112]. The origin of these myofibroblasts is uncertain; several data support a tumor origin for them, arising through epithelial-mesenchymal transition (EMT). In OSCC, EMT plays a major role in tumor invasion and metastasis; loss of E-cadherin expression and acquiring mesenchymal markers such as vimentin and smooth muscle actin characterize cells transforming from an epithelial into a mesenchymal phenotype [48]. Spheroid-derived cells (SDC) from head and neck SCC lines have stem-like features (self-renewing capacities, increased invading capacity *in vitro*, expression of putative stem cell markers, overexpression of stemness-related gene, relative quiescence) and express in part vimentin and smooth muscle actin with concomitant loss of E-cadherin (myofibroblastic immunophenotype of the tumor cells that underwent EMT) [47].

Snail is one of the molecules with major role in EMT in OSCC, and its expression is correlated with tumor aggressivity [48]. It also intervenes in inducing CSC-like cells in head and neck SCC (head and neck SCC lines transfected with Snail induce EMT and CSC-like phenotype in tumor cells; also, these Snail-transected cells present chemoresistance) [116]. Moreover, knocking down Snail in long-term nicotine-treated human oral epithelial cells of both normal gingiva and

OSCC reduces CSCs features, blocks the xenograft tumorigenesis and improves survival [117]. Likewise BMP-4 (bone morphogenetic protein-4) induces EMT with acquisition of stem cell-like phenotypes in a OSCC cell line (over-expression of ABCG2 expression of Oct-4 with concomitant Snail, Slug and vimentin over-expression and down-regulation of E-cadherin) [118]. Hypoxia also induces stem cell phenotype most likely *via* Wnt activation with subsequent induction of EMT *via* Snail [119, 120].

Treatment Resistance of CSC

CSC are resilient to chemo- and radiotherapy, due to both their intrinsic properties (apoptosis resistance by unresponsiveness to TNF alpha and anti Fas antibody, expression of ABC family transporters with multidrug resistance function, and decreased cell division) [42, 66] and also protection provided by the CSC niche [24].

Several experiments showed resistance to a variety of cytotoxic drugs of tumor cells with stem-like properties: sphere-derived cells are resistant to 5-fluorouracil; SPs are relatively resistant to platinum cytotoxic drugs and bortezomib; CD44+ cells are resistant to a panel of cytotoxic drugs (5-FU, docetaxel, paclitaxel, cisplatin, carboplatin). Moreover, chemotherapy may increase the proportion of CSCs as reported in several studies: 5-FU and paclitaxel enrich the SP and CD133+ cells; cisplatin enriches c-Met+ cells, CD133+ and c-Kit+ cells and cells with stemness markers Oct4, Nanog, Nestin, and Bmi1 [66, 96, 98, 108, 121 - 123].

In OSCC, presence of cells with CSC phenotype (CD44, Oct4 and integrin-b1) associates unsatisfactory response to radiotherapy [124]. Another study showed increased radioresistance of cells expressing CD44 and ALDH; moreover, utilization of a STAT3 signaling blocker induces tumor radiosensitivity [47].

Oral Stem Cells in Regeneration

Little is known about the role of epithelial OSCs in oral mucosa regeneration. Several studies in animals identified the presence of putative stem cells involved in oral mucosa regeneration. Intralesional administration of undifferentiated green fluorescent protein-labeled autologous bone marrow mesenchymal stem cells determines activation of the wound edge and recovering of the oral mucosa wound in a previously injured oral mucosa in an *in vivo* dog model [125]. p75NGFR positive/PCNA positive or BrdU positive cells are present in basal layers of oral epithelium during re-epitelization in irradiated mice and may indicate the presence of stem cells [126].

Several techniques of tissue engineering were developed for repairing oral mucosa, especially after surgical resection of oral cancers; to date, tissue-engineered oral mucosal models reproducing normal human oral mucosa (combination of biocompatible scaffolds seeded with cells harvested from preferably autologous stem cell reservoirs) are preferred as mucosal grafts [127, 128]. Sources of stem cells are embryonic stem cells (with all the ethical problems generated by the use of embryonic cells in therapy), induced pluripotent stem cells from oral mucosa fibroblasts and adult mesenchymal stem cells [128]. Despite the good results obtained with induced pluripotent stem cells, there are some disadvantages of using these reprogrammed cells due to three major problems: the success rate of reprogramming technology is quite low (0.01-10%); there are concerns regarding the safety of using cells in which the gene transfer was accomplished *via* modified viruses; and the capability of the reprogrammed cells to differentiate towards all three germ layers implies the danger that tumors or teratomas may arise [128 - 130].

CONCLUSIVE REMARKS

Stem cells in oral mucosa are a peculiar category of cells presenting self-renewal and differentiation properties. Their proper identification and biology are still under debate, and more basic studies are required. Very few studies about the role of stem cells in oral mucosa regeneration have been performed. More attention, although still insufficient, has been paid to the issue of CSCs in OSCC.

Several tools are used to identify CSCs in OSCC; considering the differences between density and biologic characteristics, most likely they do not highlight all the tumorigenic cells. *In vitro* methods have some limitations and sometimes they are quite laborious, whereas immunohistochemical stemness markers may identify either supplementary differentiated nontumorigenic cells that have maintained expression of that marker or may fail to identify cells with the biologic behavior of stem cells.

The role of CSCs is also not completely understood, and data about the internal collaboration within both the cellular and interstitial component of the niche offer limited insights into stem cell niche pathophysiology. As in other cancer types, presence of CSCs correlates with poor prognosis and tumor aggressivity, and most importantly, unresponsiveness to chemotherapy and radiation. This last feature is especially important since the maintenance of a quiescent state (which is an intrinsic characteristic of stem cells) technically preserves therapy-resistant tumorigenic cells, the source for recurrences. Thus, it is possible that conventional oncologic treatment (aiming to eliminate the main mass of the tumor by surgery, radiotherapy and chemotherapy) in association with CSCs targeting therapies might significantly improve the prognosis of the OSCC patients.

CONFLICT OF INTEREST

The authors confirm that they have no conflict of interest to declare for this publication.

ACKNOWLEDGEMENTS

This work was partially supported by PN-II-PT-PCCA-2013-4-1407 (Project nr. 190), PN-II-PT-PCCA-2013-4-1386 (Project nr. 185) and PN-II-ID-PCE-2-11-3-0918.

REFFERENCES

[1] Jones KB, Klein OD. Oral epithelial stem cells in tissue maintenance and disease: the first steps in a long journey. Int J Oral Sci 2013; 5(3): 121-9. [http://dx.doi.org/10.1038/ijos.2013.46] [PMID: 23887128]

[2] Van Keymeulen A, Blanpain C. Tracing epithelial stem cells during development, homeostasis, and repair. J Cell Biol 2012; 197(5): 575-84. [http://dx.doi.org/10.1083/jcb.201201041] [PMID: 22641343]

[3] Bickenbach JR. Identification and behavior of label-retaining cells in oral mucosa and skin. J Dent Res 1981; 60: 1611-20. (Spec No C)

[4] Bickenbach JR, Mackenzie IC. Identification and localization of label-retaining cells in hamster

epithelia. J Invest Dermatol 1984; 82(6): 618-22.
[http://dx.doi.org/10.1111/1523-1747.ep12261460] [PMID: 6725984]

[5] Huang YL, Tao X, Xia J, Li CY, Cheng B. Distribution and quantity of label-retaining cells in rat oral epithelia. J Oral Pathol Med 2009; 38(8): 663-7.
[http://dx.doi.org/10.1111/j.1600-0714.2009.00798.x] [PMID: 19570111]

[6] Asaka T, Akiyama M, Kitagawa Y, Shimizu H. Higher density of label-retaining cells in gingival epithelium. J Dermatol Sci 2009; 55(2): 132-4.
[http://dx.doi.org/10.1016/j.jdermsci.2009.03.006] [PMID: 19403272]

[7] Pan Q, Nicholson AM, Barr H, *et al.* Identification of lineage-uncommitted, long-lived, label-retaining cells in healthy human esophagus and stomach, and in metaplastic esophagus. Gastroenterology 2013; 144(4): 761-70.
[http://dx.doi.org/10.1053/j.gastro.2012.12.022] [PMID: 23266557]

[8] Barrandon Y, Green H. Three clonal types of keratinocyte with different capacities for multiplication. Proc Natl Acad Sci USA 1987; 84(8): 2302-6.
[http://dx.doi.org/10.1073/pnas.84.8.2302] [PMID: 2436229]

[9] Calenic B, Ishkitiev N, Yaegaki K, *et al.* Magnetic separation and characterization of keratinocyte stem cells from human gingiva. J Periodontal Res 2010; 45(6): 703-8.
[http://dx.doi.org/10.1111/j.1600-0765.2010.01284.x] [PMID: 20546107]

[10] Luo X, Okubo T, Randell S, Hogan BL. Culture of endodermal stem/progenitor cells of the mouse tongue. In Vitro Cell Dev Biol Anim 2009; 45(1-2): 44-54.
[http://dx.doi.org/10.1007/s11626-008-9149-2] [PMID: 18830772]

[11] Calenic B, Ishkitiev N, Yaegaki K, *et al.* Characterization of oral keratinocyte stem cells and prospects of its differentiation to oral epithelial equivalents. Rom J Morphol Embryol 2010; 51(4): 641-5.
[PMID: 21103620]

[12] Hirata TM, Ishkitiev N, Yaegaki K, *et al.* Expression of multiple stem cell markers in dental pulp cells cultured in serum-free media. J Endod 2010; 36(7): 1139-44.
[http://dx.doi.org/10.1016/j.joen.2010.03.002] [PMID: 20630286]

[13] Karaöz E, Doğan BN, Aksoy A, *et al.* Isolation and *in vitro* characterisation of dental pulp stem cells from natal teeth. Histochem Cell Biol 2010; 133(1): 95-112.
[http://dx.doi.org/10.1007/s00418-009-0646-5] [PMID: 19816704]

[14] Huang CY, Pelaez D, Dominguez-Bendala J, Garcia-Godoy F, Cheung HS. Plasticity of stem cells derived from adult periodontal ligament. Regen Med 2009; 4(6): 809-21.
[http://dx.doi.org/10.2217/rme.09.55] [PMID: 19903001]

[15] Dong R, Liu X, Fan M, Yang L, Peng L, Zhang L. Isolation and differentiation of nestin positive cells from rat oral mucosal lamina propria. Differentiation 2010; 79(1): 9-14.
[http://dx.doi.org/10.1016/j.diff.2009.08.010] [PMID: 19762142]

[16] Widera D, Zander C, Heidbreder M, *et al.* Adult palatum as a novel source of neural crest-related stem cells. Stem Cells 2009; 27(8): 1899-910.
[http://dx.doi.org/10.1002/stem.104] [PMID: 19544446]

[17] Kanafi M, Majumdar D, Bhonde R, Gupta P, Datta I. Midbrain cues dictate differentiation of human

dental pulp stem cells towards functional dopaminergic neurons. J Cell Physiol 2014; 229(10): 1369-77.
[http://dx.doi.org/10.1002/jcp.24570] [PMID: 24477667]

[18] Nakamura T, Endo K, Kinoshita S. Identification of human oral keratinocyte stem/progenitor cells by neurotrophin receptor p75 and the role of neurotrophin/p75 signaling. Stem Cells 2007; 25(3): 628-38.
[http://dx.doi.org/10.1634/stemcells.2006-0494] [PMID: 17110619]

[19] Pellegrini G, Dellambra E, Golisano O, *et al.* p63 identifies keratinocyte stem cells. Proc Natl Acad Sci USA 2001; 98(6): 3156-61.
[http://dx.doi.org/10.1073/pnas.061032098] [PMID: 11248048]

[20] Legg J, Jensen UB, Broad S, Leigh I, Watt FM. Role of melanoma chondroitin sulphate proteoglycan in patterning stem cells in human interfollicular epidermis. Development 2003; 130(24): 6049-63.
[http://dx.doi.org/10.1242/dev.00837] [PMID: 14573520]

[21] Sen S, Sharma S, Gupta A, *et al.* Molecular characterization of explant cultured human oral mucosal epithelial cells. Invest Ophthalmol Vis Sci 2011; 52(13): 9548-54.
[http://dx.doi.org/10.1167/iovs.11-7946] [PMID: 22064988]

[22] Tao Q, Qiao B, Lv B, Zheng C, Chen Z, Huang H. p63 and its isoforms as markers of rat oral mucosa epidermal stem cells *in vitro.* Cell Biochem Funct 2009; 27(8): 535-41.
[http://dx.doi.org/10.1002/cbf.1612] [PMID: 19918930]

[23] Thomson PJ, Potten CS, Appleton DR. *In vitro* labelling studies and the measurement of epithelial cell proliferative activity in the human oral cavity. Arch Oral Biol 2001; 46(12): 1157-64.
[http://dx.doi.org/10.1016/S0003-9969(01)00065-6] [PMID: 11684035]

[24] Borovski T, De Sousa E Melo F, Vermeulen L, Medema JP, Medema JP. Cancer stem cell niche: the place to be. Cancer Res 2011; 71(3): 634-9.
[http://dx.doi.org/10.1158/0008-5472.CAN-10-3220] [PMID: 21266356]

[25] Morrison SJ, Spradling AC. Stem cells and niches: mechanisms that promote stem cell maintenance throughout life. Cell 2008; 132(4): 598-611.
[http://dx.doi.org/10.1016/j.cell.2008.01.038] [PMID: 18295578]

[26] Lander AD, Kimble J, Clevers H, *et al.* What does the concept of the stem cell niche really mean today? BMC Biol 2012; 10: 19.
[http://dx.doi.org/10.1186/1741-7007-10-19] [PMID: 22405133]

[27] Cao YA, Wagers AJ, Beilhack A, *et al.* Shifting foci of hematopoiesis during reconstitution from single stem cells. Proc Natl Acad Sci USA 2004; 101(1): 221-6.
[http://dx.doi.org/10.1073/pnas.2637010100] [PMID: 14688412]

[28] Conti L, Pollard SM, Gorba T, *et al.* Niche-independent symmetrical self-renewal of a mammalian tissue stem cell. PLoS Biol 2005; 3(9): e283.
[http://dx.doi.org/10.1371/journal.pbio.0030283] [PMID: 16086633]

[29] Durand A, Donahue B, Peignon G, *et al.* Functional intestinal stem cells after Paneth cell ablation induced by the loss of transcription factor Math1 (Atoh1). Proc Natl Acad Sci USA 2012; 109(23): 8965-70.
[http://dx.doi.org/10.1073/pnas.1201652109] [PMID: 22586121]

[30] Brawley C, Matunis E. Regeneration of male germline stem cells by spermatogonial dedifferentiation *in vivo*. Science 2004; 304(5675): 1331-4. [http://dx.doi.org/10.1126/science.1097676] [PMID: 15143218]

[31] Kai T, Spradling A. Differentiating germ cells can revert into functional stem cells in Drosophila melanogaster ovaries. Nature 2004; 428(6982): 564-9. [http://dx.doi.org/10.1038/nature02436] [PMID: 15024390]

[32] Hsu YC, Pasolli HA, Fuchs E. Dynamics between stem cells, niche, and progeny in the hair follicle. Cell 2011; 144(1): 92-105. [http://dx.doi.org/10.1016/j.cell.2010.11.049] [PMID: 21215372]

[33] Lim X, Tan SH, Koh WL, *et al.* Interfollicular epidermal stem cells self-renew *via* autocrine Wnt signaling. Science 2013; 342(6163): 1226-30. [http://dx.doi.org/10.1126/science.1239730] [PMID: 24311688]

[34] Januschke J, Näthke I. Stem cell decisions: a twist of fate or a niche market? Semin Cell Dev Biol 2014; 34: 116-23. [http://dx.doi.org/10.1016/j.semcdb.2014.02.014] [PMID: 24613913]

[35] Scadden DT. The stem-cell niche as an entity of action. Nature 2006; 441(7097): 1075-9. [http://dx.doi.org/10.1038/nature04957] [PMID: 16810242]

[36] Yamashita YM, Mahowald AP, Perlin JR, Fuller MT. Asymmetric inheritance of mother *versus* daughter centrosome in stem cell division. Science 2007; 315(5811): 518-21. [http://dx.doi.org/10.1126/science.1134910] [PMID: 17255513]

[37] Januschke J, Llamazares S, Reina J, Gonzalez C. Drosophila neuroblasts retain the daughter centrosome. Nat Commun 2011; 2: 243-6. [http://dx.doi.org/10.1038/ncomms1245] [PMID: 21407209]

[38] Chen CT, Ettinger AW, Huttner WB, Doxsey SJ. Resurrecting remnants: the lives of post-mitotic midbodies. Trends Cell Biol 2013; 23(3): 118-28. [http://dx.doi.org/10.1016/j.tcb.2012.10.012] [PMID: 23245592]

[39] Marzesco AM, Janich P, Wilsch-Bräuninger M, *et al.* Release of extracellular membrane particles carrying the stem cell marker prominin-1 (CD133) from neural progenitors and other epithelial cells. J Cell Sci 2005; 118(Pt 13): 2849-58. [http://dx.doi.org/10.1242/jcs.02439] [PMID: 15976444]

[40] Wilcock AC, Swedlow JR, Storey KG. Mitotic spindle orientation distinguishes stem cell and terminal modes of neuron production in the early spinal cord. Development 2007; 134(10): 1943-54. [http://dx.doi.org/10.1242/dev.002519] [PMID: 17470968]

[41] Ettinger AW, Wilsch-Bräuninger M, Marzesco AM, *et al.* Proliferating *versus* differentiating stem and cancer cells exhibit distinct midbody-release behaviour. Nat Commun 2011; 2: 503. [http://dx.doi.org/10.1038/ncomms1511] [PMID: 22009035]

[42] Clarke MF, Dick JE, Dirks PB, *et al.* Cancer stem cells--perspectives on current status and future directions: AACR Workshop on cancer stem cells. Cancer Res 2006; 66(19): 9339-44. [http://dx.doi.org/10.1158/0008-5472.CAN-06-3126] [PMID: 16990346]

[43] Li F, Tiede B, Massagué J, Kang Y. Beyond tumorigenesis: cancer stem cells in metastasis. Cell Res 2007; 17(1): 3-14.
[http://dx.doi.org/10.1038/sj.cr.7310118] [PMID: 17179981]

[44] Kucia M, Ratajczak MZ. Stem cells as a two edged sword-from regeneration to tumor formation. J Physiol Pharmacol 2006; 57 (Suppl. 7): 5-16.

[45] Mani SA, Guo W, Liao MJ, *et al.* The epithelial-mesenchymal transition generates cells with properties of stem cells. Cell 2008; 133(4): 704-15.
[http://dx.doi.org/10.1016/j.cell.2008.03.027] [PMID: 18485877]

[46] Floor S, van Staveren WC, Larsimont D, Dumont JE, Maenhaut C. Cancer cells in epithelial-t--mesenchymal transition and tumor-propagating-cancer stem cells: distinct, overlapping or same populations. Oncogene 2011; 30(46): 4609-21.
[http://dx.doi.org/10.1038/onc.2011.184] [PMID: 21643013]

[47] Chen C, Wei Y, Hummel M, *et al.* Evidence for epithelial-mesenchymal transition in cancer stem cells of head and neck squamous cell carcinoma. PLoS One 2011; 6(1): e16466.
[http://dx.doi.org/10.1371/journal.pone.0016466] [PMID: 21304586]

[48] Slootweg PJ, Zurac S. Prognostic and predictive value of epithelial to mesenchymal transition associated markers in oral squamous cell carcinoma. Curr Proteomics 2013; 10(3): 218-27.
[http://dx.doi.org/10.2174/157016461131003004]

[49] Radisky DC, LaBarge MA. Epithelial-mesenchymal transition and the stem cell phenotype. Cell Stem Cell 2008; 2(6): 511-2.
[http://dx.doi.org/10.1016/j.stem.2008.05.007] [PMID: 18522839]

[50] Kalluri R, Weinberg RA. The basics of epithelial-mesenchymal transition. J Clin Invest 2009; 119(6): 1420-8.
[http://dx.doi.org/10.1172/JCI39104] [PMID: 19487818]

[51] Liu LK, Jiang XY, Zhou XX, Wang DM, Song XL, Jiang HB. Upregulation of vimentin and aberrant expression of E-cadherin/beta-catenin complex in oral squamous cell carcinomas: correlation with the clinicopathological features and patient outcome. Mod Pathol 2010; 23(2): 213-24.
[http://dx.doi.org/10.1038/modpathol.2009.160] [PMID: 19915524]

[52] Zhao D, Tang XF, Yang K, Liu JY, Ma XR. Over-expression of integrin-linked kinase correlates with aberrant expression of Snail, E-cadherin and N-cadherin in oral squamous cell carcinoma: implications in tumor progression and metastasis. Clin Exp Metastasis 2012; 29(8): 957-69.
[http://dx.doi.org/10.1007/s10585-012-9485-1] [PMID: 22638656]

[53] Scanlon CS, Van Tubergen EA, Inglehart RC, D'Silva NJ. Biomarkers of epithelial-mesenchymal transition in squamous cell carcinoma. J Dent Res 2013; 92(2): 114-21.
[http://dx.doi.org/10.1177/0022034512467352] [PMID: 23128109]

[54] Patel SS, Shah KA, Shah MJ, Kothari KC, Rawal RM. Cancer stem cells and stemness markers in oral squamous cell carcinomas. Asian Pac J Cancer Prev 2014; 15(20): 8549-56.
[http://dx.doi.org/10.7314/APJCP.2014.15.20.8549] [PMID: 25374166]

[55] Pearson H. Stem-cell tagging shows flaws. Nature 2006; 439(7076): 519.
[http://dx.doi.org/10.1038/439519a] [PMID: 16452944]

[56] Isacke CM, Yarwood H. The hyaluronan receptor, CD44. Int J Biochem Cell Biol 2002; 34(7): 718-21.
[http://dx.doi.org/10.1016/S1357-2725(01)00166-2] [PMID: 11950588]

[57] Kajita M, Itoh Y, Chiba T, *et al.* Membrane-type 1 matrix metalloproteinase cleaves CD44 and promotes cell migration. J Cell Biol 2001; 153(5): 893-904.
[http://dx.doi.org/10.1083/jcb.153.5.893] [PMID: 11381077]

[58] Weber GF, Ashkar S, Cantor H. Interaction between CD44 and osteopontin as a potential basis for metastasis formation. Proc Assoc Am Physicians 1997; 109(1): 1-9.
[PMID: 9010911]

[59] Zhang Z, Filho MS, Nör JE. The biology of head and neck cancer stem cells. Oral Oncol 2012; 48(1): 1-9.
[http://dx.doi.org/10.1016/j.oraloncology.2011.10.004] [PMID: 22070916]

[60] Sterz CM, Kulle C, Dakic B, *et al.* A basal-cell-like compartment in head and neck squamous cell carcinomas represents the invasive front of the tumor and is expressing MMP-9. Oral Oncol 2010; 46(2): 116-22.
[http://dx.doi.org/10.1016/j.oraloncology.2009.11.011] [PMID: 20036607]

[61] Desai B, Ma T, Zhu J, Chellaiah MA. Characterization of the expression of variant and standard CD44 in prostate cancer cells: identification of the possible molecular mechanism of CD44/MMP9 complex formation on the cell surface. J Cell Biochem 2009; 108(1): 272-84.
[http://dx.doi.org/10.1002/jcb.22248] [PMID: 19582779]

[62] Mishra A, Verma M. Cancer biomarkers: are we ready for the prime time? Cancers (Basel) 2010; 2(1): 190-208.
[http://dx.doi.org/10.3390/cancers2010190] [PMID: 24281040]

[63] Prince ME, Sivanandan R, Kaczorowski A, *et al.* Identification of a subpopulation of cells with cancer stem cell properties in head and neck squamous cell carcinoma. Proc Natl Acad Sci USA 2007; 104(3): 973-8.
[http://dx.doi.org/10.1073/pnas.0610117104] [PMID: 17210912]

[64] Chikamatsu K, Ishii H, Takahashi G, *et al.* Resistance to apoptosis-inducing stimuli in CD44+ head and neck squamous cell carcinoma cells. Head Neck 2012; 34(3): 336-43.
[http://dx.doi.org/10.1002/hed.21732] [PMID: 21472882]

[65] Chen YW, Chen KH, Huang PI, *et al.* Cucurbitacin I suppressed stem-like property and enhanced radiation-induced apoptosis in head and neck squamous carcinoma--derived CD44(+)ALDH1(+) cells. Mol Cancer Ther 2010; 9(11): 2879-92.
[http://dx.doi.org/10.1158/1535-7163.MCT-10-0504] [PMID: 21062915]

[66] Okamoto A, Chikamatsu K, Sakakura K, Hatsushika K, Takahashi G, Masuyama K. Expansion and characterization of cancer stem-like cells in squamous cell carcinoma of the head and neck. Oral Oncol 2009; 45(7): 633-9.
[http://dx.doi.org/10.1016/j.oraloncology.2008.10.003] [PMID: 19027347]

[67] Joshua B, Kaplan MJ, Doweck I, *et al.* Frequency of cells expressing CD44, a head and neck cancer stem cell marker: correlation with tumor aggressiveness. Head Neck 2012; 34(1): 42-9.

[http://dx.doi.org/10.1002/hed.21699] [PMID: 21322081]

[68] Chute JP, Muramoto GG, Whitesides J, *et al.* Inhibition of aldehyde dehydrogenase and retinoid signaling induces the expansion of human hematopoietic stem cells. Proc Natl Acad Sci USA 2006; 103(31): 11707-12.
[http://dx.doi.org/10.1073/pnas.0603806103] [PMID: 16857736]

[69] Chen YC, Chen YW, Hsu HS, *et al.* Aldehyde dehydrogenase 1 is a putative marker for cancer stem cells in head and neck squamous cancer. Biochem Biophys Res Commun 2009; 385(3): 307-13.
[http://dx.doi.org/10.1016/j.bbrc.2009.05.048] [PMID: 19450560]

[70] Thomasson HR, Edenberg HJ, Crabb DW, *et al.* Alcohol and aldehyde dehydrogenase genotypes and alcoholism in Chinese men. Am J Hum Genet 1991; 48(4): 677-81.
[PMID: 2014795]

[71] Ginestier C, Hur MH, Charafe-Jauffret E, *et al.* ALDH1 is a marker of normal and malignant human mammary stem cells and a predictor of poor clinical outcome. Cell Stem Cell 2007; 1(5): 555-67.
[http://dx.doi.org/10.1016/j.stem.2007.08.014] [PMID: 18371393]

[72] Rasper M, Schäfer A, Piontek G, *et al.* Aldehyde dehydrogenase 1 positive glioblastoma cells show brain tumor stem cell capacity. Neuro-oncol 2010; 12(10): 1024-33.
[http://dx.doi.org/10.1093/neuonc/noq070] [PMID: 20627895]

[73] Huang EH, Hynes MJ, Zhang T, *et al.* Aldehyde dehydrogenase 1 is a marker for normal and malignant human colonic stem cells (SC) and tracks SC overpopulation during colon tumorigenesis. Cancer Res 2009; 69(8): 3382-9.
[http://dx.doi.org/10.1158/0008-5472.CAN-08-4418] [PMID: 19336570]

[74] Jiang F, Qiu Q, Khanna A, *et al.* Aldehyde dehydrogenase 1 is a tumor stem cell-associated marker in lung cancer. Mol Cancer Res 2009; 7(3): 330-8.
[http://dx.doi.org/10.1158/1541-7786.MCR-08-0393] [PMID: 19276181]

[75] Clay MR, Tabor M, Owen JH, *et al.* Single-marker identification of head and neck squamous cell carcinoma cancer stem cells with aldehyde dehydrogenase. Head Neck 2010; 32(9): 1195-201.
[http://dx.doi.org/10.1002/hed.21315] [PMID: 20073073]

[76] Chen YC, Chang CJ, Hsu HS, *et al.* Inhibition of tumorigenicity and enhancement of radiochemosensitivity in head and neck squamous cell cancer-derived ALDH1-positive cells by knockdown of Bmi-1. Oral Oncol 2010; 46(3): 158-65.
[http://dx.doi.org/10.1016/j.oraloncology.2009.11.007] [PMID: 20036608]

[77] Yin AH, Miraglia S, Zanjani ED, *et al.* AC133, a novel marker for human hematopoietic stem and progenitor cells. Blood 1997; 90(12): 5002-12.
[PMID: 9389720]

[78] Corbeil D, Röper K, Hellwig A, *et al.* The human AC133 hematopoietic stem cell antigen is also expressed in epithelial cells and targeted to plasma membrane protrusions. J Biol Chem 2000; 275(8): 5512-20.
[http://dx.doi.org/10.1074/jbc.275.8.5512] [PMID: 10681530]

[79] Wu Y, Wu PY. CD133 as a marker for cancer stem cells: progresses and concerns. Stem Cells Dev 2009; 18(8): 1127-34.

[http://dx.doi.org/10.1089/scd.2008.0338] [PMID: 19409053]

[80] Singh SK, Hawkins C, Clarke ID, *et al.* Identification of human brain tumour initiating cells. Nature 2004; 432(7015): 396-401.
[http://dx.doi.org/10.1038/nature03128] [PMID: 15549107]

[81] Nadal R, Ortega FG, Salido M, *et al.* CD133 expression in circulating tumor cells from breast cancer patients: potential role in resistance to chemotherapy. Int J Cancer 2013; 133(10): 2398-407.
[http://dx.doi.org/10.1002/ijc.28263] [PMID: 23661576]

[82] Zhang Q, Shi S, Yen Y, Brown J, Ta JQ, Le AD. A subpopulation of CD133(+) cancer stem-like cells characterized in human oral squamous cell carcinoma confer resistance to chemotherapy. Cancer Lett 2010; 289(2): 151-60.
[http://dx.doi.org/10.1016/j.canlet.2009.08.010] [PMID: 19748175]

[83] Zhou L, Wei X, Cheng L, Tian J, Jiang JJ. CD133, one of the markers of cancer stem cells in Hep-2 cell line. Laryngoscope 2007; 117(3): 455-60.
[http://dx.doi.org/10.1097/01.mlg.0000251586.15299.35] [PMID: 17334305]

[84] Chiou SH, Yu CC, Huang CY, *et al.* Positive correlations of Oct-4 and Nanog in oral cancer stem-like cells and high-grade oral squamous cell carcinoma. Clin Cancer Res 2008; 14(13): 4085-95.
[http://dx.doi.org/10.1158/1078-0432.CCR-07-4404] [PMID: 18593985]

[85] Shmelkov SV, Butler JM, Hooper AT, *et al.* CD133 expression is not restricted to stem cells, and both CD133+ and CD133- metastatic colon cancer cells initiate tumors. J Clin Invest 2008; 118(6): 2111-20.
[PMID: 18497886]

[86] Mărgăritescu C, Pirici D, Simionescu C, Stepan A. The utility of CD44, CD117 and CD133 in identification of cancer stem cells (CSC) in oral squamous cell carcinomas (OSCC). Rom J Morphol Embryol 2011; 52(3) (Suppl.): 985-93.
[PMID: 22119814]

[87] Yu J, Vodyanik MA, Smuga-Otto K, *et al.* Induced pluripotent stem cell lines derived from human somatic cells. Science 2007; 318(5858): 1917-20.
[http://dx.doi.org/10.1126/science.1151526] [PMID: 18029452]

[88] Burdon T, Smith A, Savatier P. Signalling, cell cycle and pluripotency in embryonic stem cells. Trends Cell Biol 2002; 12(9): 432-8.
[http://dx.doi.org/10.1016/S0962-8924(02)02352-8] [PMID: 12220864]

[89] Pesce M, Wang X, Wolgemuth DJ, Schöler H. Differential expression of the Oct-4 transcription factor during mouse germ cell differentiation. Mech Dev 1998; 71(1-2): 89-98.
[http://dx.doi.org/10.1016/S0925-4773(98)00002-1] [PMID: 9507072]

[90] Chambers I, Colby D, Robertson M, *et al.* Functional expression cloning of Nanog, a pluripotency sustaining factor in embryonic stem cells. Cell 2003; 113(5): 643-55.
[http://dx.doi.org/10.1016/S0092-8674(03)00392-1] [PMID: 12787505]

[91] Boyer LA, Lee TI, Cole MF, *et al.* Core transcriptional regulatory circuitry in human embryonic stem cells. Cell 2005; 122(6): 947-56.
[http://dx.doi.org/10.1016/j.cell.2005.08.020] [PMID: 16153702]

[92] Loh YH, Wu Q, Chew JL, *et al.* The Oct4 and Nanog transcription network regulates pluripotency in mouse embryonic stem cells. Nat Genet 2006; 38(4): 431-40. [http://dx.doi.org/10.1038/ng1760] [PMID: 16518401]

[93] Marynka-Kalmani K, Treves S, Yafee M, *et al.* The lamina propria of adult human oral mucosa harbors a novel stem cell population. Stem Cells 2010; 28(5): 984-95. [PMID: 20474080]

[94] Ben-Porath I, Thomson MW, Carey VJ, *et al.* An embryonic stem cell-like gene expression signature in poorly differentiated aggressive human tumors. Nat Genet 2008; 40(5): 499-507. [http://dx.doi.org/10.1038/ng.127] [PMID: 18443585]

[95] Lim YC, Oh SY, Cha YY, Kim SH, Jin X, Kim H. Cancer stem cell traits in squamospheres derived from primary head and neck squamous cell carcinomas. Oral Oncol 2011; 47(2): 83-91. [http://dx.doi.org/10.1016/j.oraloncology.2010.11.011] [PMID: 21167769]

[96] Tsai LL, Yu CC, Chang YC, Yu CH, Chou MY. Markedly increased Oct4 and Nanog expression correlates with cisplatin resistance in oral squamous cell carcinoma. J Oral Pathol Med 2011; 40(8): 621-8. [http://dx.doi.org/10.1111/j.1600-0714.2011.01015.x] [PMID: 21342274]

[97] Shah A, Patel S, Pathak J, Swain N, Kumar S. The evolving concepts of cancer stem cells in head and neck squamous cell carcinoma. Scientific World Journal 2014 Jan; 21 2014:842491 [http://dx.doi.org/10.1155/2014/842491]

[98] Sun S, Wang Z. Head neck squamous cell carcinoma c-Met cells display cancer stem cell properties and are responsible for cisplatin-resistance and metastasis. Int J Cancer 2011; 129(10): 2337-48. [http://dx.doi.org/10.1002/ijc.25927] [PMID: 21225626]

[99] Lim YC, Han JH, Kang HJ, *et al.* Overexpression of c-Met promotes invasion and metastasis of small oral tongue carcinoma. Oral Oncol 2012; 48(11): 1114-9. [http://dx.doi.org/10.1016/j.oraloncology.2012.05.013] [PMID: 22704061]

[100] Zhao D, Wang SH, Feng Y, Hua CG, Zhao J, Tang XF. Intratumoral c-Met expression is associated with vascular endothelial growth factor C expression, lymphangiogenesis, and lymph node metastasis in oral squamous cell carcinoma: implications for use as a prognostic marker. Hum Pathol 2011; 42(10): 1514-23. [http://dx.doi.org/10.1016/j.humpath.2010.03.012] [PMID: 21531000]

[101] Hadnagy A, Gaboury L, Beaulieu R, Balicki D. SP analysis may be used to identify cancer stem cell populations. Exp Cell Res 2006; 312(19): 3701-10. [http://dx.doi.org/10.1016/j.yexcr.2006.08.030] [PMID: 17046749]

[102] Zhang P, Zhang Y, Mao L, Zhang Z, Chen W. Side population in oral squamous cell carcinoma possesses tumor stem cell phenotypes. Cancer Lett 2009; 277(2): 227-34. [http://dx.doi.org/10.1016/j.canlet.2008.12.015] [PMID: 19185988]

[103] Doyle L, Ross DD. Multidrug resistance mediated by the breast cancer resistance protein BCRP (ABCG2). Oncogene 2003; 22(47): 7340-58. [http://dx.doi.org/10.1038/sj.onc.1206938] [PMID: 14576842]

[104] Hirschmann-Jax C, Foster AE, Wulf GG, *et al.* A distinct "side population" of cells with high drug

efflux capacity in human tumor cells. Proc Natl Acad Sci USA 2004; 101(39): 14228-33. [http://dx.doi.org/10.1073/pnas.0400067101] [PMID: 15381773]

[105] Goodell MA, Brose K, Paradis G, Conner AS, Mulligan RC. Isolation and functional properties of murine hematopoietic stem cells that are replicating *in vivo.* J Exp Med 1996; 183(4): 1797-806. [http://dx.doi.org/10.1084/jem.183.4.1797] [PMID: 8666936]

[106] Wu C, Alman BA. Side population cells in human cancers. Cancer Lett 2008; 268(1): 1-9. [http://dx.doi.org/10.1016/j.canlet.2008.03.048] [PMID: 18487012]

[107] Chinn SB, Darr OA, Peters RD, Prince ME. The role of head and neck squamous cell carcinoma cancer stem cells in tumorigenesis, metastasis, and treatment failure. Front Endocrinol (Lausanne) 2012; 3: 90. [http://dx.doi.org/10.3389/fendo.2012.00090] [PMID: 22876238]

[108] Tabor MH, Clay MR, Owen JH, *et al.* Head and neck cancer stem cells: the side population. Laryngoscope 2011; 121(3): 527-33. [http://dx.doi.org/10.1002/lary.21032] [PMID: 21344428]

[109] Liu W, Feng JQ, Shen XM, Wang HY, Liu Y, Zhou ZT. Two stem cell markers, ATP-binding cassette, G2 subfamily (ABCG2) and BMI-1, predict the transformation of oral leukoplakia to cancer: a long-term follow-up study. Cancer 2012; 118(6): 1693-700. [http://dx.doi.org/10.1002/cncr.26483] [PMID: 22009787]

[110] Grimm M, Krimmel M, Polligkeit J, *et al.* ABCB5 expression and cancer stem cell hypothesis in oral squamous cell carcinoma. Eur J Cancer 2012; 48(17): 3186-97. [http://dx.doi.org/10.1016/j.ejca.2012.05.027] [PMID: 22784549]

[111] Calabrese C, Poppleton H, Kocak M, *et al.* A perivascular niche for brain tumor stem cells. Cancer Cell 2007; 11(1): 69-82. [http://dx.doi.org/10.1016/j.ccr.2006.11.020] [PMID: 17222791]

[112] Vermeulen L, De Sousa E Melo F, van der Heijden M, *et al.* Wnt activity defines colon cancer stem cells and is regulated by the microenvironment. Nat Cell Biol 2010; 12(5): 468-76. [http://dx.doi.org/10.1038/ncb2048] [PMID: 20418870]

[113] Krishnamurthy S, Dong Z, Vodopyanov D, *et al.* Endothelial cell-initiated signaling promotes the survival and self-renewal of cancer stem cells. Cancer Res 2010; 70(23): 9969-78. [http://dx.doi.org/10.1158/0008-5472.CAN-10-1712] [PMID: 21098716]

[114] Zhang Z, Neiva KG, Lingen MW, Ellis LM, Nör JE. VEGF-dependent tumor angiogenesis requires inverse and reciprocal regulation of VEGFR1 and VEGFR2. Cell Death Differ 2010; 17(3): 499-512. [http://dx.doi.org/10.1038/cdd.2009.152] [PMID: 19834490]

[115] Neiva KG, Zhang Z, Miyazawa M, Warner KA, Karl E, Nör JE. Cross talk initiated by endothelial cells enhances migration and inhibits anoikis of squamous cell carcinoma cells through STAT3/Akt/ERK signaling. Neoplasia 2009; 11(6): 583-93. [http://dx.doi.org/10.1593/neo.09266] [PMID: 19484147]

[116] Masui T, Ota I, Yook JI, *et al.* Snail-induced epithelial-mesenchymal transition promotes cancer stem cell-like phenotype in head and neck cancer cells. Int J Oncol 2014; 44(3): 693-9. [PMID: 24365974]

[117] Yu CC, Chang YC. Enhancement of cancer stem-like and epithelial-mesenchymal transdifferentiation property in oral epithelial cells with long-term nicotine exposure: reversal by targeting SNAIL. Toxicol Appl Pharmacol 2013; 266(3): 459-69.
[http://dx.doi.org/10.1016/j.taap.2012.11.023] [PMID: 23219715]

[118] Qiao B, Johnson NW, Chen X, Li R, Tao Q, Gao J. Disclosure of a stem cell phenotype in an oral squamous cell carcinoma cell line induced by BMP-4 *via* an epithelial-mesenchymal transition. Oncol Rep 2011; 26(2): 455-61.
[PMID: 21567100]

[119] Heddleston JM, Li Z, McLendon RE, Hjelmeland AB, Rich JN. The hypoxic microenvironment maintains glioblastoma stem cells and promotes reprogramming towards a cancer stem cell phenotype. Cell Cycle 2009; 8(20): 3274-84.
[http://dx.doi.org/10.4161/cc.8.20.9701] [PMID: 19770585]

[120] Cannito S, Novo E, Compagnone A, *et al.* Redox mechanisms switch on hypoxia-dependent epithelial-mesenchymal transition in cancer cells. Carcinogenesis 2008; 29(12): 2267-78.
[http://dx.doi.org/10.1093/carcin/bgn216] [PMID: 18791199]

[121] Yajima T, Ochiai H, Uchiyama T, Takano N, Shibahara T, Azuma T. Resistance to cytotoxic chemotherapy-induced apoptosis in side population cells of human oral squamous cell carcinoma cell line Ho-1-N-1. Int J Oncol 2009; 35(2): 273-80.
[PMID: 19578740]

[122] Yanamoto S, Kawasaki G, Yamada S, *et al.* Isolation and characterization of cancer stem-like side population cells in human oral cancer cells. Oral Oncol 2011; 47(9): 855-60.
[http://dx.doi.org/10.1016/j.oraloncology.2011.06.501] [PMID: 21757396]

[123] Li H, Gao Q, Guo L, Lu SH. The PTEN/PI3K/Akt pathway regulates stem-like cells in primary esophageal carcinoma cells. Cancer Biol Ther 2011; 11(11): 950-8.
[http://dx.doi.org/10.4161/cbt.11.11.15531] [PMID: 21467840]

[124] Koukourakis MI, Giatromanolaki A, Tsakmaki V, Danielidis V, Sivridis E. Cancer stem cell phenotype relates to radio-chemotherapy outcome in locally advanced squamous cell head-neck cancer. Br J Cancer 2012; 106(5): 846-53.
[http://dx.doi.org/10.1038/bjc.2012.33] [PMID: 22333601]

[125] Aly LA, El-Menoufy H, Sadeq HS, Ragae A, Sabry D. Efficiency of systemic *versus* intralesional bone marrow-derived stem cells in regeneration of oral mucosa after induction of formocresol induced ulcers in dogs. Dent Res J (Isfahan) 2014; 11(2): 212-21.
[PMID: 24932192]

[126] Ishii A, Muramatsu T, Lee JM, *et al.* Expression of p75(NGFR), a proliferative and basal cell marker, in the buccal mucosa epithelium during re-epithelialization. Acta Histochem Cytochem 2014; 47(4): 145-53.
[http://dx.doi.org/10.1267/ahc.14011] [PMID: 25392568]

[127] Abou Neel EA, Chrzanowski W, Salih VM, Kim HW, Knowles JC. Tissue engineering in dentistry. J Dent 2014; 42(8): 915-28.
[http://dx.doi.org/10.1016/j.jdent.2014.05.008] [PMID: 24880036]

[128] Nauta A, Gurtner G, Longaker MT. Wound healing and regenerative strategies. Oral Dis 2011; 17(6): 541-9.
[http://dx.doi.org/10.1111/j.1601-0825.2011.01787.x] [PMID: 21332599]

[129] Takahashi K, Tanabe K, Ohnuki M, *et al.* Induction of pluripotent stem cells from adult human fibroblasts by defined factors. Cell 2007; 131(5): 861-72.
[http://dx.doi.org/10.1016/j.cell.2007.11.019] [PMID: 18035408]

[130] Yamanaka S. Elite and stochastic models for induced pluripotent stem cell generation. Nature 2009; 460(7251): 49-52.
[http://dx.doi.org/10.1038/nature08180] [PMID: 19571877]

CHAPTER 3

Skin Stem Cells in Cutaneous Wound Healing and Tumorigenesis

Constantin Caruntu[1,2], **Daniel Boda**[3,*], **Ana Caruntu**[4] and **Caterina Longo**[5]

[1] *Immunology Department, "Victor Babes" National Institute of Pathology, Bucharest, Romania*

[2] *Department of Physiology, "Carol Davila" University of Medicine and Pharmacy, Bucharest, Romania*

[3] *Dermatology Research Laboratory, "Carol Davila" University of Medicine and Pharmacy, Bucharest, Romania*

[4] *"Dan Theodorescu" Oral and Maxillofacial Surgery Hospital, Bucharest, Romania*

[5] *Dermatology and Skin Cancer Unit, Arcispedale S Maria Nuova, Reggio Emilia, Italy*

Abstract: This chapter will highlight the main characteristics of processes like regeneration and tumorigenesis in skin. The chapter describes the skin regeneration pathways and it elaborates on interesting link regarding regeneration triggering tumorigenesis in cutaneous tissue. Stem cells that are the key player in non-melanoma and melanoma skin cancers will be described in separate sections.

Keywords: Regeneration, Skin, Stem cells, Tumorigenesis, Wound healing.

INTRODUCTION

Skin plays an essential protective role against physical, chemical or infectious, potentially harmful agents. It is also involved in thermoregulation, hemodynamic homeostasis, fluid and electrolyte balance, excretory function and has important metabolic activities. Another major role of skin is to provide information about the environment, functioning as a sense organ.

* **Corresponding author Daniel Boda:** Dermatology Research Laboratory, "Carol Davila" University of Medicine and Pharmacy, 22-24 Gr. Manolescu, 0111234, Sector 1, Bucharest, Romania; Tel: +40-757079117; E-mail: daniel.boda.umf@gmail.com.

Skin has a complex structure with various tissue types including epithelium, connective tissue, smooth muscle, fat, vessels and nerves. These are organized in separate layers with different embryonic origin, but an interdependent functionality.

Epidermis, the outer layer of the skin, is composed of stratified squamous epithelium of ectodermal origin, which functions as a protective coating. Beneath epidermis lies the dermis, a collagen-rich connective tissue of mesodermal origin, which is separated from epidermis by dermo-epidermal junction. It provides support, elasticity and nourishment for the skin. Below dermis is hypodermis, consisting of adipose tissue, is also derived from embryonic mesoderm, with important protective and metabolic functions. Hair follicles, sebaceous glands and sweat glands, known as appendages of the skin, are also components of cutaneous tissue [1, 2].

Due to its multiple functions and the permanent contact with aggressive environmental factors, skin undergoes continuous renewal and has a high regenerative potential [3], processes that depends on the numerous stem cell populations residing in various skin compartments [4]. Skin stem cells are the cells of origin for all terminally differentiated cells in cutaneous tissue [5] and are essential for preserving the genetic information and retaining the proliferative potential of the skin. They are relatively undifferentiated, have the capacity to self-renewal and the ability of differentiation into different cell types. Skin stem cells have a long life span, are slow cycling, but their quiescence may turn into a high proliferation ability when exposed to certain stimuli [6 - 14].

The skin stem cells divide asymmetrically, resulting in a daughter stem cell, and a committed progenitor cell, known as transit amplifying cell, with a limited proliferative potential. Transit amplifying cells are rapidly cycling, increasing the number of cells available for subsequent maturation. After a few division cycles, their daughter cells, called post-mitotic cells, no longer divide and undergo terminal differentiation [2, 9 - 12, 15 - 18]. Skin stem cells can also divide symmetrically, to generate two identical daughter cells [19, 20]. The proliferation and differentiation processes arising from stem cells maintain tissue homeostasis in all skin compartments [21].

Cutaneous tissue is rich in stem cells, it can be explored easily and research on skin stem cells has known a rapid development [12, 22, 23]. Skin stem cells are located in niches, which are distinct regions with specific microenvironment conditions that have a strong influence on their activities [24, 25].

In epidermis, stem cells are found in the basal layer, but their precise location varies from one region to another. In glabrous skin of palms and soles, epidermal stem cells are located in the deep part of the rete ridges, an area less exposed to aggressive environmental factors [13, 26]. In hairy skin epidermis, stem cells are found at the tips of dermal papillae; very close to the underlying blood vessel. This location favors their nourishment. [10]. Other skin stem cells populations reside in various regions such as the secretory segments of sweat glands, sebaceous glands, germinal matrix of hair follicles, and particularly the bulge region of hair follicles [10, 12 - 14, 18, 19, 27].

Dermis and hypodermis also contain multiple stem cells populations located mainly around hair follicles and blood vessels [21, 28, 29].

The heterogeneous skin stem cell pools accomplish multiple functions, being the cellular base for tissue development, renewal, and repair [3, 24]. Their activities show circadian oscillations and are controlled by complex regulatory interactions between intrinsic mechanisms, local signals from microenvironment and different extrinsic factors [30 - 32].

Interfollicular Epidermal Stem Cells

Interfollicular epidermis is the most superficial layer of skin, located between hair follicles, and is composed of keratinized stratified squamous epithelium [14]. The vast majority of epidermal cells are keratinocytes. These cells are in a continuous transition from deep regions to the surface of the epidermis. They are arranged in layers and change their differentiation degree from one layer to the next. In epidermis, the differentiation process is mainly defined by keratinization. Cells from the inner basal layer pass through the spinous and then granular layers to reach the superficial layer as flat cornified cells, full of keratin, lacking nucleus, and eventually are discarded from the surface [10, 18, 19]. Epithelial homeostasis is maintained by the fine balance between cell production in deeper layers and cell loss at the surface. Since epidermal turnover time is short, the progenitor compartment of epidermis, located in the basal layer, is in a constant intense activity [8].

The maintenance of interfollicular epidermis was explained using the epidermal proliferative units model. According to this hypothesis, epidermal cells are structured in ordered columnar units of epithelial tissue, each one derived from a single stem cell. The epidermal proliferative unit contains a central self-renewing stem cell, which gives rise to transiently amplifying cells that form a cluster of about ten basal cells and their maturing progeny, which migrate into the supra-basal layers. [9 - 1, 33 - 36]. However, recent studies have suggested alternative models, in which each progenitor basal cell can undergo both asymmetric and symmetric divisions, generating one progenitor basal cell and one differentiated progeny, two daughter progenitor cells or two differentiated progeny, depending on physiological needs of the tissue [10, 19, 20, 37 - 39].

It is assumed that the complex communication between progenitor compartment of interfollicular epidermis and basement membrane with a rich content of growth factors and extracellular matrix proteins plays an important role in these processes [10, 11, 16, 17, 40].

Different studies, using various techniques, led to a wide range of estimates for epidermal stem cells population which varied between 0.01% [41] and 10-12% of the basal layer cells [36, 42 - 44].

Recent research pointed out potential surface markers for human interfollicular epidermis stem cells that include β1 integrin extracellular matrix receptors, delta-like protein1 (DLL1), melanoma chondroitin sulfate proteoglycan (MCSP), leucine-rich repeats and immunoglobulin-like domains protein 1 (LRIG1) and cluster of differentiation 46 (CD46) [45 - 49]. Mouse interfollicular epidermis stem cells could be identified using leucine-rich repeat-containing G-protein coupled receptor 6 (LGR-6) [50, 51].

It has also been outlined the close connections between interfollicular epidermis stem cells and other skin structures. Thus, LGR-6+ stem cells that can be identified above the hair follicle bulge may participate to long-term epidermal progenitor cells in intact skin. Under certain mesenchymal signals, interfollicular epidermis stem cells are able to give rise to hair keratinocytes or sebocytes, showing their multilineage differentiation ability [18, 52].

Hair Follicle Stem Cells

Hair follicles are skin appendages with an important role in ultraviolet (UV) radiation, thermal and mechanical protection. They are present in almost entire skin surface, except palms, soles, and certain regions in genital area. The density of hair follicles varies, the highest being on scalp and face.

Hair follicles have the deepest portion located at dermo-hypodermal junction, and after an oblique path through dermis and epidermis they open on skin surface through a follicular ostia.

From surface to depth, hair follicle can be divided into several segments with distinct anatomical limits [53]:

1. acrotrichium - intraepidermal part of hair follicles, extending from follicular ostia to dermo-epidermal junction;
2. infundibulum, located between the dermo-epidermal junction and the opening of sebaceous gland;
3. isthmus, located between the opening of sebaceous duct and the bulge - insertion site of arrector pili muscle;
4. lower follicle, extending down to the hair bulb;
5. hair bulb - the deepest part of hair follicle that surrounds the dermal papilla.

Hair follicles are dynamic structures, in a constant cyclic transformation, with three distinct stages: anagen (active growing stage), catagen (regression stage) and telogen (inactive, resting stage). In the upper segments of hair follicle, forming the permanent portion, histological structure is relatively stable. However, the lower follicle and the hair bulb form the transient portion, that changes during hair follicle cycle [13, 54]. These cyclic changes are driven by stem cells of the hair follicle (see Fig. **1**), especially populations located within the bulge, a region with numerous nerve fibers, rich vascularization, that offers protection and a complex symphony of signals for residing cells [20, 25, 55]. The bulge stem cells are quiescent in most phases of follicular cycle but are activated and self-renew when a new cycle is initiated [20, 31, 56 - 58].

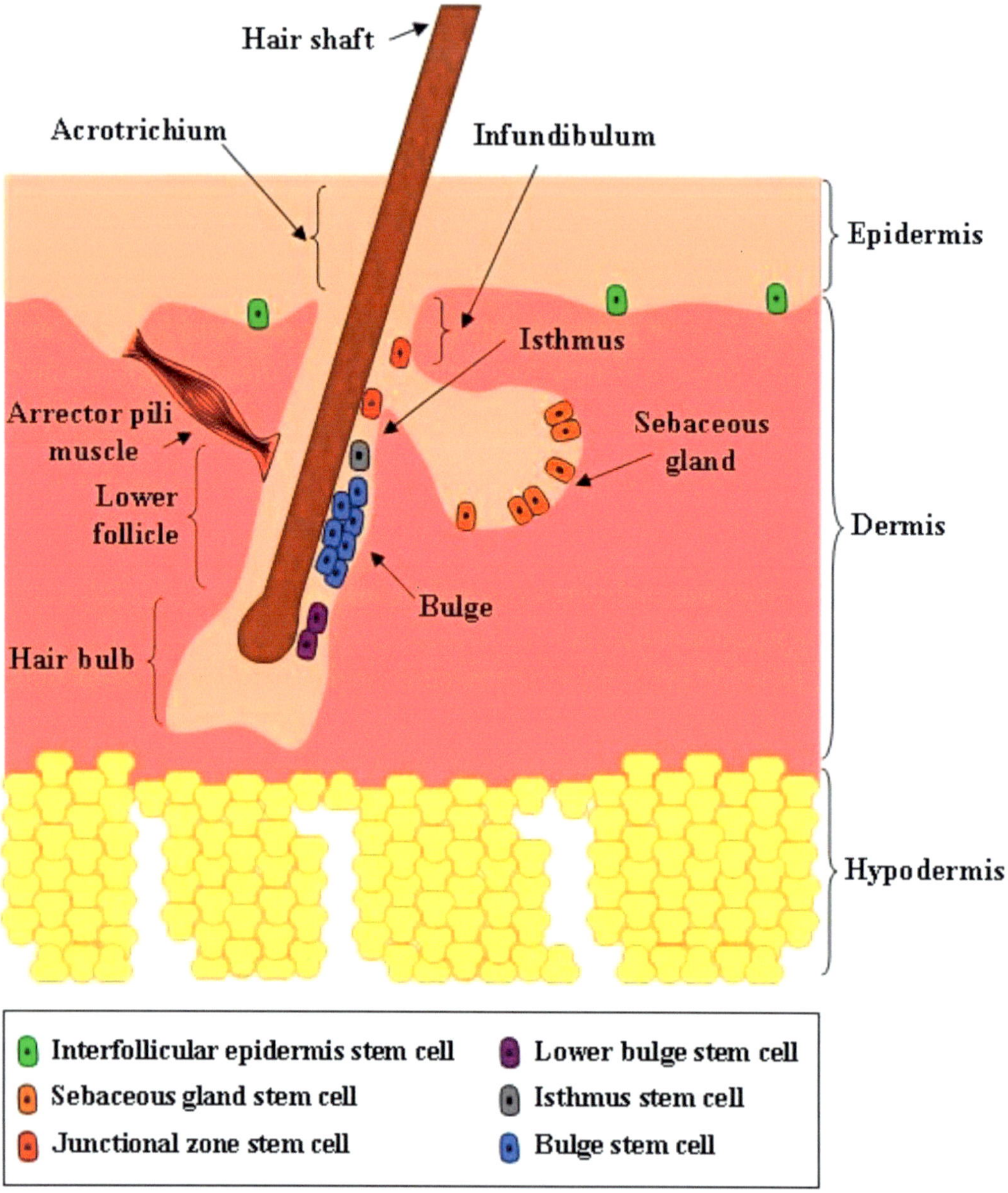

Fig. (1). Hairy skin contains several stem cell populations residing in interfollicular epidermis, sebaceous glands, isthmus and junctional zone adjacent to the infundibulum and especially in the bulge region of hair follicles.

The first type of hair follicle stem cells was identified in the bulge area [56]. Subsequently, various studies mostly using rodent experimental models, described other hair follicle stem cell populations with different proliferative potential and distinct anatomical locations, such as hair germ, isthmus and junctional zone [20, 59, 60].

Hair follicle stem cells can be identified on the basis of molecular markers, such as keratin 15 (KRT15), clusters of differentiation 34 (CD34), leucine-rich repeat-containing G-protein coupled receptor 5 (LGR5), leucine-rich repeats and immunoglobulin-like domains protein 1 (LRIG1); placenta-expressed transcript 1 protein (PLET1), leucine-rich repeat-containing G-protein coupled receptor 6 (LGR6) and glioma-associated oncogene homolog 1 (GLI1) [61 - 67].

In murine skin, CD34 and KRT15 are widely used to recognize bulge stem cells, and there is a significant overlap between expressions of these two markers [51]. KRT15+ cells are able to regenerate hair follicles and sebaceous glands [68, 69]. After skin wounding, they are able to transiently generate epidermal keratinocytes [70].

During telogen, LGR5+ cells are present in the lowermost portion of hair follicle, consisting of the lower bulge region and the hair germ, and throughout anagen they reside in the deep part of the outer root sheath. In normal mouse skin, they only participate in hair follicle regeneration, and do not seem to take part in interfollicular epidermis renewing or sebaceous glands regeneration [64].

In mouse epidermis, stem cells that co-express LRIG1 and PLET1 were identified in hair follicle isthmus and its junctional zone adjacent to the infundibulum and sebaceous gland opening [65]. LRIG1+ stem cells are multipotent and may contribute to all epidermal lineages in the skin. They participate predominantly in the maintenance of infundibulum and sebaceous gland, but upon wounding LRIG1+ stem cells make permanent contributions to interfollicular epidermis regeneration [50].

Experimental results show that in mouse skin, stem cells expressing LGR6 are located above the follicle bulge, but they can also be found in the lower follicle, sebaceous gland and germinative layer of epidermis. LGR6+ stem cells can generate all epidermal cell lineages, contributing to homeostasis of hair follicle upper portion, sebaceous gland, and interfollicular epidermis. Upon injury, LGR6+ stem cells have a long-lasting contribution to wound healing [66].

GLI1+ stem cells were identified in mouse skin, in upper and lower bulge and in the epithelial hair germ, in close proximity of sensory nerve endings. Recent results show that the activity of GLI1 expressing stem cells is modulated by sonic hedgehog (SHH) signaling from perifollicular nerves. GLI1+ stem cells are multipotent and contribute to homeostasis of hair follicle and regeneration of interfollicular epidermis in response to wounding [67].

Melanocyte stem cells are a distinct population that is found mostly in the bulge region of hair follicle. They are quiescent and function as melanocyte reservoir. Upon activation, melanocyte stem cells generate pigment-producing mature melanocytes [14, 21, 71].

Within this diversity of hair follicle stem cells populations there is certain overlapping and a high degree of plasticity in response to signals from microenvironment [60]. Future studies are required for a better understanding of hair follicle stem cells functioning.

Another important issue is that most of the initial studies on follicular stem cells were performed on murine skin, which is very different, both structurally and functionally, from human skin. Thus, future research should clarify the structural and functional details of human skin stem cells.

Stem Cells of Sebaceous Glands

Sebaceous glands have an acinar structure, with sebaceous lobes connected to an excretory duct that opens to the adjacent hair follicle. Sebocytes are predominant cells in sebaceous glands and their maintenance requires the presence of a stem cell population which gives rise to terminally differentiated cells [18, 20]. Recent research suggested the existence of sebaceous stem cells at the base of sebaceous gland. These cells express B lymphocyte maturation-induced protein 1 (BLIMP1) which acts as a transcription repressor [72]. However, other studies showed that BLIMP1 is also expressed by terminally differentiated sebocytes, and is not suitable for the identification of sebocyte progenitor cells [28].

Stem Cells of Sweat Glands

Sweat glands have a secretory unit, consisting of basal myoepithelial cells and suprabasal luminal cells, and an excretory duct, which opens on skin surface. Distinct progenitor cell populations maintain the homeostasis of sweat gland secretory component and excretory duct. Myoepithelial and luminal progenitors behave as unipotent stem cells, instead ductal progenitors may participate in epidermal wound healing [51, 73]. Experimental studies showed that stem cells from sweat glands may generate stratified epidermis [74].

Dermal Stem Cells

Dermis is the connective layer between epidermis and subcutaneous fat, which provides structural support to the skin. It incorporates skin appendages and contains a rich vascular plexus and numerous nerve fibers. It is composed of collagen, elastic fibers, extra-fibrillar matrix and contains multiple distinct cell types, such as fibroblasts, dendritic cells, macrophages, lymphocytes and mast cells. In order to maintain tissue homeostasis, dermis harbors a large pool of adult stem cells placed in perifollicular connective tissue, in dermal papillae, among pericytes surrounding blood vessels [4, 28, 75]. They are multipotent mesenchymal stem cells that have a potential to give rise to fibroblasts, smooth muscle cells, pericytes, adipocytes, tenocytes, osteoblasts, chondrocytes, hematopoietic cells, and even neurons and glia [21, 28, 76 - 80]. Thus, apart from their contribution in tissue maintenance, dermal stem cells may have a major role in wound healing and regenerative medicine.

SKIN STEM CELLS IN REGENERATION AND WOUND HEALING

Skin wound healing is a dynamic regenerative process that can be divided into three phases: inflammatory, proliferative and remodeling (see Fig. **2**). It involves a sequence of well-coordinated events with significant interactions between various local and recruited cells, leading to a substantial remodeling of all skin compartments [1, 25, 80 - 82].

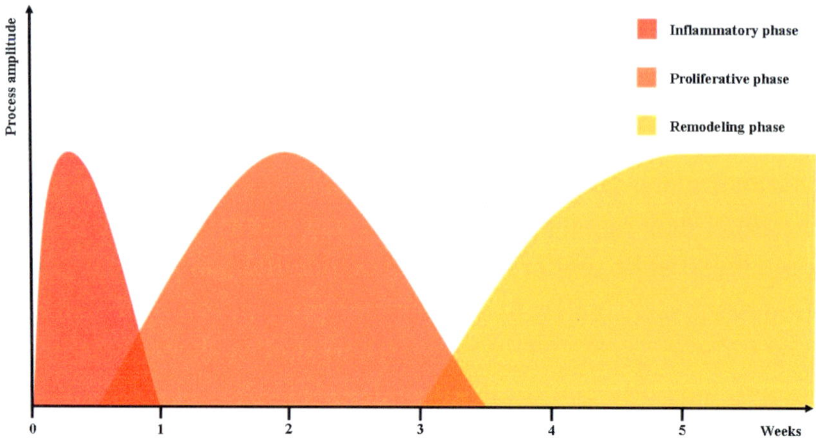

Fig. (2). Phases of skin wound healing.

Stem cells play an essential role in each of wound-healing phases (see Fig. **3**), responding to injury by activation, increased proliferation and release of biologically active factors that induce tissue regeneration and repair [1, 13, 80 - 83].

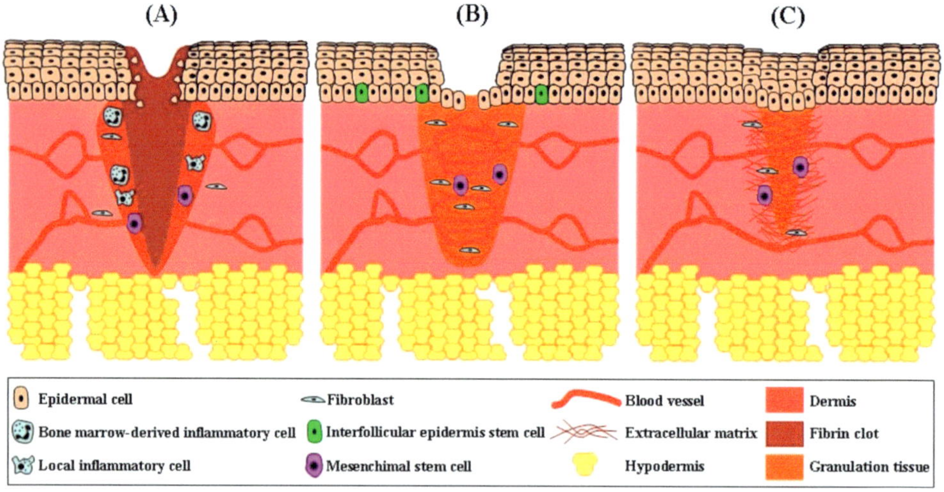

Fig. (3). Role of skin stem cells in wound healing. **(A)** In the inflammatory phase of wound healing, immune cells are recruited to the wound site; mesenchymal stem cells have immunomodulatory and antibacterial actions; (**B**) In the proliferative phase, circulating and local stem cells are recruited, promoting angiogenesis, granulation tissue formation and reepithelialization; (**C**) During the remodeling phase, mesenchymal stem cells promote an organised rearrangement of extracellular matrix.

Inflammatory Phase

The acute inflammatory phase starts very soon after lesion. Following hemostasis with platelet adhesion, activation and aggregation and formation of a fibrin clot, immune cells are attracted to the injury site in order to prevent infection and clear cell and matrix debris. They also secrete cytokines and growth factors that promote tissue formation.

Both *in vitro* studies on murine and human cell lines, and *in vivo* research using mouse models show that in this phase of wound healing, mesenchymal stem cells have immunomodulatory and antibacterial actions [80, 84, 85]. They reduce immune cells production of proinflammatory factors, such as tumor necrosis factor α (TNF-α) and interferon γ (IFN-γ), and increase their secretion of anti-inflammatory cytokines interleukin-10 (IL-10) and interleukin-4 (IL-4). They have also intrinsic antimicrobial properties induced by the production of human cathelicidin hCAP-18/LL-37 [85] and promote immune cells phagocytic properties [84]. Other studies suggested that mast cell precursors from the connective tissue sheath of hair follicles could be another population involved in inflammatory stage of wound healing [86].

Proliferative Phase

In the next proliferative phase, new blood vessels are formed, granulation tissue covers the wound and keratinocytes proliferate, migrate and produce re-epithelialization.

In this stage of wound healing circulating and local stem cells can be recruited to the wound site in response to chemotactic signals. Stem cells enhance their proliferative activity and, by paracrine signals, they stimulate keratinocytes, fibroblasts and endothelial cells migration and proliferation, promoting angiogenesis and further tissue repair [25, 80, 87 - 92].

Results from murine models suggest that the main progenitor and stem cell populations that are involved in wound healing associated neovascularization are tissue-resident endothelial precursors from dermis and adipose tissue [93], pluripotent nestin+ cells from the bulge region [94] and circulating endothelial progenitor cells [95].

Bone marrow-derived stem cells were identified in the skin and both *in vitro* research on mouse and human cell lines and *in vivo* studies using mouse models showed that they are able to participate along with dermal mesenchymal cells to reestablish the dermal fibroblast population [14, 96 - 99] and to stimulate their proliferation and migration [87].

Over the last few years, many studies have provided evidence, mostly in mouse models, that several stem cell populations from different locations can participate in the re-epithelization that occurs in the proliferative phase of wound healing. In this process, the most important role is played by interfollicular epidermis and hair follicle stem cells [81, 100 - 102].

Different hair follicle stem cells pools have distinct contributions to wound re-epithelialization [20]. Thus, KRT15+ bulge stem cells contribute especially in the early stages of wound reepithelialization, but their involvement is temporary, being replaced by other epidermal clones in several weeks [70, 86, 103, 104]. As a reaction to injury, LGR5 expressing cells migrate to the interfollicular epidermis where their progeny can still be identified after more than one year [64]. Other stem cells populations that contribute to long-term regeneration of epidermis are located in the infundibulum and isthmus of the hair follicle and express molecular markers, such as LRIG1, LGR6 and GLI1 [65 - 67, 70, 101, 104, 105].

Sweat glands ductal progenitors also contribute to wound re-epithelialization, especially around the duct openings [73].

Moreover, upon wound healing, dermal mesenchymal multipotent progenitors and bone marrow-derived stem cells are able to differentiate into epidermal cells and may also contribute to epidermal regeneration [106 - 109].

Remodeling Phase

During the phase of maturation and remodeling, new components of extracellular matrix are produced and remodeled in order to strengthen the tissue. As well, unnecessary cells and structures are removed in order to reduce the scar volume. Mesenchymal stem cells promote an organized rearrangement of extracellular matrix. They also induce an increased production of collagen type 1 by dermal fibroblasts [80, 87].

In this last phase of wound healing, sometimes new skin appendages can form and repigmentation may occur [86]. Hair follicle neogenesis was evidenced as a late event in murine skin wound healing. It involves the contribution of interfollicular epidermis and hair follicle LGR6+ stem cells and requires activation of WNT signalling [66, 81, 82, 110 - 112]. Also, recent results showed that in mouse skin sweat gland myoepithelial stem cells can reproduce the entire structure of the gland suggesting that this cell population may support sweat glands regeneration after injury [1, 73].

Finally, melanocyte stem cells residing in the hair follicle appear to be at the origin of melanocytes that induce epidermal repigmentation [86].

LINKING REGENERATION TO TUMORIGENESIS

In cutaneous tissue, there is an established link between injury, inflammation, and carcinogenesis. As described above, the fundamental distinction between these two processes is that wound healing is a self-limiting process while cancer development is a complex mechanism of constitutive pathway activation [83, 113] (see Fig. **4**). Nevertheless, we can draw some common characteristics between these processes. Thus, both processes use common molecular and signaling pathways, such as Ras, Hedgehog (HH) and WNT [82, 114, 115]. HH and WNT pathways can be impaired both in tissue repair and carcinogenesis. Hence HH signaling was connected to promotion of basal cell carcinoma (BCC), while abnormal WNT pathway was reported in several other epidermal tumours [82].

Various cell types, such as epithelial, mesenchymal and immune cells, interact intimately, in order to develop wound repair or tumor formation. All changes of cellular microenvironment induced by tissue injury can lead to the development of a tumor as well. In this complex microenvironment, stem cells interact with the influx of immune cells programmed to induce regeneration, with new blood vessel formation, with proliferating fibroblasts and remodeling of the extracellular matrix [82]. In each of these dynamic processes, stem cells are activated and may embrace a new spatial location, different from the one adopted in undamaged tissue. Henceforth during both wound healing and tumorigenesis, stem cells act outside their normal niche, they interact actively to each other and to a microenvironment that contains mesenchymal cells, bone marrow-derived cells and immune cells [25, 82].

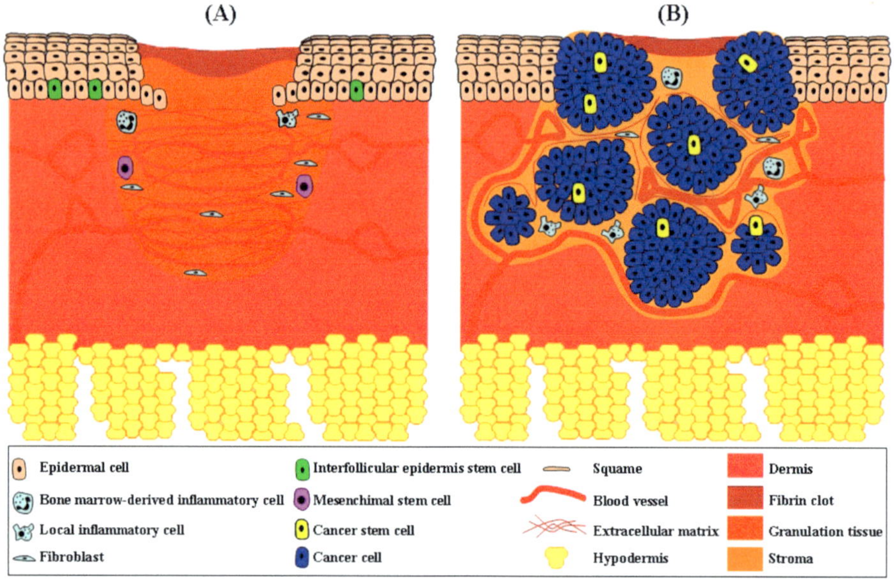

Fig. (4). Comparative representation of **(A)** skin wound healing and **(B)** development of skin cancer. In both processes, different stem cells interact intimately with epithelial, mesenchymal and immune cells, inducing keratinocyte migration and proliferation, increased number of activated fibroblasts, extracellular matrix remodeling, new blood vessel formation, in order to develop wound repair or tumor formation.

The bulge – the main follicular stem cells niche, comprises a heterogeneous population of multipotent cells with self-renewing capacity. Follicular stem cells are activated by intrinsic molecular pathways, their activity and characteristics being influenced by different extrinsic cell types. The fine balance between intrinsic and extrinsic factors guides the function of bulge cells, maintaining homeostasis of cutaneous tissue, developing skin renewal and possible carcinogenesis [116].

Epithelial-mesenchymal transition is a recently reported process in which epithelial cells lose some of their innate characteristics (cell polarity and cell-cell adhesion) and are endowed with migratory and invasive properties becoming mesenchymal stem cells [117]. Regarding this process, there have been highlighted remarkable resemblances between wound healing and tumorigenesis [81, 82]. During wound reepithelialization, epidermal keratinocytes acquire migratory phenotypes [81, 118, 119], but the migration of epithelial cells diminishes in intensity and returns to normal as the basement membrane is rebuilt. In tumors, these processes are uncontrolled. Epithelial cells can harbor oncogenic mutations that lead to immortalization, undergoing epithelial-mesenchymal transition associated with cancer stem cells properties acquisition [120]. Transdifferentiation is another recently discovered cellular process in which, a mature somatic cell changes directly into another type of mature somatic cell without passing through an intermediate state of pluripotent or progenitor cell type [121]. It is unlikely that lineage conversion from one cell type to another to be a common process in adult tissues, but upon wound healing mesenchymal stem cells may undergo transdifferentiation into epidermal, endothelial cells and pericytes [82, 107, 122].

Another link between wound healing and tumorigenesis is related to the excessive collagen synthesis by fibroblasts leading to fibrosis and scar formation. This fibrotic connective tissue is a tumor permissive microenvironment [123]. Moreover, the microenvironment shows remarkable resemblances between the multitude of growth factors, cytokines and chemokines released during tissue repair and those present in cancer development, with slight expression differences in terms of kinetics [114]. If one intends to prevent any tumor growth at wound site, the logical clinical approach is to block inflammation.

In conclusion of this section, we can draw attention on several aspects. There is a close relationship between chronic injury, inflammation and tumorigenesis [113], cancer occurrence being possible, though infrequently, at the site of chronic skin lesions [82, 124]. Skin stem cells are involved in both wound healing and tumor formation by activation, proliferation and migration of their progeny. An efficient tissue repair is crucial for skin homeostasis and even for organism survival, while a defective wound repair may lead to skin tumorigenesis [82]. Lessons learned from understanding the intimate mechanisms of carcinogenesis can also bring into focus the epigenetics of wound healing. Recent studies suggesting different ways to employ stem cells as therapeutic tools in wound repair follow the same pathways involved in the suppression of tumorigenesis [82].

CANCER STEM CELL – MAJOR PLAYER IN SKIN TUMORIGENESIS

Skin stem cell populations, as previously described above, play a key role in skin homeostasis and are essential for the regeneration of cutaneous tissue. When its normal cycle of division and death is disrupted by various genetic, epigenetic and/or proteomic deregulations, the overall physiology of skin stem cell is disturbed. Thus, the cells can be pushed down the road towards tumorigenesis. Clear experimental findings have shown that skin tumors can comprise highly tumorigenic cancer cell subpopulations with features resembling stem cells. Cancer stem cells have been identified in BCC and squamous cell carcinoma (SCC). Moreover, using these cells, skin cancer cell lines have been established [125 - 136]. Some studies have even suggested the presence of cells with cancer stem cell properties in melanoma, but this issue is still a matter of debate.

Mechanisms that Generate Non-Melanoma Skin Cancer

As previously described, the permanent renewal of epidermis is sustained by the activity of somatic stem cells. Similarly, skin cancer is triggered by the emergence of specific tumor-initiating stem cells that accumulate genetic alterations exceeding physiological mechanisms of control. Long-lived skin stem cells can accumulate oncogenic mutations over long periods of years or decades.

Even if it is a subject of extensive research, the link between epidermal stem cell, its specific microenvironment and skin tumorigenesis is still only partially elucidated [60].

Stem Cells: at the Root of Human Skin Carcinomas

In both BCCs and SCCs, the cell that initiates tumor development is likely to be a cancer stem cell, although this model requires future research. The main argument for this model is that the epidermal homeostasis is based on keratinocytes that emerge from stem cell compartment, differentiate and eventually form dead horn squames that are discarded from the skin surface. If a seldom-dividing epidermal stem cell that has a long-lived nature accumulates multiple genetic deregulations, this can induce an aberrant cell cycle and may allow neoplastic growth [60].

Recent studies have shown that accumulation of genetic mutation can reside in extensive UV radiation exposure that can trigger genetic and/or epigenetic alterations. All these genetic-related phenomena may lead to sustained changes in the activation of several signaling pathways, such as telomerase reverse transcriptase (TERT), HH, epidermal growth factor receptor (EGFR) and/or WNT/β-catenin which may affect skin stem/progenitor cells and their daughter cells.

In both BCCs and SCCs, various studies have suggested that different stem cell populations residing in hair follicles or interfollicular epidermis may be at the origin of tumor development [137 - 141]. Extensive research has shown that malignant transformation of skin multipotent epithelial stem cells or their early progenies may be at the root of certain BCC and SCC subtypes [30, 142 - 146]. The initiating cells may be found in the basal layer of interfollicular epidermis and/or in the bulge region of the hair follicles [126, 132, 133, 147 - 150].

Skin Cancer Microenvironment

Skin carcinomas are contingent on a cell-instructive and functional microenvironment [60]. Stem cells niches, deregulated upon aging and malignant transformation can generate signals that influence stem/progenitor cells located in epidermis and hair follicles. These findings have contributed to a better understanding of complex mechanisms involved in various human skin disorders and particularly in skin cancers.

Skin stem/progenitor cells and their progenies may suffer important DNA alterations and genomic instability induced by various factors, such as extensive exposure to UV radiation, increased oxidative stress, chronic inflammation, age-related telomere attrition. The induced alterations can hinder regenerative and self-renewal capacity and may impair activation of tumor suppressor gene products. Photoaging, telomerase re-activation and action of various oncogenic factors that affect skin stem/progenitor cells can lead to malignant transformation. These findings suggest that reduction of inflammation, anti-oxidant treatments, and stem cell-replacement may represent new and promising therapeutic approaches for skin cancer [133, 145 - 147, 151 - 158].

In conclusion of this section, development of skin cancers is conveyed by a broad range of alterations affecting skin stem/progenitor cells, their daughter cells and local microenvironment. An array of paracrine mitotic factors secreted by activated stromal cells may increase malignancy of skin cancer cells [159 - 161]. Targeting specific molecules as oncogenic products in skin cancer stem/progenitor cells and their niches opens new perspectives for cancer therapies [158].

Stem Cells as Triggers Of Tumorigenesis in BCC

BCC is the most common form of cancer worldwide. One of the most documented deregulations in BCC involves the HH pathway [162], a signaling system which actually has a significant contribution in the normal hair follicle cycling [82, 163].

In mice, experimental cutaneous injuries exacerbate BCC malignant progression [164, 165], this finding being consistent with the increased risk of cancer associated with chronic wounds [1]. Probably when wound microenvironment is generated, this process stimulates cancer stem cell migration and tumor promotion [137, 164, 165].

The cellular origin of BCCs has recently been described with the aid of genetic mouse models based either on lineage-tracing or on activation of oncogenic HH pathway in different cell subsets [166, 167]. The originating cell can determine the subtype of BCC, the way the tumor will appear and the incidence of tumor development [137, 168].

By expressing in adult mice epidermis a constitutively activated Smoothened (*Smo*) mutant *SmoM2* that activates HH signaling of epidermal cell populations, it was reported the spontaneous development of BCCs. The tumor-initiating cells were found among the stem/progenitor cells located in interfollicular epidermis or upper infundibulum of hair follicles [137]. Conversely, the stem cells residing in the bulge region of hair follicle do not form tumors if they are not recruited to an injury site [165]. In another study, X-ray induced BCC in mice *Ptch1(+/--)* originated from the KRT15-expressing stem cells of follicular bulge [138].

In *SmoM2* BCCs, the tumor-initiating cells located in the interfollicular epidermis are rapidly reprogrammed into an embryonic hair follicle progenitor-like fate [169], suggesting that the attempt to identify the origin of tumor cells using expression of specific differentiation markers can sometimes be misleading.

During this reprogramming process, the earliest molecular event includes the WNT/β-catenin signaling activation, pathway that is one of the main BCCs development triggers [148]. Recently, it has been revealed that BCC initiation associated with *SmoM2* is contingent on both WNT and EGFR pathways. When EGFR is pharmacological inhibited, the number and size of tumors decrease [25, 170].

Expression of the high mobility group box-comprising transcription factor SOX9 is fundamental for the formation of bulge stem cell compartment [171], and activated SOX9 is found in cutaneous neoplasms including BCC, and adnexal skin tumors such as trichoepitheliomas and trichilemmomas [2, 149].

Stem Cells and Wound-Induced Tumors

Regarding BCCs development, there is an interplay between SHH signaling, *Smo* mutation and wounding. Thus, while SHH is down regulated during wound healing in keratinocytes [101], *Smo* mutation and wounding are both necessary for tumor formation from bulge stem cells [165]. Another study revealed that the progeny of LGR5+ cells can be involved in the development of BCC independently of wounding, but it was also emphasized that wounding is required for the initiation of BCC in the interfollicular epidermis [164]. The stem cell translocation into the interfollicular epidermis associated with activation of HH signaling was revealed as a possible mechanism of BCC formation induced by tumor promoter 12-O-tetradecanoyl-phorbol-13-acetate (TPA) [164, 165].

It can be concluded that wounding is one of multiple causes of BCC, promoting BCC formation through stem cells recruitment into the interfollicular epidermis [70, 164, 165].

Perturbation of Stem Cell Behavior in BCC

As previously mentioned, translocation of bulge stem cells carrying BCC-promoting mutations into the interfollicular epidermis could lead to a rapid tumor induction [164, 165]. This process is sustained by the interfollicular epidermis ability to allow an increased cellular proliferation linked to oncogenes expression [50, 137, 139]. In the molecular pattern of BCCs development, the HH signaling pathway deregulations are highly incriminated [82, 163, 172].

In most human BCCs can occur point mutation of genes in SHH signaling pathway [21]. The pathogenesis of some inherited or sporadic human BCC subtypes involves an unrelenting activation of SHH signaling, probably owing to an excessive expression of SHH ligand. This HH signaling pathway, in cooperation with WNT/β-catenin cascade, has been associated with BCCs onset [146, 150, 158, 173 - 177]. Studying the early molecular events following tumor induction, it has been shown that the development of BCC summarizes early pilosebaceous unit morphogenesis in a very similar WNT/β-catenin flow dependent pattern [169].

The involvement of HH pathway in BCCs development was underlined *in vivo* in murine models which associate an altered expression of several key effectors, such as SHH ligand, GLI1 or 2 in skin stem/progenitor cells from hair follicles or germinative layer of epidermis. Disturbance of signaling pathways may lead to cellular hyper-proliferation and BCC formation [126, 148, 163, 178 - 183]. Intensive research has focused on several transcription factors associated with HH pathway in an attempt to define the genetic background of BCCs. Activation of GLI2 transcription factor - the major mediator of HH pathway in bulge cells, results in BCC development [168]. Moreover, activated SMO receptors in bulge cells alongside with activation of HH pathway in the interfollicular epidermis can induce BCC [137]. It was also suggested that BCC formation relies more on the magnitude of HH pathway activation than on cell types that could generate BCC.

In a mouse model of NOTCH1 receptor ablation, a sustained expression of GLI2 was emphasized in epidermis, associated in early stages with epidermal hyperplasia and further with the development of BCC-like tumors [184]. Thus, it could be suggested that NOTCH1 could operate as an epidermal tumor-suppressor gene factor, counterbalancing the pro-tumoral character of HH pathway deregulation. In addition, it has been highlighted that SOX9 transcription factor, a possible target of HH signaling pathway, is strongly expressed in BCCs [149].

Other genetic alterations responsible for constitutive activation of HH pathway, such as loss-of-function mutations in PTCH1 receptor or activation in *Smo* gene have been associated with rare diseases such as Gorlin syndrome or accelerated aging disorder both of them prone to a high BCC incidence [158, 176, 178, 185 - 188].

As a preliminary conclusion, tumor-associated mutations affecting stem cells of the follicular bulge can induce tissue hyperplasia. The presence of stem cells with similar mutations in the interfollicular epidermis can potentially generate invasive carcinomas such as BCC [4, 137].

Stem Cell –Related Tumorigenesis in SCC

SCC is the second most common skin cancer after BCC and may show an aggressive behavior, being capable of metastasizing [189]. Many SCCs are proven to originate from interfollicular epidermis and follicular bulge. It is also highly probable that SCCs arise from stem/progenitor cells. As shown in the previous section, stratified epithelia undergo permanent regeneration and have an intrinsic rapid turnover. Thus, mutations that trigger SCC initiation can affect the resident stem cells that maintain epithelial homeostasis [190]. It was reported by several groups that SCCs are associated with mutations of *Hras*, *Tp53*, *Tgfb1b* and *Fos* [21, 191]. Moreover, it was demonstrated that oncogenic *Ras* transformation associated with a reduced action of tumor suppressors, such as p53, BRCA1 or transforming growth factor β (TGF-β) can be the molecular triggers for SCC [191].

Hair follicle stem cells can be at the root of cancer stem cell emergence in certain subsets of SCC. In humans, SCCs arising from the hair follicle comprise around 1% of all primary SCCs [192]. During the resting stage of the hair follicle cycle, oncogenic factors cannot trigger tumors since the mechanisms that induce dormancy of hair follicle stem cells dominate over the gain of oncogenic transformations or the reduction of tumor suppressors. Phosphatase and tensin homolog (PTEN) activity maintains the dormancy state, as its removal decreases tumor suppression without influencing proliferation [190]. Stem cell quiescence maintains tumor suppression in hair follicle stem cells, while PTEN probably has a major role in maintaining this resistance to carcinogenic stimuli [193]. It was reported that oncogenic *KRasG12D* is expressed in specific regions of the hair follicle [139, 140]. When the mutant *KRas* is expressed in stem cells of follicular bulge, hair germ and outer root sheath, but lacks in matrix cells, benign squamous skin tumors arise. When *KRasiG12D* expression is combined with p53 deletion invasive SCCs are initiated [25]. Another group showed in *K15.KrasG12D. Smad4–/–* mice models that the loss of *Smad4*, combined with the activation of *KrasG12D* in un-committed multipotent stem cells or in their progenitors, induces tumor development. All these cells can subsequently initiate tumor lineages expected to develop from bulge stem cells. This study proves that metastatic behavior is likely to be influenced by certain mutations arising in multipotent skin stem cells [116, 190].

Experimental models have shown that the ablation of β-catenin gene in CD34+ cancer stem/progenitor cells induces a complete tumor regression [132]. The suppression of p53 induces activation of EGFR/c-Jun pathway in keratinocytes, down-regulating Notch1-related differentiation pathway and promoting SCC development [144, 194]. However, specific involvement of epidermal stem cells that reside in the germinal layer of epidermis, near the basement membrane, in the initiation of various BCC or SCC subtypes is still a subject of investigation [158].

Epigenetic regulations such as the one performed by microRNA-9 (miR-9) has a contribution in side population proliferation and metastasis. The target of miR-9 is mainly α-catenin but not E-cadherin. Indeed, an increased miR-9 correlated with α-catenin loss was noticed in human metastatic SCC and similar results were obtained in primary tumors. Switching "on-off" miR-9 induced an increase or, on the contrary, a reduction of SCC invasiveness. [190].

Like in other cancer types, tumor-initiating cells are highly dependent on their niches, furthermore, the microenvironment is influenced by blood vessels and immune cells infiltration. Recently, cancer stem cells have been isolated from both mouse skin papillomas and SCCs [132, 195, 196]. Transplanted into a host recipient mouse, these cancer stem cells are able to induce formation of a new SCC [195].

In vivo lineage tracing has shown that cancer stem cells are resident in the tumor-stroma interface, expressing hence high levels of integrins [195 - 197]. Cancer stem cells are regulated by their microenvironment, TGF-β, signals from integrin and focal adhesion kinase orchestrating their proliferation [195]. Angiogenesis, an essential process in SCC development, can also be favored by cancer stem cells that express vascular endothelial growth factor (VEGF). Moreover, VEGF modulates the activity of cancer stem cells in an autocrine fashion [198]. Like in normal stem cells, the self-renewal and differentiation of cancer stem cells depend on autocrine and paracrine microenvironmental signals [1].

Regarding the extracellular trigger for SCC in humans, one causal carcinogen can be UV radiation. Like in other cutaneous tumors, UV radiation induces the initial strike that creates a favorable milieu for the accumulation of genetic alterations (*e.g.* p53 mutations array). Indeed, SCC occurs mainly on sun-exposed sites [199] and, as UVB radiation can reach the basal layer, interfollicular epidermis is the preferred target for UVB deleterious effect in human skin [60]. Interestingly, it was reported that mice having their interfollicular epidermis removed and further subjected to experimental UV-irradiation also developed SCCs. It is highly probable that in this case SCC originated from hair follicles, probably from bulge stem cells. This experimental finding showed that, at least in mice skin, UV penetration was sufficient to affect hair follicle cells [200].

Two-Stage Skin Carcinogenesis Mouse Model – Clues for Cancer Stem Cells Involvement in SCC Initiation

As in the experimental mice with interfollicular epidermis removed, two-stage skin carcinogenesis model is also suggesting a hair follicle origin of SCCs [60].

Chemically induced carcinogenesis, a widely used SCC model in mice [201, 202], has a first stage of "initiating" carcinogen application (usually 7,12-dimethylben--alpha-anthracene, DMBA) followed by chronic application of the "promoting" agent (usually TPA). This model, allowed us to reveal several important aspects regarding cancer stem cells involvement in SCC induction. One, already mentioned, is that SCC may arise from interfollicular epidermis [203], while interfollicular epidermis progenitors expressing mutant *KRas* or *HRas* can form papillomas [140, 204]. Thus, oncogenic Ras is a trigger for skin tumors but dependent on the targeted cell type [25]. Another interesting finding is that chemical carcinogens induce a significant increase of cells expressing the bulge marker CD34 [132, 82], cells that after transplantation can initiate secondary tumors exhibiting similar characteristics as the primary ones [25].

Several other molecular events were highlighted using this model, hence the identification of several direct SOX2 target genes that are involved in modulation of tumor cells stemness, survival, proliferative capacity, adhesion, invasiveness. SOX2 is a key transcription factor that favors tumor initiation and progression [205]. SOX4 probably operates as a transcriptional regulator important in adult tissue homeostasis which is also involved in maintaining cancer stem cells functions, hence its *knock-down* creates a carcinogenesis resistance pattern [206]. Another molecular player recently reported is myeloid ecotropic insertion site1 (Meis1) and its expression increases in relation to the benign-malign switch. During malignant transformation of skin cells, a functional shift occurs in Meis1 [207].

In conclusion, identification of *in situ* cancer stem cells may contribute to understanding of intrinsic and extrinsic mechanisms of skin carcinogenesis. These processes are involved in the regulation of cancer stem cells as tumor initiating cells, as inductors of tumor progression and probably as main cells related to therapy resistance of tumors. Large genomic studies can identify gene expression signatures of cancer stem cells involved in SCC induction and can bring new evidence regarding gene expression alteration that are contributory to SCC development. Moreover, genetic manipulation of cancer stem cells along with their niches could represent future therapeutic approaches [1, 195, 198, 208, 209].

Potential Role of Cancer Stem Cells in Melanoma – Still a Matter of Debate

Several studies suggested that cutaneous melanoma, a skin cancer with high metastatic capacity and resistance to current therapies, can have its origin in cells with cancer stem cell properties [127 - 130, 134, 135, 210 - 214]. Cancer stem/progenitor cells have been identified in small amounts and further isolated from primary and secondary tumors. In cell lines like human primary WM115, metastatic WM239A and murine B16F10 melanoma were also identified subpopulations with cancer stem cell characteristics [127 - 130, 134 - 136, 210 - 212]. Moreover, this cancer stem cells and progenitor cells had a specific biomarkers phenotype comprising CD133, CD44, CD20, TERT, nestin, NANOG and/or ABC transporters (MDR-1, ABCG2 and ABCB5) [158].

Traditionally, it was assumed that cancer cells arose from melanocytes [215] but, the hypothesis that melanoma originates from extrafollicular stem cells altered by factors such as UVA and UVB radiation has recently been drawn [215, 216]. Nevertheless, further experiments are necessary to elucidate the mechanisms that damage stem cell's DNA to sustain this hypothesis [215, 216]. Future research must establish if the root of skin melanoma resides in the malignant transformation of stem/progenitor cells from the follicular bulge, such as epidermal neural crest stem cells, melanocyte stem cells and/or their early progeny. Moreover, further studies may have a great therapeutic impact and could open new possibilities for multitargeted therapies aimed against the initiating cells and their progenies in all stages of the disease [158].

CONCLUSIVE REMARKS

Skin stem cells can regenerate any cutaneous structure and are key players in skin renewal. Impaired stem cells function and perturbation of signaling from their niches may affect the regenerative capacity of the skin, inducing aging and carcinogenesis.

The study of skin stem cells is a rapidly emerging field and a better understanding of the functioning mechanisms of stem cells and their interactions with the microenvironment opens new pathways for therapy of skin cancer, wound healing and regenerative medicine.

CONFLICT OF INTEREST

The authors confirm that they have no conflict of interest to declare for this publication.

ACKNOWLEDGEMENTS

This work was partially supported by PN-II-PT-PCCA-2013-4-1407 (Project nr. 190), PN-II-PT-PCCA-2013-4-1386 (Project nr. 185). Author CC has a post-doctoral fellowship supported by the Sectorial Operational Programme Human Resources Development (SOPHRD), financed by the European Social Fund and the Romanian Government under the contract number POSDRU 141531/2014.

REFERENCES

[1] Hsu YC, Li L, Fuchs E. Emerging interactions between skin stem cells and their niches. Nat Med 2014; 20(8): 847-56.
[http://dx.doi.org/10.1038/nm.3643] [PMID: 25100530]

[2] Sun X, Fu X, Han W, Zhao M, Chalmers L. Epidermal stem cells: an update on their potential in regenerative medicine. Expert Opin Biol Ther 2013; 13(6): 901-10.
[http://dx.doi.org/10.1517/14712598.2013.776036] [PMID: 23517553]

[3] Forni MF, Trombetta-Lima M, Sogayar MC. Stem cells in embryonic skin development. Biol Res 2012; 45(3): 215-22.
[http://dx.doi.org/10.4067/S0716-97602012000300003] [PMID: 23283431]

[4] Schepeler T, Page ME, Jensen KB. Heterogeneity and plasticity of epidermal stem cells. Development 2014; 141(13): 2559-67.
[http://dx.doi.org/10.1242/dev.104588] [PMID: 24961797]

[5] Cotsarelis G. Epithelial stem cells: a folliculocentric view. J Invest Dermatol 2006; 126(7): 1459-68.
[http://dx.doi.org/10.1038/sj.jid.5700376] [PMID: 16778814]

[6] Bickenbach JR, Mackenzie IC. Identification and localization of label-retaining cells in hamster epithelia. J Invest Dermatol 1984; 82(6): 618-22.
[http://dx.doi.org/10.1111/1523-1747.ep12261460] [PMID: 6725984]

[7] Morris RJ, Fischer SM, Slaga TJ. Evidence that a slowly cycling subpopulation of adult murine epidermal cells retains carcinogen. Cancer Res 1986; 46(6): 3061-6.
[PMID: 3698024]

[8] Cotsarelis G, Kaur P, Dhouailly D, Hengge U, Bickenbach J. Epithelial stem cells in the skin: definition, markers, localization and functions. Exp Dermatol 1999; 8(1): 80-8.
[http://dx.doi.org/10.1111/j.1600-0625.1999.tb00351.x] [PMID: 10206725]

[9] Fuchs E. Skin stem cells: rising to the surface. J Cell Biol 2008; 180(2): 273-84.
[http://dx.doi.org/10.1083/jcb.200708185] [PMID: 18209104]

[10] Schreder A, Pierard GE, Paquet P, Reginster MA, Pierard-Franchimont C, Quatresooz P. Facing towards epidermal stem cells (Review). Int J Mol Med 2010; 26(2): 171-4. [Review].
[PMID: 20596595]

[11] Barker N, Bartfeld S, Clevers H. Tissue-resident adult stem cell populations of rapidly self-renewing organs. Cell Stem Cell 2010; 7(6): 656-70.

[http://dx.doi.org/10.1016/j.stem.2010.11.016] [PMID: 21112561]

[12] Staniszewska M, Słuczanowska-Głąbowska S, Drukała J. Stem cells and skin regeneration. Folia Histochem Cytobiol 2011; 49(3): 375-80.
[http://dx.doi.org/10.5603/FHC.2011.0053] [PMID: 22038214]

[13] Terskikh VV, Vasiliev AV, Vorotelyak EA. Label retaining cells and cutaneous stem cells. Stem Cell Rev 2012; 8(2): 414-25.
[http://dx.doi.org/10.1007/s12015-011-9299-6] [PMID: 21744048]

[14] Eckert RL, Adhikary G, Balasubramanian S, *et al.* Biochemistry of epidermal stem cells. Biochim Biophys Acta 2013; 1830(2): 2427-34.
[http://dx.doi.org/10.1016/j.bbagen.2012.07.002] [PMID: 22820019]

[15] Janes SM, Lowell S, Hutter C. Epidermal stem cells. J Pathol 2002; 197(4): 479-91.
[http://dx.doi.org/10.1002/path.1156] [PMID: 12115864]

[16] Alonso L, Fuchs E. Stem cells in the skin: waste not, Wnt not. Genes Dev 2003; 17(10): 1189-200.
[http://dx.doi.org/10.1101/gad.1086903] [PMID: 12756224]

[17] Alonso L, Fuchs E. Stem cells of the skin epithelium. Proc Natl Acad Sci USA 2003; 100 (Suppl. 1): 11830-5.
[http://dx.doi.org/10.1073/pnas.1734203100] [PMID: 12913119]

[18] Zouboulis CC, Adjaye J, Akamatsu H, Moe-Behrens G, Niemann C. Human skin stem cells and the ageing process. Exp Gerontol 2008; 43(11): 986-97.
[http://dx.doi.org/10.1016/j.exger.2008.09.001] [PMID: 18809487]

[19] Shen Q, Jin H, Wang X. Epidermal stem cells and their epigenetic regulation. Int J Mol Sci 2013; 14(9): 17861-80.
[http://dx.doi.org/10.3390/ijms140917861] [PMID: 23999591]

[20] Plikus MV, Gay DL, Treffeisen E, Wang A, Supapannachart RJ, Cotsarelis G. Epithelial stem cells and implications for wound repair. Semin Cell Dev Biol 2012; 23(9): 946-53.
[http://dx.doi.org/10.1016/j.semcdb.2012.10.001] [PMID: 23085626]

[21] Dahl MV. Stem cells and the skin. J Cosmet Dermatol 2012; 11(4): 297-306.
[http://dx.doi.org/10.1111/jocd.12008] [PMID: 23174053]

[22] Barthel R, Aberdam D. Epidermal stem cells. J Eur Acad Dermatol Venereol 2005; 19(4): 405-13.
[http://dx.doi.org/10.1111/j.1468-3083.2005.01279.x] [PMID: 15987283]

[23] Gola M, Czajkowski R, Bajek A, Dura A, Drewa T. Melanocyte stem cells: biology and current aspects. Med Sci Monit 2012; 18(10): RA155-9.
[http://dx.doi.org/10.12659/MSM.883475] [PMID: 23018363]

[24] Clevers H, Loh KM, Nusse R. Stem cell signaling. An integral program for tissue renewal and regeneration: Wnt signaling and stem cell control. Science 2014; 346(6205): 1248012.
[http://dx.doi.org/10.1126/science.1248012] [PMID: 25278615]

[25] Doma E, Rupp C, Baccarini M. EGFR-ras-raf signaling in epidermal stem cells: roles in hair follicle development, regeneration, tissue remodeling and epidermal cancers. Int J Mol Sci 2013; 14(10): 19361-84.
[http://dx.doi.org/10.3390/ijms141019361] [PMID: 24071938]

[26] Kaur P. Interfollicular epidermal stem cells: identification, challenges, potential. J Invest Dermatol 2006; 126(7): 1450-8.
[http://dx.doi.org/10.1038/sj.jid.5700184] [PMID: 16543901]

[27] Fuchs E, Horsley V. More than one way to skin. Genes Dev 2008; 22(8): 976-85.
[http://dx.doi.org/10.1101/gad.1645908] [PMID: 18413712]

[28] Sellheyer K, Krahl D. Skin mesenchymal stem cells: prospects for clinical dermatology. J Am Acad Dermatol 2010; 63(5): 859-65.
[http://dx.doi.org/10.1016/j.jaad.2009.09.022] [PMID: 20471137]

[29] Lindroos B, Suuronen R, Miettinen S. The potential of adipose stem cells in regenerative medicine. Stem Cell Rev 2011; 7(2): 269-91.
[http://dx.doi.org/10.1007/s12015-010-9193-7] [PMID: 20853072]

[30] Petersson M, Niemann C. Stem cell dynamics and heterogeneity: implications for epidermal regeneration and skin cancer. Curr Med Chem 2012; 19(35): 5984-92.
[http://dx.doi.org/10.2174/092986731209065984] [PMID: 22963565]

[31] Fuchs E, Chen T. A matter of life and death: self-renewal in stem cells. EMBO Rep 2013; 14(1): 39-48.
[http://dx.doi.org/10.1038/embor.2012.197] [PMID: 23229591]

[32] Janich P, Toufighi K, Solanas G, *et al.* Human epidermal stem cell function is regulated by circadian oscillations. Cell Stem Cell 2013; 13(6): 745-53.
[http://dx.doi.org/10.1016/j.stem.2013.09.004] [PMID: 24120744]

[33] Mackenzie IC. Relationship between mitosis and the ordered structure of the stratum corneum in mouse epidermis. Nature 1970; 226(5246): 653-5.
[http://dx.doi.org/10.1038/226653a0] [PMID: 5444930]

[34] Potten CS. The epidermal proliferative unit: the possible role of the central basal cell. Cell Tissue Kinet 1974; 7(1): 77-88.
[PMID: 4129708]

[35] Mackenzie IC, Zimmerman K, Peterson L. The pattern of cellular organization of human epidermis. J Invest Dermatol 1981; 76(6): 459-61.
[http://dx.doi.org/10.1111/1523-1747.ep12521107] [PMID: 7240792]

[36] Mackenzie IC. Retroviral transduction of murine epidermal stem cells demonstrates clonal units of epidermal structure. J Invest Dermatol 1997; 109(3): 377-83.
[http://dx.doi.org/10.1111/1523-1747.ep12336255] [PMID: 9284108]

[37] Clayton E, Doupé DP, Klein AM, Winton DJ, Simons BD, Jones PH. A single type of progenitor cell maintains normal epidermis. Nature 2007; 446(7132): 185-9.
[http://dx.doi.org/10.1038/nature05574] [PMID: 17330052]

[38] Doupé DP, Klein AM, Simons BD, Jones PH. The ordered architecture of murine ear epidermis is maintained by progenitor cells with random fate. Dev Cell 2010; 18(2): 317-23.
[http://dx.doi.org/10.1016/j.devcel.2009.12.016] [PMID: 20159601]

[39] Doupé DP, Alcolea MP, Roshan A, *et al.* A single progenitor population switches behavior to maintain and repair esophageal epithelium. Science 2012; 337(6098): 1091-3.

[http://dx.doi.org/10.1126/science.1218835] [PMID: 22821983]

[40] Watt FM, Lo Celso C, Silva-Vargas V. Epidermal stem cells: an update. Curr Opin Genet Dev 2006; 16(5): 518-24.
[http://dx.doi.org/10.1016/j.gde.2006.08.006] [PMID: 16919447]

[41] Schneider TE, Barland C, Alex AM, *et al.* Measuring stem cell frequency in epidermis: a quantitative *in vivo* functional assay for long-term repopulating cells. Proc Natl Acad Sci USA 2003; 100(20): 11412-7.
[http://dx.doi.org/10.1073/pnas.2034935100] [PMID: 13679571]

[42] Mackenzie IC, Mackenzie SL, Rittman GA. Isolation of subpopulations of murine epidermal cells using monoclonal antibodies against differentiation-related cell surface molecules. Differentiation 1989; 41(2): 127-38.
[http://dx.doi.org/10.1111/j.1432-0436.1989.tb00740.x] [PMID: 2612763]

[43] Jones PH, Harper S, Watt FM. Stem cell patterning and fate in human epidermis. Cell 1995; 80(1): 83-93.
[http://dx.doi.org/10.1016/0092-8674(95)90453-0] [PMID: 7813021]

[44] Kolodka TM, Garlick JA, Taichman LB. Evidence for keratinocyte stem cells *in vitro*: long term engraftment and persistence of transgene expression from retrovirus-transduced keratinocytes. Proc Natl Acad Sci USA 1998; 95(8): 4356-61.
[http://dx.doi.org/10.1073/pnas.95.8.4356] [PMID: 9539741]

[45] Jones PH, Watt FM. Separation of human epidermal stem cells from transit amplifying cells on the basis of differences in integrin function and expression. Cell 1993; 73(4): 713-24.
[http://dx.doi.org/10.1016/0092-8674(93)90251-K] [PMID: 8500165]

[46] Lowell S, Jones P, Le Roux I, Dunne J, Watt FM. Stimulation of human epidermal differentiation by delta-notch signalling at the boundaries of stem-cell clusters. Curr Biol 2000; 10(9): 491-500.
[http://dx.doi.org/10.1016/S0960-9822(00)00451-6] [PMID: 10801437]

[47] Legg J, Jensen UB, Broad S, Leigh I, Watt FM. Role of melanoma chondroitin sulphate proteoglycan in patterning stem cells in human interfollicular epidermis. Development 2003; 130(24): 6049-63.
[http://dx.doi.org/10.1242/dev.00837] [PMID: 14573520]

[48] Jensen KB, Watt FM. Single-cell expression profiling of human epidermal stem and transit-amplifying cells: Lrig1 is a regulator of stem cell quiescence. Proc Natl Acad Sci USA 2006; 103(32): 11958-63.
[http://dx.doi.org/10.1073/pnas.0601886103] [PMID: 16877544]

[49] Tan DW, Jensen KB, Trotter MW, Connelly JT, Broad S, Watt FM. Single-cell gene expression profiling reveals functional heterogeneity of undifferentiated human epidermal cells. Development 2013; 140(7): 1433-44.
[http://dx.doi.org/10.1242/dev.087551] [PMID: 23482486]

[50] Page ME, Lombard P, Ng F, Göttgens B, Jensen KB. The epidermis comprises autonomous compartments maintained by distinct stem cell populations. Cell Stem Cell 2013; 13(4): 471-82.
[http://dx.doi.org/10.1016/j.stem.2013.07.010] [PMID: 23954751]

[51] Kretzschmar K, Watt FM. Markers of epidermal stem cell subpopulations in adult mammalian skin. Cold Spring Harb Perspect Med 2014; 4(10): a013631.

[http://dx.doi.org/10.1101/cshperspect.a013631] [PMID: 24993676]

[52] Ferraris C, Bernard BA, Dhouailly D. Adult epidermal keratinocytes are endowed with pilosebaceous forming abilities. Int J Dev Biol 1997; 41(3): 491-8.
[PMID: 9240566]

[53] Kanitakis J. Anatomy, histology and immunohistochemistry of normal human skin. Eur J Dermatol 2002; 12(4): 390-9.
[PMID: 12095893]

[54] Schneider MR, Schmidt-Ullrich R, Paus R. The hair follicle as a dynamic miniorgan. Curr Biol 2009; 19(3): R132-42.
[http://dx.doi.org/10.1016/j.cub.2008.12.005] [PMID: 19211055]

[55] Hsu YC, Fuchs E. A family business: stem cell progeny join the niche to regulate homeostasis. Nat Rev Mol Cell Biol 2012; 13(2): 103-14.
[http://dx.doi.org/10.1038/nrm3272] [PMID: 22266760]

[56] Cotsarelis G, Sun TT, Lavker RM. Label-retaining cells reside in the bulge area of pilosebaceous unit: implications for follicular stem cells, hair cycle, and skin carcinogenesis. Cell 1990; 61(7): 1329-37.
[http://dx.doi.org/10.1016/0092-8674(90)90696-C] [PMID: 2364430]

[57] Tumbar T, Guasch G, Greco V, *et al.* Defining the epithelial stem cell niche in skin. Science 2004; 303(5656): 359-63.
[http://dx.doi.org/10.1126/science.1092436] [PMID: 14671312]

[58] Hsu YC, Pasolli HA, Fuchs E. Dynamics between stem cells, niche, and progeny in the hair follicle. Cell 2011; 144(1): 92-105.
[http://dx.doi.org/10.1016/j.cell.2010.11.049] [PMID: 21215372]

[59] Blanpain C, Lowry WE, Geoghegan A, Polak L, Fuchs E. Self-renewal, multipotency, and the existence of two cell populations within an epithelial stem cell niche. Cell 2004; 118(5): 635-48.
[http://dx.doi.org/10.1016/j.cell.2004.08.012] [PMID: 15339667]

[60] Boehnke K, Falkowska-Hansen B, Stark HJ, Boukamp P. Stem cells of the human epidermis and their niche: composition and function in epidermal regeneration and carcinogenesis. Carcinogenesis 2012; 33(7): 1247-58.
[http://dx.doi.org/10.1093/carcin/bgs136] [PMID: 22461521]

[61] Liu Y, Lyle S, Yang Z, Cotsarelis G. Keratin 15 promoter targets putative epithelial stem cells in the hair follicle bulge. J Invest Dermatol 2003; 121(5): 963-8.
[http://dx.doi.org/10.1046/j.1523-1747.2003.12600.x] [PMID: 14708593]

[62] Trempus CS, Morris RJ, Bortner CD, *et al.* Enrichment for living murine keratinocytes from the hair follicle bulge with the cell surface marker CD34. J Invest Dermatol 2003; 120(4): 501-11.
[http://dx.doi.org/10.1046/j.1523-1747.2003.12088.x] [PMID: 12648211]

[63] Nijhof JG, Braun KM, Giangreco A, *et al.* The cell-surface marker MTS24 identifies a novel population of follicular keratinocytes with characteristics of progenitor cells. Development 2006; 133(15): 3027-37.
[http://dx.doi.org/10.1242/dev.02443] [PMID: 16818453]

[64] Jaks V, Barker N, Kasper M, *et al.* Lgr5 marks cycling, yet long-lived, hair follicle stem cells. Nat

Genet 2008; 40(11): 1291-9.
[http://dx.doi.org/10.1038/ng.239] [PMID: 18849992]

[65] Jensen KB, Collins CA, Nascimento E, *et al.* Lrig1 expression defines a distinct multipotent stem cell population in mammalian epidermis. Cell Stem Cell 2009; 4(5): 427-39.
[http://dx.doi.org/10.1016/j.stem.2009.04.014] [PMID: 19427292]

[66] Snippert HJ, Haegebarth A, Kasper M, *et al.* Lgr6 marks stem cells in the hair follicle that generate all cell lineages of the skin. Science 2010; 327(5971): 1385-9.
[http://dx.doi.org/10.1126/science.1184733] [PMID: 20223988]

[67] Brownell I, Guevara E, Bai CB, Loomis CA, Joyner AL. Nerve-derived sonic hedgehog defines a niche for hair follicle stem cells capable of becoming epidermal stem cells. Cell Stem Cell 2011; 8(5): 552-65.
[http://dx.doi.org/10.1016/j.stem.2011.02.021] [PMID: 21549329]

[68] Morris RJ, Liu Y, Marles L, *et al.* Capturing and profiling adult hair follicle stem cells. Nat Biotechnol 2004; 22(4): 411-7.
[http://dx.doi.org/10.1038/nbt950] [PMID: 15024388]

[69] Petersson M, Brylka H, Kraus A, *et al.* TCF/Lef1 activity controls establishment of diverse stem and progenitor cell compartments in mouse epidermis. EMBO J 2011; 30(15): 3004-18.
[http://dx.doi.org/10.1038/emboj.2011.199] [PMID: 21694721]

[70] Ito M, Liu Y, Yang Z, *et al.* Stem cells in the hair follicle bulge contribute to wound repair but not to homeostasis of the epidermis. Nat Med 2005; 11(12): 1351-4.
[http://dx.doi.org/10.1038/nm1328] [PMID: 16288281]

[71] Lee JH, Fisher DE. Melanocyte stem cells as potential therapeutics in skin disorders. Expert Opin Biol Ther 2014; 14(11): 1569-79.
[http://dx.doi.org/10.1517/14712598.2014.935331] [PMID: 25104310]

[72] Horsley V, O'Carroll D, Tooze R, *et al.* Blimp1 defines a progenitor population that governs cellular input to the sebaceous gland. Cell 2006; 126(3): 597-609.
[http://dx.doi.org/10.1016/j.cell.2006.06.048] [PMID: 16901790]

[73] Lu CP, Polak L, Rocha AS, *et al.* Identification of stem cell populations in sweat glands and ducts reveals roles in homeostasis and wound repair. Cell 2012; 150(1): 136-50.
[http://dx.doi.org/10.1016/j.cell.2012.04.045] [PMID: 22770217]

[74] Biedermann T, Pontiggia L, Böttcher-Haberzeth S, *et al.* Human eccrine sweat gland cells can reconstitute a stratified epidermis. J Invest Dermatol 2010; 130(8): 1996-2009.
[http://dx.doi.org/10.1038/jid.2010.83] [PMID: 20376062]

[75] Driskell RR, Lichtenberger BM, Hoste E, *et al.* Distinct fibroblast lineages determine dermal architecture in skin development and repair. Nature 2013; 504(7479): 277-81.
[http://dx.doi.org/10.1038/nature12783] [PMID: 24336287]

[76] Biernaskie J, Paris M, Morozova O, *et al.* SKPs derive from hair follicle precursors and exhibit properties of adult dermal stem cells. Cell Stem Cell 2009; 5(6): 610-23.
[http://dx.doi.org/10.1016/j.stem.2009.10.019] [PMID: 19951689]

[77] McKenzie IA, Biernaskie J, Toma JG, Midha R, Miller FD. Skin-derived precursors generate

myelinating Schwann cells for the injured and dysmyelinated nervous system. J Neurosci 2006; 26(24): 6651-60.
[http://dx.doi.org/10.1523/JNEUROSCI.1007-06.2006] [PMID: 16775154]

[78] Hu MS, Rennert RC, McArdle A, *et al.* The Role of Stem Cells During Scarless Skin Wound Healing. Adv Wound Care (New Rochelle) 2014; 3(4): 304-14.
[http://dx.doi.org/10.1089/wound.2013.0471] [PMID: 24761362]

[79] Kuhn NZ, Tuan RS. Regulation of stemness and stem cell niche of mesenchymal stem cells: implications in tumorigenesis and metastasis. J Cell Physiol 2010; 222(2): 268-77.
[http://dx.doi.org/10.1002/jcp.21940] [PMID: 19847802]

[80] Maxson S, Lopez EA, Yoo D, Danilkovitch-Miagkova A, Leroux MA. Concise review: role of mesenchymal stem cells in wound repair. Stem Cells Transl Med 2012; 1(2): 142-9.
[http://dx.doi.org/10.5966/sctm.2011-0018] [PMID: 23197761]

[81] Plikus MV, Guerrero-Juarez CF, Treffeisen E, Gay DL. Epigenetic control of skin and hair regeneration after wounding. Exp Dermatol 2015; 24(3): 167-70.
[http://dx.doi.org/10.1111/exd.12488] [PMID: 25039994]

[82] Arwert EN, Hoste E, Watt FM. Epithelial stem cells, wound healing and cancer. Nat Rev Cancer 2012; 12(3): 170-80.
[http://dx.doi.org/10.1038/nrc3217] [PMID: 22362215]

[83] Watt FM, Driskell RR. The therapeutic potential of stem cells. Philos Trans R Soc Lond B Biol Sci 2010; 365(1537): 155-63.

[84] Mei SH, Haitsma JJ, Dos Santos CC, *et al.* Mesenchymal stem cells reduce inflammation while enhancing bacterial clearance and improving survival in sepsis. Am J Respir Crit Care Med 2010; 182(8): 1047-57.
[http://dx.doi.org/10.1164/rccm.201001-0010OC] [PMID: 20558630]

[85] Krasnodembskaya A, Song Y, Fang X, *et al.* Antibacterial effect of human mesenchymal stem cells is mediated in part from secretion of the antimicrobial peptide LL-37. Stem Cells 2010; 28(12): 2229-38.
[http://dx.doi.org/10.1002/stem.544] [PMID: 20945332]

[86] Lau K, Paus R, Tiede S, Day P, Bayat A. Exploring the role of stem cells in cutaneous wound healing. Exp Dermatol 2009; 18(11): 921-33.
[http://dx.doi.org/10.1111/j.1600-0625.2009.00942.x] [PMID: 19719838]

[87] Smith AN, Willis E, Chan VT, *et al.* Mesenchymal stem cells induce dermal fibroblast responses to injury. Exp Cell Res 2010; 316(1): 48-54.
[http://dx.doi.org/10.1016/j.yexcr.2009.08.001] [PMID: 19666021]

[88] Lee EY, Xia Y, Kim WS, *et al.* Hypoxia-enhanced wound-healing function of adipose-derived stem cells: increase in stem cell proliferation and up-regulation of VEGF and bFGF. Wound Repair Regen 2009; 17(4): 540-7.
[http://dx.doi.org/10.1111/j.1524-475X.2009.00499.x] [PMID: 19614919]

[89] Chen L, Tredget EE, Wu PY, Wu Y. Paracrine factors of mesenchymal stem cells recruit macrophages and endothelial lineage cells and enhance wound healing. PLoS One 2008; 3(4) e1886

[90] Gurtner GC, Werner S, Barrandon Y, Longaker MT. Wound repair and regeneration. Nature 2008;

453(7193): 314-21.
[http://dx.doi.org/10.1038/nature07039] [PMID: 18480812]

[91] Gnecchi M, Zhang Z, Ni A, Dzau VJ. Paracrine mechanisms in adult stem cell signaling and therapy. Circ Res 2008; 103(11): 1204-19.
[http://dx.doi.org/10.1161/CIRCRESAHA.108.176826] [PMID: 19028920]

[92] Hocking AM, Gibran NS. Mesenchymal stem cells: paracrine signaling and differentiation during cutaneous wound repair. Exp Cell Res 2010; 316(14): 2213-9.
[http://dx.doi.org/10.1016/j.yexcr.2010.05.009] [PMID: 20471978]

[93] Grenier G, Scimè A, Le Grand F, *et al.* Resident endothelial precursors in muscle, adipose, and dermis contribute to postnatal vasculogenesis. Stem Cells 2007; 25(12): 3101-10.
[http://dx.doi.org/10.1634/stemcells.2006-0795] [PMID: 17823241]

[94] Amoh Y, Li L, Yang M, *et al.* Nascent blood vessels in the skin arise from nestin-expressing hair-follicle cells. Proc Natl Acad Sci USA 2004; 101(36): 13291-5.
[http://dx.doi.org/10.1073/pnas.0405250101] [PMID: 15331785]

[95] Asahara T, Masuda H, Takahashi T, *et al.* Bone marrow origin of endothelial progenitor cells responsible for postnatal vasculogenesis in physiological and pathological neovascularization. Circ Res 1999; 85(3): 221-8.
[http://dx.doi.org/10.1161/01.RES.85.3.221] [PMID: 10436164]

[96] Jahoda CA, Reynolds AJ. Hair follicle dermal sheath cells: unsung participants in wound healing. Lancet 2001; 358(9291): 1445-8.
[http://dx.doi.org/10.1016/S0140-6736(01)06532-1] [PMID: 11705511]

[97] Badiavas EV, Abedi M, Butmarc J, Falanga V, Quesenberry P. Participation of bone marrow derived cells in cutaneous wound healing. J Cell Physiol 2003; 196(2): 245-50.
[http://dx.doi.org/10.1002/jcp.10260] [PMID: 12811816]

[98] Fathke C, Wilson L, Hutter J, *et al.* Contribution of bone marrow-derived cells to skin: collagen deposition and wound repair. Stem Cells 2004; 22(5): 812-22.
[http://dx.doi.org/10.1634/stemcells.22-5-812] [PMID: 15342945]

[99] Lorenz K, Sicker M, Schmelzer E, *et al.* Multilineage differentiation potential of human dermal skin-derived fibroblasts. Exp Dermatol 2008; 17(11): 925-32.
[http://dx.doi.org/10.1111/j.1600-0625.2008.00724.x] [PMID: 18557932]

[100] Liang L, Bickenbach JR. Somatic epidermal stem cells can produce multiple cell lineages during development. Stem Cells 2002, 20(1): 21-31.
[http://dx.doi.org/10.1634/stemcells.20-1-21] [PMID: 11796919]

[101] Levy V, Lindon C, Zheng Y, Harfe BD, Morgan BA. Epidermal stem cells arise from the hair follicle after wounding. FASEB J 2007; 21(7): 1358-66.
[http://dx.doi.org/10.1096/fj.06-6926com] [PMID: 17255473]

[102] Mascré G, Dekoninck S, Drogat B, *et al.* Distinct contribution of stem and progenitor cells to epidermal maintenance. Nature 2012; 489(7415): 257-62.
[http://dx.doi.org/10.1038/nature11393] [PMID: 22940863]

[103] Langton AK, Herrick SE, Headon DJ. An extended epidermal response heals cutaneous wounds in the

absence of a hair follicle stem cell contribution. J Invest Dermatol 2008; 128(5): 1311-8. [http://dx.doi.org/10.1038/sj.jid.5701178] [PMID: 18037901]

[104] Nowak JA, Polak L, Pasolli HA, Fuchs E. Hair follicle stem cells are specified and function in early skin morphogenesis. Cell Stem Cell 2008; 3(1): 33-43. [http://dx.doi.org/10.1016/j.stem.2008.05.009] [PMID: 18593557]

[105] Jensen UB, Yan X, Triel C, Woo SH, Christensen R, Owens DM. A distinct population of clonogenic and multipotent murine follicular keratinocytes residing in the upper isthmus. J Cell Sci 2008; 121(Pt 5): 609-17. [http://dx.doi.org/10.1242/jcs.025502] [PMID: 18252795]

[106] Borue X, Lee S, Grove J, *et al.* Bone marrow-derived cells contribute to epithelial engraftment during wound healing. Am J Pathol 2004; 165(5): 1767-72. [http://dx.doi.org/10.1016/S0002-9440(10)63431-1] [PMID: 15509544]

[107] Brittan M, Braun KM, Reynolds LE, *et al.* Bone marrow cells engraft within the epidermis and proliferate *in vivo* with no evidence of cell fusion. J Pathol 2005; 205(1): 1-13. [http://dx.doi.org/10.1002/path.1682] [PMID: 15546160]

[108] Deng W, Han Q, Liao L, *et al.* Engrafted bone marrow-derived flk-(1+) mesenchymal stem cells regenerate skin tissue. Tissue Eng 2005; 11(1-2): 110-9. [http://dx.doi.org/10.1089/ten.2005.11.110] [PMID: 15738666]

[109] Crigler L, Kazhanie A, Yoon TJ, *et al.* Isolation of a mesenchymal cell population from murine dermis that contains progenitors of multiple cell lineages. FASEB J 2007; 21(9): 2050-63. [http://dx.doi.org/10.1096/fj.06-5880com] [PMID: 17384147]

[110] Silva-Vargas V, Lo Celso C, Giangreco A, *et al.* Beta-catenin and Hedgehog signal strength can specify number and location of hair follicles in adult epidermis without recruitment of bulge stem cells. Dev Cell 2005; 9(1): 121-31. [http://dx.doi.org/10.1016/j.devcel.2005.04.013] [PMID: 15992546]

[111] Ito M, Yang Z, Andl T, *et al.* Wnt-dependent de novo hair follicle regeneration in adult mouse skin after wounding. Nature 2007; 447(7142): 316-20. [http://dx.doi.org/10.1038/nature05766] [PMID: 17507982]

[112] Nguyen H, Merrill BJ, Polak L, *et al.* Tcf3 and Tcf4 are essential for long-term homeostasis of skin epithelia. Nat Genet 2009; 41(10): 1068-75. [http://dx.doi.org/10.1038/ng.431] [PMID: 19718027]

[113] Gonda TA, Tu S, Wang TC. Chronic inflammation, the tumor microenvironment and carcinogenesis. Cell Cycle 2009; 8(13): 2005-13. [http://dx.doi.org/10.4161/cc.8.13.8985] [PMID: 19550141]

[114] Pedersen TX, Leethanakul C, Patel V, *et al.* Laser capture microdissection-based *in vivo* genomic profiling of wound keratinocytes identifies similarities and differences to squamous cell carcinoma. Oncogene 2003; 22(25): 3964-76. [http://dx.doi.org/10.1038/sj.onc.1206614] [PMID: 12813470]

[115] Schäfer M, Werner S. Cancer as an overhealing wound: an old hypothesis revisited. Nat Rev Mol Cell Biol 2008; 9(8): 628-38.

[http://dx.doi.org/10.1038/nrm2455] [PMID: 18628784]

[116] Goldstein J, Horsley V. Home sweet home: skin stem cell niches. Cell Mol Life Sci 2012; 69(15): 2573-82.
[http://dx.doi.org/10.1007/s00018-012-0943-3] [PMID: 22410738]

[117] Kong D, Li Y, Wang Z, Sarkar FH. Cancer stem cells and epithelial-to-mesenchymal transition (EMT)-phenotypic cells: are they cousins or twins? Cancers (Basel) 2011; 3(1): 716-29.
[http://dx.doi.org/10.3390/cancers30100716] [PMID: 21643534]

[118] Yan C, Grimm WA, Garner WL, *et al.* Epithelial to mesenchymal transition in human skin wound healing is induced by tumor necrosis factor-alpha through bone morphogenic protein-2. Am J Pathol 2010; 176(5): 2247-58.
[http://dx.doi.org/10.2353/ajpath.2010.090048] [PMID: 20304956]

[119] Leopold PL, Vincent J, Wang H. A comparison of epithelial-to-mesenchymal transition and re-epithelialization. Semin Cancer Biol 2012; 22(5-6): 471-83.
[http://dx.doi.org/10.1016/j.semcancer.2012.07.003] [PMID: 22863788]

[120] Mani SA, Guo W, Liao MJ, *et al.* The epithelial-mesenchymal transition generates cells with properties of stem cells. Cell 2008; 133(4): 704-15.
[http://dx.doi.org/10.1016/j.cell.2008.03.027] [PMID: 18485877]

[121] Graf T, Enver T. Forcing cells to change lineages. Nature 2009; 462(7273): 587-94.
[http://dx.doi.org/10.1038/nature08533] [PMID: 19956253]

[122] Sasaki M, Abe R, Fujita Y, Ando S, Inokuma D, Shimizu H. Mesenchymal stem cells are recruited into wounded skin and contribute to wound repair by transdifferentiation into multiple skin cell type. J Immunol 2008; 180(4): 2581-7.
[http://dx.doi.org/10.4049/jimmunol.180.4.2581] [PMID: 18250469]

[123] Egeblad M, Nakasone ES, Werb Z. Tumors as organs: complex tissues that interface with the entire organism. Dev Cell 2010; 18(6): 884-901.
[http://dx.doi.org/10.1016/j.devcel.2010.05.012] [PMID: 20627072]

[124] Dunham LJ. Cancer in man at site of prior benign lesion of skin or mucous membrane: a review. Cancer Res 1972; 32(7): 1359-74.
[PMID: 4555382]

[125] Ackermann J, Frutschi M, Kaloulis K, McKee T, Trumpp A, Beermann F. Metastasizing melanoma formation caused by expression of activated N-RasQ61K on an INK4a-deficient background. Cancer Res 2005; 65(10): 4005-11.
[http://dx.doi.org/10.1158/0008-5472.CAN-04-2970] [PMID: 15899789]

[126] Hutchin ME, Kariapper MS, Grachtchouk M, *et al.* Sustained Hedgehog signaling is required for basal cell carcinoma proliferation and survival: conditional skin tumorigenesis recapitulates the hair growth cycle. Genes Dev 2005; 19(2): 214-23.
[http://dx.doi.org/10.1101/gad.1258705] [PMID: 15625189]

[127] Frank NY, Margaryan A, Huang Y, *et al.* ABCB5-mediated doxorubicin transport and chemoresistance in human malignant melanoma. Cancer Res 2005; 65(10): 4320-33.
[http://dx.doi.org/10.1158/0008-5472.CAN-04-3327] [PMID: 15899824]

[128] Fang D, Nguyen TK, Leishear K, *et al.* A tumorigenic subpopulation with stem cell properties in melanomas. Cancer Res 2005; 65(20): 9328-37.
[http://dx.doi.org/10.1158/0008-5472.CAN-05-1343] [PMID: 16230395]

[129] Dou J, Pan M, Wen P, *et al.* Isolation and identification of cancer stem-like cells from murine melanoma cell lines. Cell Mol Immunol 2007; 4(6): 467-72.
[PMID: 18163959]

[130] Monzani E, Facchetti F, Galmozzi E, *et al.* Melanoma contains CD133 and ABCG2 positive cells with enhanced tumourigenic potential. Eur J Cancer 2007; 43(5): 935-46.
[http://dx.doi.org/10.1016/j.ejca.2007.01.017] [PMID: 17320377]

[131] Grichnik JM, Burch JA, Schulteis RD, *et al.* Melanoma, a tumor based on a mutant stem cell? J Invest Dermatol 2006; 126(1): 142-53.
[http://dx.doi.org/10.1038/sj.jid.5700017] [PMID: 16417230]

[132] Malanchi I, Peinado H, Kassen D, *et al.* Cutaneous cancer stem cell maintenance is dependent on beta-catenin signalling. Nature 2008; 452(7187): 650-3.
[http://dx.doi.org/10.1038/nature06835] [PMID: 18385740]

[133] Chen S, Takahara M, Kido M, *et al.* Increased expression of an epidermal stem cell marker, cytokeratin 19, in cutaneous squamous cell carcinoma. Br J Dermatol 2008; 159(4): 952-5.
[http://dx.doi.org/10.1111/j.1365-2133.2008.08731.x] [PMID: 18647309]

[134] Keshet GI, Goldstein I, Itzhaki O, *et al.* MDR1 expression identifies human melanoma stem cells. Biochem Biophys Res Commun 2008; 368(4): 930-6.
[http://dx.doi.org/10.1016/j.bbrc.2008.02.022] [PMID: 18279661]

[135] Schatton T, Murphy GF, Frank NY, *et al.* Identification of cells initiating human melanomas. Nature 2008; 451(7176): 345-9.
[http://dx.doi.org/10.1038/nature06489] [PMID: 18202660]

[136] Rappa G, Fodstad O, Lorico A. The stem cell-associated antigen CD133 (Prominin-1) is a molecular therapeutic target for metastatic melanoma. Stem Cells 2008; 26(12): 3008-17.
[http://dx.doi.org/10.1634/stemcells.2008-0601] [PMID: 18802032]

[137] Youssef KK, Van Keymeulen A, Lapouge G, *et al.* Identification of the cell lineage at the origin of basal cell carcinoma. Nat Cell Biol 2010; 12(3): 299-305.
[PMID: 20154679]

[138] Wang GY, Wang J, Mancianti ML, Epstein EH Jr. Basal cell carcinomas arise from hair follicle stem cells in Ptch1(+/-) mice. Cancer Cell 2011; 19(1): 114-24.
[http://dx.doi.org/10.1016/j.ccr.2010.11.007] [PMID: 21215705]

[139] White AC, Tran K, Khuu J, *et al.* Defining the origins of Ras/p53-mediated squamous cell carcinoma. Proc Natl Acad Sci USA 2011; 108(18): 7425-30.
[http://dx.doi.org/10.1073/pnas.1012670108] [PMID: 21502519]

[140] Lapouge G, Youssef KK, Vokaer B, *et al.* Identifying the cellular origin of squamous skin tumors. Proc Natl Acad Sci USA 2011; 108(18): 7431-6.
[http://dx.doi.org/10.1073/pnas.1012720108] [PMID: 21502497]

[141] Kyrgidis A, Tzellos TG, Triaridis S. Melanoma: Stem cells, sun exposure and hallmarks for carcinogenesis, molecular concepts and future clinical implications. J Carcinog 2010; 9: 3. [http://dx.doi.org/10.4103/1477-3163.62141] [PMID: 20442802]

[142] Shimizu T, Izumi H, Oga A, *et al.* Epidermal growth factor receptor overexpression and genetic aberrations in metastatic squamous-cell carcinoma of the skin. Dermatology (Basel) 2001; 202(3): 203-6.
[http://dx.doi.org/10.1159/000051637] [PMID: 11385224]

[143] Lu J, Rho O, Wilker E, Beltran L, Digiovanni J. Activation of epidermal akt by diverse mouse skin tumor promoters. Mol Cancer Res 2007; 5(12): 1342-52.
[http://dx.doi.org/10.1158/1541-7786.MCR-07-0115] [PMID: 18171992]

[144] Kolev V, Mandinova A, Guinea-Viniegra J, *et al.* EGFR signalling as a negative regulator of Notch1 gene transcription and function in proliferating keratinocytes and cancer. Nat Cell Biol 2008; 10(8): 902-11.
[http://dx.doi.org/10.1038/ncb1750] [PMID: 18604200]

[145] Pacifico A, Goldberg LH, Peris K, Chimenti S, Leone G, Ananthaswamy HN. Loss of CDKN2A and p14ARF expression occurs frequently in human nonmelanoma skin cancers. Br J Dermatol 2008; 158(2): 291-7.
[http://dx.doi.org/10.1111/j.1365-2133.2007.08360.x] [PMID: 18070208]

[146] Rass K, Reichrath J. UV damage and DNA repair in malignant melanoma and nonmelanoma skin cancer. Adv Exp Med Biol 2008; 624: 162-78.
[http://dx.doi.org/10.1007/978-0-387-77574-6_13] [PMID: 18348455]

[147] Athar M, Tang X, Lee JL, Kopelovich L, Kim AL. Hedgehog signalling in skin development and cancer. Exp Dermatol 2006; 15(9): 667-77.
[http://dx.doi.org/10.1111/j.1600-0625.2006.00473.x] [PMID: 16881963]

[148] Yang SH, Andl T, Grachtchouk V, *et al.* Pathological responses to oncogenic Hedgehog signaling in skin are dependent on canonical Wnt/beta3-catenin signaling. Nat Genet 2008; 40(9): 1130-5.
[http://dx.doi.org/10.1038/ng.192] [PMID: 19165927]

[149] Vidal VP, Ortonne N, Schedl A. SOX9 expression is a general marker of basal cell carcinoma and adnexal-related neoplasms. J Cutan Pathol 2008; 35(4): 373-9.
[http://dx.doi.org/10.1111/j.1600-0560.2007.00815.x] [PMID: 18333897]

[150] Tanese K, Fukuma M, Yamada T, *et al.* G-protein-coupled receptor GPR49 is up-regulated in basal cell carcinoma and promotes cell proliferation and tumor formation. Am J Pathol 2008; 173(3): 835-43.
[http://dx.doi.org/10.2353/ajpath.2008.071091] [PMID: 18688030]

[151] Hussein MR. Ultraviolet radiation and skin cancer: molecular mechanisms. J Cutan Pathol 2005; 32(3): 191-205.
[http://dx.doi.org/10.1111/j.0303-6987.2005.00281.x] [PMID: 15701081]

[152] Pons M, Quintanilla M. Molecular biology of malignant melanoma and other cutaneous tumors. Clin Transl Oncol 2006; 8(7): 466-74.
[http://dx.doi.org/10.1007/s12094-006-0046-4] [PMID: 16870533]

[153] Blanpain C, Fuchs E. Epidermal stem cells of the skin. Annu Rev Cell Dev Biol 2006; 22: 339-73. [http://dx.doi.org/10.1146/annurev.cellbio.22.010305.104357] [PMID: 16824012]

[154] Lupi O. Correlations between the Sonic Hedgehog pathway and basal cell carcinoma. Int J Dermatol 2007; 46(11): 1113-7. [http://dx.doi.org/10.1111/j.1365-4632.2007.03391.x] [PMID: 17988327]

[155] Benitah SA. Epidermal stem cells in skin homeostasis and cutaneous carcinomas. Clin Transl Oncol 2007; 9(12): 760-6. [http://dx.doi.org/10.1007/s12094-007-0137-x] [PMID: 18158979]

[156] Müller EJ, Williamson L, Kolly C, Suter MM. Outside-in signaling through integrins and cadherins: a central mechanism to control epidermal growth and differentiation? J Invest Dermatol 2008; 128(3): 501-16. [http://dx.doi.org/10.1038/sj.jid.5701248] [PMID: 18268536]

[157] Bachmann IM, Puntervoll HE, Otte AP, Akslen LA. Loss of BMI-1 expression is associated with clinical progress of malignant melanoma. Mod Pathol 2008; 21(5): 583-90. [http://dx.doi.org/10.1038/modpathol.2008.17] [PMID: 18246051]

[158] Mimeault M, Batra SK. Recent advances on skin-resident stem/progenitor cell functions in skin regeneration, aging and cancers and novel anti-aging and cancer therapies. J Cell Mol Med 2010; 14(1-2): 116-34. [http://dx.doi.org/10.1111/j.1582-4934.2009.00885.x] [PMID: 19725922]

[159] Lacina L, Dvoránkova B, Smetana K Jr, *et al.* Marker profiling of normal keratinocytes identifies the stroma from squamous cell carcinoma of the oral cavity as a modulatory microenvironment in co-culture. Int J Radiat Biol 2007; 83(11-12): 837-48. [http://dx.doi.org/10.1080/09553000701694343] [PMID: 17952768]

[160] Marsh D, Dickinson S, Neill GW, Marshall JF, Hart IR, Thomas GJ. alpha vbeta 6 Integrin promotes the invasion of morphoeic basal cell carcinoma through stromal modulation. Cancer Res 2008; 68(9): 3295-303. [http://dx.doi.org/10.1158/0008-5472.CAN-08-0174] [PMID: 18451156]

[161] Ruiter D, Bogenrieder T, Elder D, Herlyn M. Melanoma-stroma interactions: structural and functional aspects. Lancet Oncol 2002; 3(1): 35-43. [http://dx.doi.org/10.1016/S1470-2045(01)00620-9] [PMID: 11905603]

[162] Epstein EH. Basal cell carcinomas: attack of the hedgehog. Nat Rev Cancer 2008; 8(10): 743-54. [http://dx.doi.org/10.1038/nrc2503] [PMID: 18813320]

[163] Oro AE, Higgins K. Hair cycle regulation of Hedgehog signal reception. Dev Biol 2003; 255(2): 238-48. [http://dx.doi.org/10.1016/S0012-1606(02)00042-8] [PMID: 12648487]

[164] Kasper M, Jaks V, Are A, *et al.* Wounding enhances epidermal tumorigenesis by recruiting hair follicle keratinocytes. Proc Natl Acad Sci USA 2011; 108(10): 4099-104. [http://dx.doi.org/10.1073/pnas.1014489108] [PMID: 21321199]

[165] Wong SY, Reiter JF. Wounding mobilizes hair follicle stem cells to form tumors. Proc Natl Acad Sci USA 2011; 108(10): 4093-8.

[http://dx.doi.org/10.1073/pnas.1013098108] [PMID: 21321207]

[166] Lomas A, Leonardi-Bee J, Bath-Hextall F. A systematic review of worldwide incidence of nonmelanoma skin cancer. Br J Dermatol 2012; 166(5): 1069-80. [http://dx.doi.org/10.1111/j.1365-2133.2012.10830.x] [PMID: 22251204]

[167] Blanpain C. Tracing the cellular origin of cancer. Nat Cell Biol 2013; 15(2): 126-34. [http://dx.doi.org/10.1038/ncb2657] [PMID: 23334500]

[168] Grachtchouk M, Pero J, Yang SH, *et al.* Basal cell carcinomas in mice arise from hair follicle stem cells and multiple epithelial progenitor populations. J Clin Invest 2011; 121(5): 1768-81. [http://dx.doi.org/10.1172/JCI46307] [PMID: 21519145]

[169] Youssef KK, Lapouge G, Bouvrée K, *et al.* Adult interfollicular tumour-initiating cells are reprogrammed into an embryonic hair follicle progenitor-like fate during basal cell carcinoma initiation. Nat Cell Biol 2012; 14(12): 1282-94. [http://dx.doi.org/10.1038/ncb2628] [PMID: 23178882]

[170] Eberl M, Klingler S, Mangelberger D, *et al.* Hedgehog-EGFR cooperation response genes determine the oncogenic phenotype of basal cell carcinoma and tumour-initiating pancreatic cancer cells. EMBO Mol Med 2012; 4(3): 218-33. [http://dx.doi.org/10.1002/emmm.201100201] [PMID: 22294553]

[171] Vidal VP, Chaboissier MC, Lützkendorf S, *et al.* Sox9 is essential for outer root sheath differentiation and the formation of the hair stem cell compartment. Curr Biol 2005; 15(15): 1340-51. [http://dx.doi.org/10.1016/j.cub.2005.06.064] [PMID: 16085486]

[172] Hahn H, Wicking C, Zaphiropoulous PG, *et al.* Mutations of the human homolog of Drosophila patched in the nevoid basal cell carcinoma syndrome. Cell 1996; 85(6): 841-51. [http://dx.doi.org/10.1016/S0092-8674(00)81268-4] [PMID: 8681379]

[173] Ghali L, Wong ST, Green J, Tidman N, Quinn AG. Gli1 protein is expressed in basal cell carcinomas, outer root sheath keratinocytes and a subpopulation of mesenchymal cells in normal human skin. J Invest Dermatol 1999; 113(4): 595-9. [http://dx.doi.org/10.1046/j.1523-1747.1999.00729.x] [PMID: 10504446]

[174] Aszterbaum M, Rothman A, Johnson RL, *et al.* Identification of mutations in the human PATCHED gene in sporadic basal cell carcinomas and in patients with the basal cell nevus syndrome. J Invest Dermatol 1998; 110(6): 885-8. [http://dx.doi.org/10.1046/j.1523-1747.1998.00222.x] [PMID: 9620294]

[175] Beachy PA, Karhadkar SS, Berman DM. Tissue repair and stem cell renewal in carcinogenesis. Nature 2004; 432(7015): 324-31. [http://dx.doi.org/10.1038/nature03100] [PMID: 15549094]

[176] Crowson AN. Basal cell carcinoma: biology, morphology and clinical implications. Mod Pathol 2006; 19 (Suppl. 2): S127-47. [http://dx.doi.org/10.1038/modpathol.3800512] [PMID: 16446711]

[177] Reinisch CM, Uthman A, Erovic BM, Pammer J. Expression of BMI-1 in normal skin and inflammatory and neoplastic skin lesions. J Cutan Pathol 2007; 34(2): 174-80. [http://dx.doi.org/10.1111/j.1600-0560.2006.00587.x] [PMID: 17244030]

[178] Mancuso M, Pazzaglia S, Tanori M, *et al.* Basal cell carcinoma and its development: insights from radiation-induced tumors in Ptch1-deficient mice. Cancer Res 2004; 64(3): 934-41. [http://dx.doi.org/10.1158/0008-5472.CAN-03-2460] [PMID: 14871823]

[179] Dahmane N, Lee J, Robins P, Heller P, Ruiz i Altaba A. Activation of the transcription factor Gli1 and the Sonic hedgehog signalling pathway in skin tumours. Nature 1997; 389(6653): 876-81. [http://dx.doi.org/10.1038/39918] [PMID: 9349822]

[180] Fan H, Oro AE, Scott MP, Khavari PA. Induction of basal cell carcinoma features in transgenic human skin expressing Sonic Hedgehog. Nat Med 1997; 3(7): 788-92. [http://dx.doi.org/10.1038/nm0797-788] [PMID: 9212109]

[181] Grachtchouk V, Grachtchouk M, Lowe L, *et al.* The magnitude of hedgehog signaling activity defines skin tumor phenotype. EMBO J 2003; 22(11): 2741-51. [http://dx.doi.org/10.1093/emboj/cdg271] [PMID: 12773389]

[182] Nilsson M, Undèn AB, Krause D, *et al.* Induction of basal cell carcinomas and trichoepitheliomas in mice overexpressing GLI-1. Proc Natl Acad Sci USA 2000; 97(7): 3438-43. [http://dx.doi.org/10.1073/pnas.97.7.3438] [PMID: 10725363]

[183] Athar M, Li C, Tang X, *et al.* Inhibition of smoothened signaling prevents ultraviolet B-induced basal cell carcinomas through regulation of Fas expression and apoptosis. Cancer Res 2004; 64(20): 7545-52. [http://dx.doi.org/10.1158/0008-5472.CAN-04-1393] [PMID: 15492281]

[184] Nicolas M, Wolfer A, Raj K, *et al.* Notch1 functions as a tumor suppressor in mouse skin. Nat Genet 2003; 33(3): 416-21. [http://dx.doi.org/10.1038/ng1099] [PMID: 12590261]

[185] de Gruijl FR, van Kranen HJ, Mullenders LH. UV-induced DNA damage, repair, mutations and oncogenic pathways in skin cancer. J Photochem Photobiol B 2001; 63(1-3): 19-27. [http://dx.doi.org/10.1016/S1011-1344(01)00199-3] [PMID: 11684448]

[186] Couvé-Privat S, Bouadjar B, Avril MF, Sarasin A, Daya-Grosjean L. Significantly high levels of ultraviolet-specific mutations in the smoothened gene in basal cell carcinomas from DNA repair-deficient xeroderma pigmentosum patients. Cancer Res 2002; 62(24): 7186-9. [PMID: 12499255]

[187] Gailani MR, Bale AE. Developmental genes and cancer: role of patched in basal cell carcinoma of the skin. J Natl Cancer Inst 1997; 89(15): 1103-9. [http://dx.doi.org/10.1093/jnci/89.15.1103] [PMID: 9262247]

[188] Daya-Grosjean L, Sarasin A. UV-specific mutations of the human patched gene in basal cell carcinomas from normal individuals and xeroderma pigmentosum patients. Mutat Res 2000; 450(1-2): 193-9. [http://dx.doi.org/10.1016/S0027-5107(00)00025-7] [PMID: 10838143]

[189] Krediet JT, Beyer M, Lenz K, *et al.* Sentinel lymph node biopsy and risk factors for predicting metastasis in cutaneous squamous cell carcinoma. Br J Dermatol 2015; 172(4): 1029-36. [http://dx.doi.org/10.1111/bjd.13508] [PMID: 25362868]

[190] White RA, Neiman JM, Reddi A, *et al.* Epithelial stem cell mutations that promote squamous cell

carcinoma metastasis. J Clin Invest 2013; 123(10): 4390-404.
[http://dx.doi.org/10.1172/JCI65856] [PMID: 23999427]

[191] Ratushny V, Gober MD, Hick R, Ridky TW, Seykora JT. From keratinocyte to cancer: the pathogenesis and modeling of cutaneous squamous cell carcinoma. J Clin Invest 2012; 122(2): 464-72.
[http://dx.doi.org/10.1172/JCI57415] [PMID: 22293185]

[192] Shendrik I, Crowson AN, Magro CM. Follicular cutaneous squamous cell carcinoma: an under-recognized neoplasm arising from hair appendage structures. Br J Dermatol 2013; 169(2): 384-8.
[http://dx.doi.org/10.1111/bjd.12374] [PMID: 23574514]

[193] White AC, Khuu JK, Dang CY, *et al.* Stem cell quiescence acts as a tumour suppressor in squamous tumours. Nat Cell Biol 2014; 16(1): 99-107.
[http://dx.doi.org/10.1038/ncb2889] [PMID: 24335650]

[194] Lefort K, Mandinova A, Ostano P, *et al.* Notch1 is a p53 target gene involved in human keratinocyte tumor suppression through negative regulation of ROCK1/2 and MRCKalpha kinases. Genes Dev 2007; 21(5): 562-77.
[http://dx.doi.org/10.1101/gad.1484707] [PMID: 17344417]

[195] Schober M, Fuchs E. Tumor-initiating stem cells of squamous cell carcinomas and their control by TGF-β and integrin/focal adhesion kinase (FAK) signaling. Proc Natl Acad Sci USA 2011; 108(26): 10544-9.
[http://dx.doi.org/10.1073/pnas.1107807108] [PMID: 21670270]

[196] Lapouge G, Beck B, Nassar D, Dubois C, Dekoninck S, Blanpain C. Skin squamous cell carcinoma propagating cells increase with tumour progression and invasiveness. EMBO J 2012; 31(24): 4563-75.
[http://dx.doi.org/10.1038/emboj.2012.312] [PMID: 23188079]

[197] Driessens G, Beck B, Caauwe A, Simons BD, Blanpain C. Defining the mode of tumour growth by clonal analysis. Nature 2012; 488(7412): 527-30.
[http://dx.doi.org/10.1038/nature11344] [PMID: 22854777]

[198] Beck B, Driessens G, Goossens S, *et al.* A vascular niche and a VEGF-Nrp1 loop regulate the initiation and stemness of skin tumours. Nature 2011; 478(7369): 399-403.
[http://dx.doi.org/10.1038/nature10525] [PMID: 22012397]

[199] Boukamp P. UV-induced skin cancer: similarities--variations. J Dtsch Dermatol Ges 2005; 3(7): 493-503.
[http://dx.doi.org/10.1111/j.1610-0387.2005.05037.x] [PMID: 15967008]

[200] Faurschou A, Haedersdal M, Poulsen T, Wulf HC. Squamous cell carcinoma induced by ultraviolet radiation originates from cells of the hair follicle in mice. Exp Dermatol 2007; 16(6): 485-9.
[http://dx.doi.org/10.1111/j.1600-0625.2007.00551.x] [PMID: 17518988]

[201] Quintanilla M, Brown K, Ramsden M, Balmain A. Carcinogen-specific mutation and amplification of Ha-ras during mouse skin carcinogenesis. Nature 1986; 322(6074): 78-80.
[http://dx.doi.org/10.1038/322078a0] [PMID: 3014349]

[202] Abel EL, Angel JM, Kiguchi K, DiGiovanni J. Multi-stage chemical carcinogenesis in mouse skin: fundamentals and applications. Nat Protoc 2009; 4(9): 1350-62.
[http://dx.doi.org/10.1038/nprot.2009.120] [PMID: 19713956]

[203] Argyris TS, Slaga TJ. Promotion of carcinomas by repeated abrasion in initiated skin of mice. Cancer Res 1981; 41(12 Pt 1): 5193-5.
[PMID: 7307015]

[204] Brown K, Strathdee D, Bryson S, Lambie W, Balmain A. The malignant capacity of skin tumours induced by expression of a mutant H-ras transgene depends on the cell type targeted. Curr Biol 1998; 8(9): 516-24.
[http://dx.doi.org/10.1016/S0960-9822(98)70203-9] [PMID: 9560338]

[205] Boumahdi S, Driessens G, Lapouge G, *et al.* SOX2 controls tumour initiation and cancer stem-cell functions in squamous-cell carcinoma. Nature 2014; 511(7508): 246-50.
[http://dx.doi.org/10.1038/nature13305] [PMID: 24909994]

[206] Foronda M, Martínez P, Schoeftner S, *et al.* Sox4 links tumor suppression to accelerated aging in mice by modulating stem cell activation. Cell Reports 2014; 8(2): 487-500.
[http://dx.doi.org/10.1016/j.celrep.2014.06.031] [PMID: 25043184]

[207] Okumura K, Saito M, Isogai E, *et al.* Meis1 regulates epidermal stem cells and is required for skin tumorigenesis. PLoS One 2014; 9(7): e102111.
[http://dx.doi.org/10.1371/journal.pone.0102111] [PMID: 25013928]

[208] Beronja S, Janki P, Heller E, *et al.* RNAi screens in mice identify physiological regulators of oncogenic growth. Nature 2013; 501(7466): 185-90.
[http://dx.doi.org/10.1038/nature12464] [PMID: 23945586]

[209] Schramek D, Sendoel A, Segal JP, *et al.* Direct *in vivo* RNAi screen unveils myosin IIa as a tumor suppressor of squamous cell carcinomas. Science 2014; 343(6168): 309-13.
[http://dx.doi.org/10.1126/science.1248627] [PMID: 24436421]

[210] Bongiorno MR, Doukaki S, Malleo F, Aricò M. Identification of progenitor cancer stem cell in lentigo maligna melanoma. Dermatol Ther (Heidelb) 2008; 21 (Suppl. 1): S1-5.
[http://dx.doi.org/10.1111/j.1529-8019.2008.00193.x] [PMID: 18727808]

[211] Chen KG, Szakács G, Annereau JP, *et al.* Principal expression of two mRNA isoforms (ABCB 5alpha and ABCB 5beta) of the ATP-binding cassette transporter gene ABCB 5 in melanoma cells and melanocytes. Pigment Cell Res 2005; 18(2): 102-12.
[http://dx.doi.org/10.1111/j.1600-0749.2005.00214.x] [PMID: 15760339]

[212] Hadnagy A, Gaboury L, Beaulieu R, Balicki D. SP analysis may be used to identify cancer stem cell populations. Exp Cell Res 2006; 312(19): 3701-10.
[http://dx.doi.org/10.1016/j.yexcr.2006.08.030] [PMID: 17046749]

[213] Chin L, Garraway LA, Fisher DE. Malignant melanoma: genetics and therapeutics in the genomic era. Genes Dev 2006; 20(16): 2149-82.
[http://dx.doi.org/10.1101/gad.1437206] [PMID: 16912270]

[214] Klein WM, Wu BP, Zhao S, Wu H, Klein-Szanto AJ, Tahan SR. Increased expression of stem cell markers in malignant melanoma. Mod Pathol 2007; 20(1): 102-7.
[http://dx.doi.org/10.1038/modpathol.3800720] [PMID: 17143262]

[215] Ogliari KS, Marinowic D, Brum DE, Loth F. Stem cells in dermatology. An Bras Dermatol 2014; 89(2): 286-91.

[http://dx.doi.org/10.1590/abd1806-4841.20142530] [PMID: 24770506]

[216] Hoerter JD, Bradley P, Casillas A, *et al.* Extrafollicular dermal melanocyte stem cells and melanoma. Stem Cells Int 2012.
[http://dx.doi.org/10.1155/2012/407079]

CHAPTER 4

Stem Cells in Neurodegeneration

Ana-Maria Enciu[1,2,*]

[1] *Department of Cellular and Molecular Biology and Histology, Carol Davila University of Medicine and Pharmacy, Bucharest, Romania*

[2] *V.Babes National Institute of Pathology, Bucharest, Romania*

Abstract: The neural stem cell niche in the adult brain is a complex environment formed by several types of cells, a particular type of extracellular matrix and specific molecular cues. Ageing is characterized by reduced neurogenesis, due to altered neurotrophin signaling, activation of senescence programmes within the niche, imbalanced growth factor signalling. A recent concept is that blood-born ageing factors signal or cross the blood brain barrier to negatively influence the neurogenesis. Neurodegenerative diseases are an appealing target for regenerative medicine and stem cell transplant was held for quite some time in high consideration and brought about high hopes for the replacement of lost cells and tissue function restoration. One of the major drawbacks is that most transplanted cells differentiate on glial line; therefore, new strategies were implemented. From transplantation of committed neuronal progenitors, to neurosphere transplant or autologous mesenchymal stem cells, these strategies are presented and some of the drawbacks highlighted in the present chapter.

Keywords: Alzheimer's disease, Hippocampus, Neural precursors, Neurodegeneration, Parkinson's disease, Stem cells, Subventricular zone.

INTRODUCTION

Paradoxically, age is a risk factor for both neurodegenerative diseases and brain tumors. Taken separately, it makes sense that, on one hand, loss of neuronal stem cells (NSC), neurons, neuronal plasticity and synapses would lead to neuro-degeneration. On the other hand, an increase in oxidative stress, accumulation of

* **Corresponding author Ana-Maria Enciu:** No. 8 BdulEroilorSanitari, 050474 sect 5 Bucharest, Romania; Tel: +4021 318 0762; Fax: +40213194528; E-mail: ana.enciu@umf.ro.

Cristiana Tanase & Monica Neagu (Eds.)

mutations and deficiency of DNA repair mechanisms with age would accumulate in generation of malignancy. The paradox rises when both processes are asserted from the stem cell perspective: neurodegeneration is thought to be accompanied by low regenerative reserve, whereas brain tumors "benefit" from the proliferative advantage of cancer stem cells (CSCs). The transplant of stem cells for neurodegenerative disorders has been considered for the restoration of neural networks supporting the cognitive processes. The evaluation of the faith of transplanted cells, as well as the accumulation of knowledge regarding adult stem cells form other tissues, highlighted interesting behaviours of stem cells, such as migration out of transplanted site or the niche into chemotactive areas. Such areas could be stroke area, traumatic area or even a tumor. There is also a concern that transplanted stem cells could harbour a latent potential of malignant transformation.

NEURAL STEM CELL NICHE ORGANIZATION IN ADULT MAMMALIAN BRAIN

Neural stem cell niche of mammalian brain is mainly found in the sub ventricular zone, responsible for the generation of new neurons for the olfactory bulb and hippocampus dentate gyrus. It generates small granular neurons that are included into neuronal networks involved in learning and memory [1]. Additional areas of adult neurogenesis have been described in hypothalamus [2], spinal central canal ependymal zone [3], substantia nigra, striatum, amygdala, neocortex [4] and leptomeninges [5]. Literature abounds in morphological and molecular characterization of mammalian neural stem cell niche, mostly in animal models. Same as most stem niches in the adult mammalian organisms, the neural niche has a cellular compartment with several subpopulations: quiescent stem cells, transit amplifying cells, precursor cells and adult cells. Each of those subpopulations is characterized by the expression of specific markers [4] and phenotypic features [6] (Fig **1**). From sub ventricular zone, newly formed neuroblasts migrate towards the olfactory bulb *via* the rostral migratory stream. But unlike other mammalian brains, human sub ventricular neurogenic zone is tri-layered: Layer I - the ependymal lining of lateral ventricle, Layer II - cellular elongations and rare neuroblasts and Layer III, or "astrocytic ribbon", - mitotic active cells [7]. The rostral migratory stream is prominent and populated with migrating neuroblasts in humans, mostly during foetal period. However, during adulthood, this olfactory tract seems to contain a very low number of cells that double stain positive for doublecortin and polysialylated-neural cell adhesion molecule, indicating either a low generation rate of neuroblasts, or a very low migratory capacity [8].

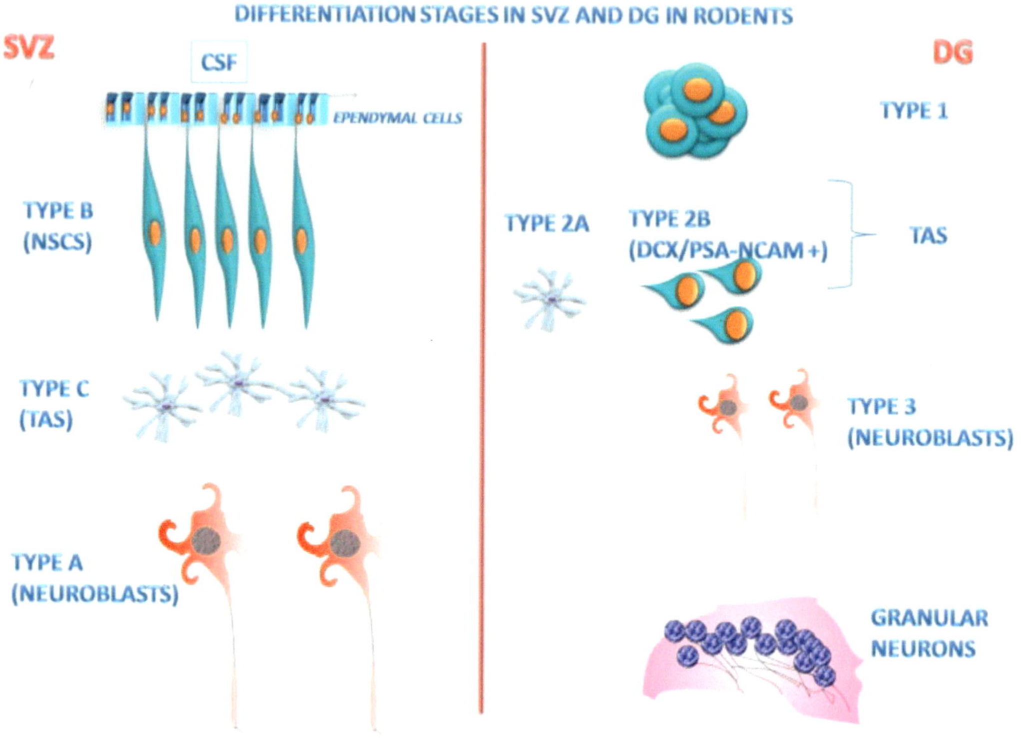

Fig. (1). Models of neuroregenerations. Abbreviations: SVZ – subventricular zone; NSCS – neural stem cells; TAS – transit amplifying cells; DG – dentate gyrus; DCX – doublecortin; PSA-NCAM - Polysialylated-neural cell adhesion molecule.

Increasing evidence had revealed the importance of brain endothelial cells of the vascular niche in the regulation of neurogenesis [9]. In adult rodent brain, neural progenitor cells are located next to endothelial cells in both SVZ and the dentate gyrus and express genes involved in angiogenesis [10]. Teng *et al.* showed that there is a dual relationship between NSC and angiogenic niche: activated NSC release angiogenic factors and in turn, activated endothelial cells enhance neural progenitor cells proliferation and differentiation [10].

Although acknowledged only recently, the extracellular matrix seems to play an important role in the maintenance of the neural niche. Adult neurogenic zones are associated with the polysialylated form of neural cell adhesion molecule [11], N-sulphated heparan sulphate proteoglycans [12] laminin beta1 and gamma1, collagen IV, nidogen, and perlecan [13]. It is believed that the pattern of sulfation of proteoglycans is determinant for the selective recognition of growth factors and cytokines [14]. The pial basal lamina of the subventricular zone also provides essential signals for NSC function, through cell-to-matrix junctions. These contacts are kept during asymmetric division of stem cells, thus providing both daughter cells with survival signals. The loss of integrin (the main protein organizing these junctions) led to the detachment of stem cells and apoptosis [15].

The NSCs are not, however, the only type of stem cells found in the brain, which influence adult neurogenesis. The perivascular mesenchymal stem cell, which differs from adult NSC by location, morphology and surface markers, is able to differentiate into neuronal progenitors. Situated at vascular branching points, these cells express mesenchymal (CD105, CD13) and pericyte markers (PDGFR-β, α-SMA, CD13, CD73, CD90, CD106, CD49d, RGS5, NG2) [16]. They are multipotent, able to differentiate *in vitro* under appropriated stimuli into adipocytes, chondroblasts and osteoblasts. When exposed to glial induction medium, they differentiate into oligodendrocytes or astrocytes, as well as into an immature neuronal phenotype [16].

MAINTENANCE OF NEURAL STEM CELL NICHE IN THE OLD BRAIN

Aged brain is apparently less able to learn than young brain, ability impaired (among other things) by decreased neurogenesis, decreased neural plasticity and/or deficit of neurotransmitters. Numerous hypothesis and mechanisms have been proposed to sustain this paradigm. Surprisingly, studies revealed that cohorts of both aged animals and aged humans without any deleterious neurodegenerative disorders, contain both "bad" and "good learners" [17], not correlated with subsequent anatomopathological findings, or number of hippocampal dividing cells [18].

The general consensus, repeatedly sustained by experimental results, is that neurogenesis is reduced in aged hippocampus and SVZ. Ageing leads to ventral stenosis of the walls of the lateral ventricle; therefore, most ventral SVZ deteriorates, thus restricting the neurogenic region to only the dorsolateral region of the lateral ventricle [7]. The evaluation of SVZ in aged laboratory animals, as well as of the output of newly formed neuroblasts to olfactory bulbs, revealed that, although the total NSC number and discreet regenerative subunits decline with age, the percentage of actively, mitotic NSCs increases [19].The output is, nevertheless, poor, either due to transition of progenitor cells to a quiescent state [20], or a consequence of low neurotrophin signaling. Neurotrophins are the most prominent factors of adult neurogenesis. Neurotrophin family is formed by closely related secreted proteins involved in neuronal survival, neurite outgrowth and migration, hippocampal learning. Neurotrophic factors family includes nerve-growth factor (NGF), brain-derived nerve growth factor (BDNF) and neurotrophins 3 to 7 (NT3-7). Recently, glial-derived neurotrophic factor (GDNF) family ligands (GDNF, artemin, neurturin, persephin), as well as cerebral dopamine neurotrophic factor –CDNF- and mesencephalic astrocyte-derived neurotrophic factor – MANF - have also been included in this family, based on their neurotrophic effects [21, 22]. In aged rats, the BDNF system is affected at different levels, from reduced gene transcription, impaired protein synthesis and processing, to decreased activation of the TrkB receptors [23]. Mature form of NGF was shown to be decreased in the cortex of aged rats, but not its precursor, proNGF [24], which signals through p75 receptor, releasing proapoptic signals.

Low telomerase activity, a hallmark of ageing, has also been described in the SVZ. Reactivation of telomerase in deficient mice was shown to restore neurogenesis [25].

Relevant signaling patterns also change with age within the niche, with senescence programmes activated through inhibition of cycle-dependant kinases [26] or down regulation of family transcriptional repressor Bmi-1 [27]. TGF-b pathway activation was persistently increased in the SVZ niches of aged mice and selective inhibition of this pathway significantly improved neurogenesis]9].

Impaired antioxidative protection is a feature of ageing and it has also been described in the aged SVZ [28]. Furthermore, a double hit, ageing and exposure to environmental toxins *in vivo*- can synergize in a negative fashion to reduce NPC survival, proliferation, and neuron differentiation capacity [28].

Adult neurogenesis is also modulated by neurotransmitter input and the fate of stem cells is influenced by the balance between inhibitory and excitatory synapse transmission. Recently, serotonin was shown to positively modulate proliferation of adult SVZ cells, in basal conditions [29] as well as indirectly, as a result of physical exercise [30]. On the other hand, the γ-aminobutyric acid was demonstrated to induce quiescence of nestin-expressing radial glia-like neural stem cells [31]. A more detailed analysis revealed that the cellular effect is mediated by the type of neurotransmitter receptor, rather than the class of neurotransmitter: 5-HT1a receptors are involved in the self-renewal of precursor cells, whereas 5-HT2 receptors induce proliferation and promote neuronal differentiation [32]. Vasoactive intestinal polypeptide (VIP) receptors shift the fate of symmetrically dividing NSPCs towards granule cell neurogenesis or a nestin-only phenotype, respectively [33]. Fate shifting in disfavour of neuronal line is another proposed mechanism of neural senescence [34]. Growth factors, such as IGF-1, FGF-2 and VEGF, have a proliferative effect that is counterbalanced during ageing by decreased expression and/or decreased receptor expression on hippocampal cells [35]. Age decreases expression of VEGF and its receptor Flk-1, as well as FGF-2 receptors FGFR 1 and 2 on GFAP positive cells of neurogenic zones [36]. A significant increase in TGF-β1 binding was observed on cycling SVZ cells during ageing, along with subsequent signaling activation and strong Smad3 phosphorylation in NSCs [9]. Both angiogenic and neurogenic response to VEGF stimulation is attenuated in the aged mouse brain [37] and ageing is well known to be associated with increased vasculopathy burden and oxidative stress. Further on, ageing seems to be associated with a significant decrease in superficial cortical vessels [38].

When transplanted *in vitro*, aged SVZ cells are still capable of responding to extrinsic growth factors, albeit with a lower efficiency compared to younger precursor cells [28]. Furthermore, supporting cells, notably astrocytes, also present differences between young and aged brain, which may influence their supporting and trophic activity in the neural stem cell niche.

Aged brain seems also less capable of coping with injury than young brain, as focal hippocampal injury or acute seizures revealed substantial age-related impairment in the ability of hippocampal NSCs to augment neurogenesis in response to injury [39].

Interestingly, although the concept of ageing is almost inconceivable without the idea of decline, there seem to be organisms with remarkable neurogenic potential and no significant age-related decline in the absolute levels of stem/progenitor cell proliferation, neuronal and glial differentiation, or long-term survival of newly generated cells [40].

Last but not the least, we must consider the fact that although separated from peripheral blood by an endothelial barrier, the brain consistently exchanges nutrients and information with the rest of the body. Therefore, circulating ageing-related factors would negatively influence the adult neurogenesis, as it has been elegantly demonstrated in animal experiments [41]. These systemic regulators of adult neurogenesis were identified as haptoglobin and β2-microglobulin, along with several chemokines (CCLs), out of which CCL11/ eotaxin seems to exert a potent inhibitory effect [42]. Hormones and circulating growth factors also influence neurogenesis and modify the neurogenic niche of aged brain. Data collected in rodents demonstrated a negative effect of corticosterone on hippocampal cell proliferation; furthermore, basal level of this steroid hormone is increased in aged animals and its receptors up regulated in aged NSCs and immature neurons [34].

NEUROREGENERATION RESERVE IN NEURODEGENERATIVE DISEASES

The term ''neurodegeneration" harbours under its umbrella a plethora of pathologies with common nominator ''loss of neurons" and subsequent neural deficits spanning from cognitive to motor impairments. However, at least in terms of cognition, it is admitted more that the correlation with neuropathological lesions is the modest and accounts for about a quarter of the variance in cognition of older adults [43]. Currently, hippocampus MRI measurement is used to assess dementia risk and several visual rating scales existing for the evaluation of atrophy in patients with cognitive impairment [44]. Neurogenesis has been investigated in AD patients by screening of cerebrospinal fluid for neurogenesis-related factors or post-mortem immunohistochemistry for immature neuronal proteins [45]. Most data indicated that, although neurogenesis is present, maturation of neuroblasts is hindered. Leonard *et al.* reported in 2009 that neural stem cells and progenitor cells can be cultured from rapid autopsy samples of SVZ from elderly human subjects, including patients with age-related neurologic disorders and generate neurospheres [46]. Interestingly, the percentages of β-tubulin (III)$^{+}$ cells differentiated from AD and control SVZ neurospheres using a neuron-promoting differentiation protocol were roughly equivalent, supporting the idea that NSCs are not intrinsically modified by the course of the disease.

Several valuable insights on the neurogenesis reserve in the pathologically aged brain are provided by mouse models. The P8 strain of senescence-accelerated mice (SAMP8) shows at 5 months old an increased number of BrdU-retaining cells (characteristic for slow dividing stem cells) in the dentate gyrus, as well as rising numbers of doublecortin-positive (DCX(+)) immature neurons. However, 10 months old animals display a lower number of BrdU+ cells than controls, as well as a decreased number of newly formed astrocytes [47]. Same tendency to an increased number of BrdU positive cells was also reported for a transgenic AD model [48]. However, data on neurogenesis and neuroblast formation in adult life are dependent on the animal model and the cell marker used for quantization [49]. One of AD's central pathogenic player, the fibrillar Aβ-42 resulted from enzymatic cleavage of amyloid beta protein precursor, plays a role in hindering neurogenesis either by generating an inappropriate environment for neuroblasts to mature, or by inducing senescence of neuronal stem or progenitor cells [50]. Exposure of NSCs to amyloid-β oligomers lowers their dividing potential, favours gliogenesis and attenuates mobility [51]. Another central feature of AD, loss of cholinergic neurons, has also been recently shown to negatively influence neurogenesis, triggering a compensatory increase in cell proliferation, which is not accompanied by an increased number of migratory neuroblasts and fully differentiated neurons.

USE OF STEM CELLS AS THERAPEUTIC APPROACHES IN NEURODEGENERATIVE DISORDERS

Neurodegenerative diseases are an appealing target for regenerative medicine, as stem cell transplant was held for quite some time in high consideration and brought about high hopes for the replacement of lost cells and tissue function restoration. Stem cell transplant has been attempted for several central nervous system pathologies, but most reports are still at animal model level. Replacement of lost neurons by transplant has been mostly experimented and reported in humans for PD, focused on the delivery of dopaminergic neurons, either allografts of foetal ventral mesencephalic tissue, or stem cells (Table **1**).

Table 1. Examples of cell-based therapies used in research for neurodegenerative disorders.

Central Nervous System Pathology	Type of Cellsused for Transplant	Ref.
Alzheimer's Disease	neural stem cells	[52]
	mesenchymal stem cells	[53, 54]
	VEGF overexpressing bone marrow mesenchymal stem cells	[55]
	placenta-derived mesenchymal stem cells	[56]
	adipose-derived mesenchymal stem cells	[57]
	human umbilical cord mesenchymal stem cells	[58]
Parkinson's Disease	allografts of foetal ventral mesencephalic tissue	[59, 60]
	human embryonic stem cells-derived dopaminergic neurons	[61]
	induced pluripotent stem cells-derived mesodiencephalic dopaminergic neurons	[62]
Huntington's Disease	foetal striatal allografts	[63]
	spontaneously derived and noggin-primed human neural precursors derived from embryonic stem cells	[64]
	adult neural sub ventricular cells	[65]
	human embryonic stem cells	[66]
	human embryonic stem cells-derived neural precursors	[67]
	patient-derived adipose stem cells	[68]

(Table 1) contd.....

Central Nervous System Pathology	Type of Cellsused for Transplant	Ref.
Amyotrophic lateral sclerosis	human motor neurons derived from neural stem cells	[69]
	human neural stem cells overexpressing growth factors	[70, 71]
	human glial-restricted progenitors	[72]
	adipose tissue-derived stem cells	[73]
	human amniotic mesenchymal stem cells	[74]
Multiple sclerosis	mesenchymal stem cells, originating from various sources (bone-marrow, adipose tissue, placenta)	reviewed in [75]
	neural precursor cells	[76]
	autologous haematopoietic stem cell transplantation	[77]
	iPSC-derived neural precursors	[78]
Stroke	mesenchymal stem cells	[79]
	preconditioned bone marrow mesenchymal stem cells	[80]
	neural stem cells	[81]
	embryonic stem cells-derived neuronal precursors	[82]
	induced pluripotent stem cells	[83]

Numerous studies have demonstrated that different types of stem cells transplanted into the mammalian brain preferentially migrate towards pathologically burdened cerebral regions, regardless of administration method, distance to affected site, or pathology (reviewed in [84]). This astonishing homing ability is related to cytokine production at injury site and activation of various cellular signalling pathways such as stem cell factor - c-kit [85], VEGF- VEGFRs [86] and downstream receptor tyrosine-kinases signaling. Migratory capacity of stem cells has been increasingly used in research in order to facilitate the actual transplant manoeuvres, with an increased number of studies reporting systemic administration of stem cells for regenerative purposes [74, 87, 88].

Neuronal Stem Cells and Committed Progenitors - Based Therapies

As in any other field of regenerative medicine, neurorestoration through stem cell transplant aims at replacing lost or damaged cells, in this case, neurons. Depending on the type of degenerative disease, the neuronal phenotype required differs, as, for example, AD is characterized by the loss of cholinergic neurons, PD is related to dopaminergic neurons depletion and motor neurons are required in amyotrophic lateral sclerosis. Not only that transplanted cells must acquire the desired phenotype, but they also need to integrate in specific neural circuitries. This drawback can be avoided by therapies that enhance endogenous neurogenesis, taking advantage of the existent environmental neurogenic niches. Furthermore, exogenous NSC implantation seems to promote regeneration in ageing organisms not so much through niche occupancy and differentiation, but through stimulation of endogenous neurogenesis [89]. One of the most reliable sources of neural stem cells is the embryonic hippocampus, that can generate *in vitro* neurospheres further used for grafting. The major drawback is that most precursors differentiate in glial cells [90]. Neonatal sub ventricular zone can also, under specific culturing conditions, be used as a reliable source of precursor cells to be further differentiated into specific neuronal types [91].

Another fact to be taken into consideration is the need to generate from the transplanted stem cells mostly neuroblasts to further differentiate into functional neurons, rather than glial cells. Thus, neural-committed progenitor cells are a transplantation alternative to stem cells. There are several reports that describe successful transplantation of neural progenitor cells in the rat or human hippocampus (reviewed in [92]). These progenitors can be differentiated from either neural stem cells, mesenchymal stem cells or human embryonic stem cells [93] and they were shown to modulate neural network activity through functional synaptic integration [94]. Although indeed neural stem cells differentiate predominantly into glia, they can still be useful for the stimulation of endogenous neural stem cells in the SGZ to differentiate into new dentate granule cells [95].

One of the main fields of neurodegenerative disorders that intensely explored the regenerative transplantation therapies is Parkinson's disease (PD). The degenerative process affects a specific subset of neurons (A9 midbrain dopaminergic neurons) which have been intensively characterized for *in vitro* efficient generation from stem cells. This goal has apparently been attained during the last few years [61, 96, 97]. Some of the strategies used in PD animal model research are:i) nigratalembryonic tissue transplantation, ii) striatal transplantation of dopaminergic neurons, obtained *in vitro* by differentiation from embryonic stem cells, iii) transplantation of neural stem cells, iv) transplantation of neurospheres, thought to expose neural progenitor cells to an environment reminiscent of the stem cell niche [98], v) transplantation of mesenchymal stem cells, that were shown to have *in vitro* a neuroprotective action against the presence of a neurotoxin [99], vi) use of stem cells as delivery vectors for growth factors [100] and, more recently, vii) transplantation of a unique class of astrocytes, GDAs(BMP), generated *in vitro* by directed differentiation of glial precursors [101], used for their quality to produce several neurotrophic factors. This particular strategy, apart from delivering neurotrophic factors without involvement of vectors or epigenetic altering, also has a beneficial effect on several types of adult neurons, such as tyrosine hydroxylase positive neurons and GABAergic neurons. And, of course, there is the alternative of combinatorial therapy [102].

However, if animal testing results were encouraging, human clinical trials, although successful from the surgical point of view of viable cells delivery , showed dismal results. The major drawbacks were limited survival of transplanted cells, limited integration in neural networks or lack of clinical response. Still, lessons were learned and new tactics were developed to overcome poor results in NSC transplantation in PD models, such as enhancement of polysialic acid expression – a cell surface glycan that apparently serves to modulate distance between cells, with immediate involvement in cell migration, axonal growth and neuronal plasticity [103]. A very recent report on transplanted dopaminergic neurons in long-time survivors with PD stated that "the vast majority of transplanted neurons remain healthy for the long term in PD patients, consistent with clinical findings that foetal dopamine neuron transplants maintain function for up to 15-18 years in patients" [104], which is only encouraging news and favours further development of stem-cell-based dopamine neuronal replacement therapies for PD.

Induced Pluripotent Stem Cells Based Therapies

Even before induced pluripotent stem cells (iPS) were reported and recognized as a useful tool of biology research, in 2005 a phase 1 clinical trial announced the use of genetically modified autologous fibroblasts as NGF delivery cargo to the forebrain of mild AD patients [105]. Several proofs of principle results have already been published regarding iPS-derived neural progenitors transplantation for ALS [106] and Parkinson [107]. Recent developments in reprogramming technology have allowed for more selective cell fate interconversion events, *e.g.* from skin fibroblasts directly to diverse induced neuron (iN) subtypes [108]. Depending on the targeted disease, the transplanted cells could be dopaminergic neurons (for PD), GABAergic neurons (for Huntington disease) or cholinergic neurons (for AD).

The generation of neuronal precursors through reprogramming greatly facilitates gene therapy for neurodegenerative diseases. However, their utility is limited in genetic diseases, such as familial forms of AD (FAD), in which the mature cells originating from IPS, carriers of the mutations, recapitulate the characteristics of the disease. In turn, this makes them useful *in vitro* models, especially for high throughput screening practices, which will ensure rapid analysis of thousands of compounds. Such an early identification of toxic or inefficient drugs would prevent animal studies and subsequent clinical failures [109].

Mesenchymal Stem Cell Based Therapies

Mesenchymal stem cells are interstitial cells firstly identified in hematopoietic bone marrow as a distinct population of cells able to recreate a hematopoietic microenvironment in subcutaneous transplants, but are also found in adipose tissue, umbilical cord blood, tendons, synovial membrane, skeletal muscle, placenta and the dental pulp. Although they are all named mesenchymal, it is not clear though to what extent these subpopulations are equivalent to one another in terms of gene expression and phenotype [110].

Classically, these cells are known to differentiate under controlled conditions into adipocytes, chondrocytes and osteocytes, but recently MSCs proved to be more versatile and generate cells from neuronal lineage under specific cues, such as adhesion proteins enriched 3D culturing matrix [111] or TNFα treatment [112]. A quick browsing of recent literature finds adipose-tissue derived stem cells to be a commonly used source for either cell transplant, or *in vitro* progenitor generation. They are readily accessible, as opposed to bone-marrow derived mesenchymal stem cells, rather well characterized and pluripotent features that makes adipocyte tissue-derived stem cells a new powerful tool for regenerative medicine research.

In AD models, MSC transplantation significantly augments hippocampal neurogenesis and enhances the differentiation of neuronal progenitor cells into mature neurons, by up regulating the Wnt signaling pathway [54]. Another study reported that bone marrow-derived mesenchymal stem cells, transplanted in the hippocampus of an AD mouse model, were shown to increase microglial-related Aβ clearance, to reduce tau hyperphosphorylation and to improve spatial learning and memory impairments. One putative effect would be activation of microglia to an alternative phenotype, which is considered part of the repair process and extracellular matrix reorganization and is not accompanied by production of pro-inflammatory cytokines [53]. Interestingly, although transplantation of neural stem cells in the hippocampus of an AD mouse model improved cognitive function, it didn't modify the amyloid pathology [52].

MSC were proven to be involved in autophagy regulatory digestion pathway for aggregated proteins and organelles that cause various neurodegenerative diseases, such as α synuclein in PD or tau deposits in tauopathies and AD. *In vitro,* MSCs increase neuronal viability and attenuate expression of α-synuclein and *in vivo*, in an animal model of Parkinson's disease, MSC administration significantly increased final maturation of late autophagic vacuoles, along with the reduction of the α-synucleinlevel in dopaminergic neurons [113].

Same as neuronal stem cells, mesenchymal stem cells can be used as delivery vectors for VEGF [55], BDNF [114], in order to alleviate symptomatology in animal models of neurodegenerative diseases. Also, reported results of MSCs transplantation in stroke animal models were encouraging, leading to improved neuronal recovery and no significant adverse effects [115].

MSCs are currently used in human trials for a plethora of pathologies (available at http://www.clinicaltrials.gov). In neurodegenerative diseases, the most addressed pathology for MSCs transplant is multiple sclerosis, in which MSCs may prove beneficial through their immunosuppressive and tissue repair abilities [116], but they are also used for spinal cord injury, ischemic stroke, ALS and PD. There is one report of a completed clinical trial regarding administration of human umbilical venous cord blood derived mesenchymal stem cells in phase one of a clinical trial for AD (NEUROSTEM®-AD).

Cell-Free Extracts of Stem Cells as Alternative Therapy to Stem Cell Transplant in Neurodegenerative Diseases

It has been long proposed and confirmed that certain soluble molecules (cytokines, neurotrophins, growth factors) exert a protective effect in neurodegenerative diseases, such as, for example, BDNF [117], NGF [118] growth factors like granulocyte colony stimulating factor [119] or erythropoietin [120] in Alzheimers'disease, neurotrophic factors in Parkinson's disease [121] or leukemia inhibitory factor in demyelinating diseases [122]. Stem cells are also involved, along with mature cells, in the production of paracrine-acting secretory molecules, possibly partially responsible for their beneficial effects. Insulin-like growth factor-1 and -2 ,VEGF and stromal- derived factor-1 mediate interaction between stem cells and their microenvironment and are important for neuronal survival, neurogenesis, and mitochondrial activation *via* a bystander-like mechanism [123]. Recent results have pointed out that adipose-derived stem cells secrete anti-apoptotic growth factors, anti-inflammatory cytokines, useful for the survival of neurons in neurodegenerative diseases and thus, cell-free extracts of ASCs (ASCs-E) could be a potential candidate therapy in Huntington disease [124] or stroke [125].

CONCLUSION

In the field of regenerative medicine through cell transplant, neurodegenerative diseases continue to occupy a leading position, mostly because of high prevalence, lack of clinically efficient therapeutic interventions and high social burden. Although animal testing and even clinical trials with stem cells transplants began more than 2 decades ago, little or no remarkable clinical progress has been attained for patients with neurodegenerative diseases. However, the results accumulated during these years of tremendous efforts yielded significant landmarks of knowledge that granted acceleration of fundamental research findings, further boosted by major breakthroughs like discovery of iPS. Perfecting the technology of iPS generation from autologous tissues offered an easily accessible source of stem-like transplantable, self-compatible cells for clinical trials. On the same level, enhanced knowledge on mesenchymal stem cells revealed them as universal substitutes for tissue-committed stem cells, a very important issue for central nervous tissue, which is scarce in neural stem cells and more so difficult to harvest and highly improbable for autologous transplants.

Higher transplant efficacy is also to be expected from *in vitro* generation of high purity subpopulations of differentiated cells, such as A9 dopaminergic neurons for Parkinson's disease therapy or motor neurons for ALS. Overall, progress translated into clinical practice is to be expected during following years and, most likely, a new era in modern regenerative medicine.

CONFLICT OF INTEREST

The author confirms that author has no conflict of interest to declare for this publication.

ACKNOWLEDGEMENTS

The authors would like to thank Irina Radu, certified translator in Medicine – Pharmacy, certificate credentials: series E number 0048, for professional linguistic assistance. This work has been partially supported by and by a grant of the Romanian National Authority for Scientific Research and Innovation CNCS-UEFISCDI, project no.PN-II-RU-TE-2014-4-1534 the Sectorial Operational Programme Human Resources Development (SOPHRD), financed by the European Social Fund and the Romanian Government under the contract number POSDRU 141531.

REFERENCES

[1] Kee N, Teixeira CM, Wang AH, Frankland PW. Preferential incorporation of adult-generated granule cells into spatial memory networks in the dentate gyrus. Nat Neurosci 2007; 10(3): 355-62. [http://dx.doi.org/10.1038/nn1847] [PMID: 17277773]

[2] Yuan TF, Arias-Carrión O. Adult neurogenesis in the hypothalamus: evidence, functions, and implications. CNS Neurol Disord Drug Targets 2011; 10(4): 433-9. [http://dx.doi.org/10.2174/187152711795563985] [PMID: 21495965]

[3] Hamilton LK, Truong MK, Bednarczyk MR, Aumont A, Fernandes KJ. Cellular organization of the central canal ependymal zone, a niche of latent neural stem cells in the adult mammalian spinal cord. Neuroscience 2009; 164(3): 1044-56. [http://dx.doi.org/10.1016/j.neuroscience.2009.09.006] [PMID: 19747531]

[4] Decimo I, Bifari F, Krampera M, Fumagalli G. Neural stem cell niches in health and diseases. Curr Pharm Des 2012; 18(13): 1755-83.

[http://dx.doi.org/10.2174/138161212799859611] [PMID: 22394166]

[5] Bifari F, Decimo I, Chiamulera C, *et al.* Novel stem/progenitor cells with neuronal differentiation potential reside in the leptomeningeal niche. J Cell Mol Med 2009; 13(9B): 3195-208. [http://dx.doi.org/10.1111/j.1582-4934.2009.00706.x] [PMID: 19228261]

[6] Quiñones-Hinojosa A, Sanai N, Soriano-Navarro M, *et al.* Cellular composition and cytoarchitecture of the adult human subventricular zone: a niche of neural stem cells. J Comp Neurol 2006; 494(3): 415-34. [http://dx.doi.org/10.1002/cne.20798] [PMID: 16320258]

[7] Conover JC, Shook BA. Aging of the subventricular zone neural stem cell niche. Aging Dis 2011; 2(1): 49-63. [PMID: 22396866]

[8] Wang C, Liu F, Liu YY, *et al.* Identification and characterization of neuroblasts in the subventricular zone and rostral migratory stream of the adult human brain. Cell Res 2011; 21(11): 1534-50. [http://dx.doi.org/10.1038/cr.2011.83] [PMID: 21577236]

[9] Pineda JR, Daynac M, Chicheportiche A, *et al.* Vascular-derived TGF-β increases in the stem cell niche and perturbs neurogenesis during aging and following irradiation in the adult mouse brain. EMBO Mol Med 2013; 5(4): 548-62. [http://dx.doi.org/10.1002/emmm.201202197] [PMID: 23526803]

[10] Teng H, Zhang ZG, Wang L, *et al.* Coupling of angiogenesis and neurogenesis in cultured endothelial cells and neural progenitor cells after stroke. J Cereb Blood Flow metabol: Official J Int Soc Cereb Blood Flow and Metabol 2008 Apr.; 28(4): 764-1. [http://dx.doi.org/10.1038/sj.jcbfm.9600573]

[11] Seki T. Hippocampal adult neurogenesis occurs in a microenvironment provided by PSA-NCA--expressing immature neurons. J Neurosci Res 2002; 69(6): 772-83. [http://dx.doi.org/10.1002/jnr.10366] [PMID: 12205671]

[12] Mercier F, Arikawa-Hirasawa E. Heparan sulfate niche for cell proliferation in the adult brain. Neurosci Lett 2012; 510(2): 67-72. [http://dx.doi.org/10.1016/j.neulet.2011.12.046] [PMID: 22230891]

[13] Kerever A, Schnack J, Vellinga D, *et al.* Novel extracellular matrix structures in the neural stem cell niche capture the neurogenic factor fibroblast growth factor 2 from the extracellular milieu. Stem Cells 2007; 25(9): 2146-57. [http://dx.doi.org/10.1634/stemcells.2007-0082] [PMID: 17569787]

[14] Mercier F, Schnack J, Chaumet M. Fractones: Home and Conductors of the Neural Stem Cell Niche. In: Seki T, Sawamoto K, Parent J, Alvarez-Buylla A, Eds. Neurogenesis in the Adult Brain I: . Japan: Springer 2011; pp. 109-33. [http://dx.doi.org/10.1007/978-4-431-53933-9_4]

[15] Marthiens V, Kazanis I, Moss L, Long K, Ffrench-Constant C. Adhesion molecules in the stem cell niche--more than just staying in shape? J Cell Sci 2010; 123(Pt 10): 1613-22. [http://dx.doi.org/10.1242/jcs.054312] [PMID: 20445012]

[16] Ozen I, Boix J, Paul G. Perivascular mesenchymal stem cells in the adult human brain: a future target

for neuroregeneration? Clin Transl Med 2012; 1(1): 30.
[http://dx.doi.org/10.1186/2001-1326-1-30] [PMID: 23369339]

[17] Couillard-Despres S, Iglseder B, Aigner L. Neurogenesis, cellular plasticity and cognition: the impact of stem cells in the adult and aging brain--a mini-review. Gerontology 2011; 57(6): 559-64.
[http://dx.doi.org/10.1159/000323481] [PMID: 21311170]

[18] Bizon JL, Lee HJ, Gallagher M. Neurogenesis in a rat model of age-related cognitive decline. Aging Cell 2004; 3(4): 227-34.
[http://dx.doi.org/10.1111/j.1474-9728.2004.00099.x] [PMID: 15268756]

[19] Shook BA, Manz DH, Peters JJ, Kang S, Conover JC. Spatiotemporal changes to the subventricular zone stem cell pool through aging. J Neurosci 2012; 32(20): 6947-56.
[http://dx.doi.org/10.1523/JNEUROSCI.5987-11.2012] [PMID: 22593063]

[20] Lugert S, Basak O, Knuckles P, *et al.* Quiescent and active hippocampal neural stem cells with distinct morphologies respond selectively to physiological and pathological stimuli and aging. Cell Stem Cell 2010; 6(5): 445-56.
[http://dx.doi.org/10.1016/j.stem.2010.03.017] [PMID: 20452319]

[21] Zihlmann KB, Ducray AD, Schaller B, *et al.* The GDNF family members neurturin, artemin and persephin promote the morphological differentiation of cultured ventral mesencephalic dopaminergic neurons. Brain Res Bull 2005; 68(1-2): 42-53.
[http://dx.doi.org/10.1016/j.brainresbull.2004.10.012] [PMID: 16325003]

[22] Lindholm P, Saarma M. Novel CDNF/MANF family of neurotrophic factors. Dev Neurobiol 2010; 70(5): 360-71.
[PMID: 20186704]

[23] Calabrese F, Guidotti G, Racagni G, Riva MA. Reduced neuroplasticity in aged rats: a role for the neurotrophin brain-derived neurotrophic factor. Neurobiol Aging 2013; 34(12): 2768-76.
[http://dx.doi.org/10.1016/j.neurobiolaging.2013.06.014] [PMID: 23870838]

[24] Perovic M, Tesic V, Mladenovic Djordjevic A, *et al.* BDNF transcripts, proBDNF and proNGF, in the cortex and hippocampus throughout the life span of the rat. Age (Dordr) 2013; 35(6): 2057-70.
[http://dx.doi.org/10.1007/s11357-012-9495-6] [PMID: 23255148]

[25] Jaskelioff M, Muller FL, Paik JH, *et al.* Telomerase reactivation reverses tissue degeneration in aged telomerase-deficient mice. Nature 2011; 469(7328): 102-6.
[http://dx.doi.org/10.1038/nature09603] [PMID: 21113150]

[26] Molofsky AV, Slutsky SG, Joseph NM, *et al.* Increasing p16INK4a expression decreases forebrain progenitors and neurogenesis during ageing. Nature 2006; 443(7110): 448-52.
[http://dx.doi.org/10.1038/nature05091] [PMID: 16957738]

[27] Wang Y, Guan Y, Wang F, Huang A, Wang S, Zhang YA. Bmi-1 regulates self-renewal, proliferation and senescence of human fetal neural stem cells *in vitro.* Neurosci Lett 2010; 476(2): 74-8.
[http://dx.doi.org/10.1016/j.neulet.2010.04.006] [PMID: 20385206]

[28] L'Episcopo F, Tirolo C, Testa N, *et al.* Aging-induced Nrf2-ARE pathway disruption in the subventricular zone drives neurogenic impairment in parkinsonian mice *via* PI3K-Wnt/β-catenin dysregulation. J Neurosci 2013; 33(4): 1462-85.

[http://dx.doi.org/10.1523/JNEUROSCI.3206-12.2013] [PMID: 23345222]

[29] Tong CK, Chen J, Cebrián-Silla A, *et al.* Axonal control of the adult neural stem cell niche. Cell Stem Cell 2014; 14(4): 500-11.
[http://dx.doi.org/10.1016/j.stem.2014.01.014] [PMID: 24561083]

[30] Klempin F, Beis D, Mosienko V, Kempermann G, Bader M, Alenina N. Serotonin is required for exercise-induced adult hippocampal neurogenesis. J Neurosci 2013; 33(19): 8270-5.
[http://dx.doi.org/10.1523/JNEUROSCI.5855-12.2013] [PMID: 23658167]

[31] Song J, Zhong C, Bonaguidi MA, *et al.* Neuronal circuitry mechanism regulating adult quiescent neural stem-cell fate decision. Nature 2012; 489(7414): 150-4.
[http://dx.doi.org/10.1038/nature11306] [PMID: 22842902]

[32] Klempin F, Babu H, De Pietri Tonelli D, Alarcon E, Fabel K, Kempermann G. Oppositional effects of serotonin receptors 5-HT1a, 2, and 2c in the regulation of adult hippocampal neurogenesis. Front Mol Neurosci 2010; 3: 3.
[PMID: 20721314]

[33] Zaben M, Sheward WJ, Shtaya A, *et al.* The neurotransmitter VIP expands the pool of symmetrically dividing postnatal dentate gyrus precursors *via* VPAC2 receptors or directs them toward a neuronal fate *via* VPAC1 receptors. Stem Cells 2009; 27(10): 2539-51.
[http://dx.doi.org/10.1002/stem.184] [PMID: 19650041]

[34] Artegiani B, Calegari F. Age-related cognitive decline: can neural stem cells help us? Aging (Albany, NY) 2012; 4(3): 176-86.
[http://dx.doi.org/10.18632/aging.100446] [PMID: 22466406]

[35] Shetty AK, Hattiangady B, Shetty GA. Stem/progenitor cell proliferation factors FGF-2, IGF-1, and VEGF exhibit early decline during the course of aging in the hippocampus: role of astrocytes. Glia 2005; 51(3): 173-86.
[http://dx.doi.org/10.1002/glia.20187] [PMID: 15800930]

[36] Bernal GM, Peterson DA. Phenotypic and gene expression modification with normal brain aging in GFAP-positive astrocytes and neural stem cells. Aging Cell 2011; 10(3): 466-82.
[http://dx.doi.org/10.1111/j.1474-9726.2011.00694.x] [PMID: 21385309]

[37] Gao P, Shen F, Gabriel RA, *et al.* Attenuation of brain response to vascular endothelial growth factor-mediated angiogenesis and neurogenesis in aged mice. Stroke 2009; 40(11): 3596-600.
[http://dx.doi.org/10.1161/STROKEAHA.109.561050] [PMID: 19745179]

[38] Petcu EB, Smith RA, Miroiu RI, Opris MM. Angiogenesis in old-aged subjects after ischemic stroke: a cautionary note for investigators. J Angiogenes Res 2010; 2: 26.
[http://dx.doi.org/10.1186/2040-2384-2-26] [PMID: 21110846]

[39] Shetty AK, Hattiangady B, Rao MS, Shuai B. Deafferentation enhances neurogenesis in the young and middle aged hippocampus but not in the aged hippocampus. Hippocampus 2011; 21(6): 631-46.
[http://dx.doi.org/10.1002/hipo.20776] [PMID: 20333732]

[40] Traniello IM, Sirbulescu RF, Ilies I, *et al.* Age-related changes in stem cell dynamics, neurogenesis, apoptosis, and gliosis in the adult brain: A novel teleost fish model of negligible senescence. Dev Neurobiol 2013.

[PMID: 24293183]

[41] Villeda SA, Luo J, Mosher KI, *et al.* The ageing systemic milieu negatively regulates neurogenesis and cognitive function. Nature 2011; 477(7362): 90-4.
[http://dx.doi.org/10.1038/nature10357] [PMID: 21886162]

[42] Mendelsohn AR, Larrick JW. Overcoming the aging systemic milieu to restore neural stem cell function. Rejuvenation Res 2011; 14(6): 681-4.
[http://dx.doi.org/10.1089/rej.2011.1301] [PMID: 22175510]

[43] Jellinger KA, Attems J. Neuropathological approaches to cerebral aging and neuroplasticity. Dialogues Clin Neurosci 2013; 15(1): 29-43.
[PMID: 23576887]

[44] Szabo K, Förster A, Gass A. Conventional and diffusion-weighted MRI of the hippocampus. Front Neurol Neurosci 2014; 34: 71-84.
[http://dx.doi.org/10.1159/000357925] [PMID: 24777132]

[45] Lazarov O, Marr RA. Neurogenesis and Alzheimer's disease: at the crossroads. Exp Neurol 2010; 223(2): 267-81.
[http://dx.doi.org/10.1016/j.expneurol.2009.08.009] [PMID: 19699201]

[46] Leonard BW, Mastroeni D, Grover A, *et al.* Subventricular zone neural progenitors from rapid brain autopsies of elderly subjects with and without neurodegenerative disease. J Comp Neurol 2009; 515(3): 269-94.
[http://dx.doi.org/10.1002/cne.22040] [PMID: 19425077]

[47] Gang B, Yue C, Han N, *et al.* Limited hippocampal neurogenesis in SAMP8 mouse model of Alzheimer's disease. Brain Res 2011; 1389: 183-93.
[http://dx.doi.org/10.1016/j.brainres.2011.03.039] [PMID: 21439270]

[48] Yu Y, He J, Zhang Y, *et al.* Increased hippocampal neurogenesis in the progressive stage of Alzheimer's disease phenotype in an APP/PS1 double transgenic mouse model. Hippocampus 2009; 19(12): 1247-53.
[http://dx.doi.org/10.1002/hipo.20587] [PMID: 19309037]

[49] Zhang C, McNeil E, Dressler L, Siman R. Long-lasting impairment in hippocampal neurogenesis associated with amyloid deposition in a knock-in mouse model of familial Alzheimer's disease. Exp Neurol 2007; 204(1): 77-87.
[http://dx.doi.org/10.1016/j.expneurol.2006.09.018] [PMID: 17070803]

[50] He N, Jin WL, Lok KH, Wang Y, Yin M, Wang ZJ. Amyloid-β(1-42) oligomer accelerates senescence in adult hippocampal neural stem/progenitor cells *via* formylpeptide receptor 2. Cell Death Dis 2013; 4: e924.
[http://dx.doi.org/10.1038/cddis.2013.437] [PMID: 24263098]

[51] Lee IS, Jung K, Kim IS, Park KI. Amyloid-β oligomers regulate the properties of human neural stem cells through GSK-3β signaling. Exp Mol Med 2013; 45: e60.
[http://dx.doi.org/10.1038/emm.2013.125] [PMID: 24232259]

[52] Zhang W, Wang PJ, Sha HY, Ni J, Li MH, Gu GJ. Neural stem cell transplants improve cognitive function without altering amyloid pathology in an APP/PS1 double transgenic model of Alzheimer's

disease. Mol Neurobiol 2014; 50(2): 423-37.
[http://dx.doi.org/10.1007/s12035-014-8640-x] [PMID: 24481678]

[53] Lee JK, Jin HK, Endo S, Schuchman EH, Carter JE, Bae JS. Intracerebral transplantation of bone marrow-derived mesenchymal stem cells reduces amyloid-beta deposition and rescues memory deficits in Alzheimer's disease mice by modulation of immune responses. Stem Cells 2010; 28(2): 329-43.
[PMID: 20014009]

[54] Oh SH, Kim HN, Park HJ, *et al.* Mesenchymal stem cells increase hippocampal neurogenesis and neuronal differentiation by enhancing the Wnt signaling pathway in Alzheimer's disease model. Cell Transplant 2014.
[PMID: 24612635]

[55] Garcia KO, Ornellas FL, Martin PK, *et al.* Therapeutic effects of the transplantation of VEGF overexpressing bone marrow mesenchymal stem cells in the hippocampus of murine model of Alzheimer's disease. Front Aging Neurosci 2014; 6: 30.
[http://dx.doi.org/10.3389/fnagi.2014.00030] [PMID: 24639647]

[56] Yun HM, Kim HS, Park KR, *et al.* Placenta-derived mesenchymal stem cells improve memory dysfunction in an Aβ1-42-infused mouse model of Alzheimer's disease. Cell Death Dis 2013; 4: e958.
[http://dx.doi.org/10.1038/cddis.2013.490] [PMID: 24336078]

[57] Ma T, Gong K, Ao Q, *et al.* Intracerebral transplantation of adipose-derived mesenchymal stem cells alternatively activates microglia and ameliorates neuropathological deficits in Alzheimer's disease mice. Cell Transplant 2013; 22 (Suppl. 1): S113-26.
[http://dx.doi.org/10.3727/096368913X672181] [PMID: 24070198]

[58] Yang H, Yue C, Yang H, *et al.* Intravenous Administration of Human Umbilical Cord Mesenchymal Stem Cells Improves Cognitive Impairments and Reduces Amyloid-Beta Deposition in an AβPP/PS1 Transgenic Mouse Model. Neurochem Res 2013.
[http://dx.doi.org/10.1007/s11064-013-1161-6] [PMID: 24078266]

[59] Hoffer BJ, Leenders KL, Young D, *et al.* Eighteen-month course of two patients with grafts of fetal dopamine neurons for severe Parkinson's disease. Exp Neurol 1992; 118(3): 243-52.
[http://dx.doi.org/10.1016/0014-4886(92)90181-O] [PMID: 1306484]

[60] Peschanski M, Defer G, N'Guyen JP, *et al.* Bilateral motor improvement and alteration of L-dopa effect in two patients with Parkinson's disease following intrastriatal transplantation of foetal ventral mesencephalon. Brain : a journal of neurology 1994 Jun.; 117(Pt 3): 487-99.

[61] Sundberg M, Bogetofte H, Lawson T, *et al.* Improved cell therapy protocols for Parkinson's disease based on differentiation efficiency and safety of hESC-, hiPSC-, and non-human primate iPSC-derived dopaminergic neurons. Stem Cells 2013; 31(8): 1548-62.
[http://dx.doi.org/10.1002/stem.1415] [PMID: 23666606]

[62] Roessler R, Smallwood SA, Veenvliet JV, *et al.* Detailed analysis of the genetic and epigenetic signatures of iPSC-derived mesodiencephalic dopaminergic neurons. Stem Cell Rep 2014; 2(4): 520-33.
[http://dx.doi.org/10.1016/j.stemcr.2014.03.001] [PMID: 24749075]

[63] Cicchetti F, Saporta S, Hauser RA, *et al.* Neural transplants in patients with Huntington's disease

undergo disease-like neuronal degeneration. Proc Natl Acad Sci USA 2009; 106(30): 12483-8. [http://dx.doi.org/10.1073/pnas.0904239106] [PMID: 19620721]

[64] Vazey EM, Dottori M, Jamshidi P, *et al.* Comparison of transplant efficiency between spontaneously derived and noggin-primed human embryonic stem cell neural precursors in the quinolinic acid rat model of Huntington's disease. Cell Transplant 2010; 19(8): 1055-62. [http://dx.doi.org/10.3727/096368910X494632] [PMID: 20350346]

[65] Vazey EM, Chen K, Hughes SM, Connor B. Transplanted adult neural progenitor cells survive, differentiate and reduce motor function impairment in a rodent model of Huntington's disease. Exp Neurol 2006; 199(2): 384-96. [http://dx.doi.org/10.1016/j.expneurol.2006.01.034] [PMID: 16626705]

[66] Lee ST, Chu K, Park JE, *et al.* Intravenous administration of human neural stem cells induces functional recovery in Huntington's disease rat model. Neurosci Res 2005; 52(3): 243-9. [http://dx.doi.org/10.1016/j.neures.2005.03.016] [PMID: 15896865]

[67] Song J, Lee ST, Kang W, *et al.* Human embryonic stem cell-derived neural precursor transplants attenuate apomorphine-induced rotational behavior in rats with unilateral quinolinic acid lesions. Neurosci Lett 2007; 423(1): 58-61. [http://dx.doi.org/10.1016/j.neulet.2007.05.066] [PMID: 17669593]

[68] Im W, Lee ST, Park JE, *et al.* Transplantation of patient-derived adipose stem cells in YAC128 Huntington's disease transgenic mice. PLoS Curr 2010; 2: 2. [http://dx.doi.org/10.1371/currents.RRN1183] [PMID: 20890444]

[69] Lee HJ, Kim KS, Ahn J, Bae HM, Lim I, Kim SU. Human motor neurons generated from neural stem cells delay clinical onset and prolong life in ALS mouse model. PLoS One 2014; 9(5): e97518. [http://dx.doi.org/10.1371/journal.pone.0097518] [PMID: 24844281]

[70] Hwang DH, Lee HJ, Park IH, *et al.* Intrathecal transplantation of human neural stem cells overexpressing VEGF provide behavioral improvement, disease onset delay and survival extension in transgenic ALS mice. Gene Ther 2009; 16(10): 1234-44. [http://dx.doi.org/10.1038/gt.2009.80] [PMID: 19626053]

[71] Park S, Kim HT, Yun S, *et al.* Growth factor-expressing human neural progenitor cell grafts protect motor neurons but do not ameliorate motor performance and survival in ALS mice. Exp Mol Med 2009; 41(7): 487-500. [http://dx.doi.org/10.3858/emm.2009.41.7.054] [PMID: 19322031]

[72] Lepore AC, O'Donnell J, Kim AS, *et al.* Human glial-restricted progenitor transplantation into cervical spinal cord of the SOD1 mouse model of ALS. PLoS One 2011; 6(10): e25968. [http://dx.doi.org/10.1371/journal.pone.0025968] [PMID: 21998733]

[73] Kim KS, Lee HJ, An J, *et al.* Transplantation of human adipose tissue-derived stem cells delays clinical onset and prolongs life span in ALS mouse model. Cell Transplant 2013. [PMID: 24070071]

[74] Sun H, Hou Z, Yang H, *et al.* Multiple systemic transplantations of human amniotic mesenchymal stem cells exert therapeutic effects in an ALS mouse model. Cell Tissue Res 2014; 357(3): 571-82. [http://dx.doi.org/10.1007/s00441-014-1903-z] [PMID: 24906288]

[75] Ardeshiry Lajimi A, Hagh MF, Saki N, Mortaz E, Soleimani M, Rahim F. Feasibility of cell therapy in multiple sclerosis: a systematic review of 83 studies. Int J Hematol Oncol Stem Cell Res 2013; 7(1): 15-33.
[PMID: 24505515]

[76] Einstein O, Friedman-Levi Y, Grigoriadis N, Ben-Hur T. Transplanted neural precursors enhance host brain-derived myelin regeneration. J Neurosci 2009; 29(50): 15694-702.
[http://dx.doi.org/10.1523/JNEUROSCI.3364-09.2009] [PMID: 20016084]

[77] Burman J, Iacobaeus E, Svenningsson A, *et al.* Autologous haematopoietic stem cell transplantation for aggressive multiple sclerosis: the Swedish experience. J Neurol Neurosurg Psychiatry 2014; 85(10): 1116-21.
[http://dx.doi.org/10.1136/jnnp-2013-307207] [PMID: 24554104]

[78] Laterza C, Merlini A, De Feo D, *et al.* iPSC-derived neural precursors exert a neuroprotective role in immune-mediated demyelination *via* the secretion of LIF. Nat Commun 2013; 4: 2597.
[http://dx.doi.org/10.1038/ncomms3597] [PMID: 24169527]

[79] Boltze J, Lukomska B, Jolkkonen J. Mesenchymal stromal cells in stroke: improvement of motor recovery or functional compensation? J Cereb Blood Flow Metabol: Official J Int Soc Cereb Blood Flow Metabol 2014 May; 21
[http://dx.doi.org/10.1038/jcbfm.2014.94]

[80] Wei N, Yu SP, Gu X, *et al.* Delayed intranasal delivery of hypoxic-preconditioned bone marrow mesenchymal stem cells enhanced cell homing and therapeutic benefits after ischemic stroke in mice. Cell Transplant 2013; 22(6): 977-91.
[http://dx.doi.org/10.3727/096368912X657251] [PMID: 23031629]

[81] Tang Y, Wang J, Lin X, *et al.* Neural stem cell protects aged rat brain from ischemia-reperfusion injury through neurogenesis and angiogenesis. Journal of cerebral blood flow and metabolism : official journal of the International Society of Cerebral Blood Flow and Metabolism 2014 Apr.; 9
[http://dx.doi.org/10.1038/jcbfm.2014.61]

[82] Kim DY, Park SH, Lee SU, *et al.* Effect of human embryonic stem cell-derived neuronal precursor cell transplantation into the cerebral infarct model of rat with exercise. Neurosci Res 2007; 58(2): 164-75.
[http://dx.doi.org/10.1016/j.neures.2007.02.016] [PMID: 17408791]

[83] Jiang M, Lv L, Ji H, *et al.* Induction of pluripotent stem cells transplantation therapy for ischemic stroke. Mol Cell Biochem 2011; 354(1-2): 67-75.
[http://dx.doi.org/10.1007/s11010-011-0806-5] [PMID: 21465238]

[84] Carney BJ, Shah K. Migration and fate of therapeutic stem cells in different brain disease models. Neuroscience 2011; 197: 37-47.
[http://dx.doi.org/10.1016/j.neuroscience.2011.08.063] [PMID: 21946010]

[85] Bantubungi K, Blum D, Cuvelier L, *et al.* Stem cell factor and mesenchymal and neural stem cell transplantation in a rat model of Huntington's disease. Mol Cell Neurosci 2008; 37(3): 454-70.
[http://dx.doi.org/10.1016/j.mcn.2007.11.001] [PMID: 18083596]

[86] Zhao D, Najbauer J, Garcia E, *et al.* Neural stem cell tropism to glioma: critical role of tumor hypoxia. Mol Cancer Res 2008; 6(12): 1819-29.

[http://dx.doi.org/10.1158/1541-7786.MCR-08-0146] [PMID: 19074827]

[87] Marconi S, Castiglione G, Turano E, *et al.* Human adipose-derived mesenchymal stem cells systemically injected promote peripheral nerve regeneration in the mouse model of sciatic crush. Tissue Eng Part A 2012; 18(11-12): 1264-72.
[http://dx.doi.org/10.1089/ten.tea.2011.0491] [PMID: 22332955]

[88] Marconi S, Bonaconsa M, Scambi I, *et al.* Systemic treatment with adipose-derived mesenchymal stem cells ameliorates clinical and pathological features in the amyotrophic lateral sclerosis murine model. Neuroscience 2013; 248: 333-43.
[http://dx.doi.org/10.1016/j.neuroscience.2013.05.034] [PMID: 23727509]

[89] Park DH, Eve DJ, Sanberg PR, *et al.* Increased neuronal proliferation in the dentate gyrus of aged rats following neural stem cell implantation. Stem Cells Dev 2010; 19(2): 175-80.
[http://dx.doi.org/10.1089/scd.2009.0172] [PMID: 19569872]

[90] Shetty AK, Rao MS, Hattiangady B. Behavior of hippocampal stem/progenitor cells following grafting into the injured aged hippocampus. J Neurosci Res 2008; 86(14): 3062-74.
[http://dx.doi.org/10.1002/jnr.21764] [PMID: 18618674]

[91] Chen X, Lepier A, Berninger B, Tolkovsky AM, Herbert J. Cultured subventricular zone progenitor cells transduced with neurogenin-2 become mature glutamatergic neurons and integrate into the dentate gyrus. PLoS One 2012; 7(2): e31547.
[http://dx.doi.org/10.1371/journal.pone.0031547] [PMID: 22348101]

[92] Lazarov O, Mattson MP, Peterson DA, Pimplikar SW, van Praag H. When neurogenesis encounters aging and disease. Trends Neurosci 2010; 33(12): 569-79.
[http://dx.doi.org/10.1016/j.tins.2010.09.003] [PMID: 20961627]

[93] Jung YW, Hysolli E, Kim KY, Tanaka Y, Park IH. Human induced pluripotent stem cells and neurodegenerative disease: prospects for novel therapies. Curr Opin Neurol 2012; 25(2): 125-30.
[http://dx.doi.org/10.1097/WCO.0b013e3283518226] [PMID: 22357218]

[94] Weick JP, Liu Y, Zhang SC. Human embryonic stem cell-derived neurons adopt and regulate the activity of an established neural network. Proc Natl Acad Sci USA 2011; 108(50): 20189-94.
[http://dx.doi.org/10.1073/pnas.1108487108] [PMID: 22106298]

[95] Hattiangady B, Shuai B, Cai J, Coksaygan T, Rao MS, Shetty AK. Increased dentate neurogenesis after grafting of glial restricted progenitors or neural stem cells in the aging hippocampus. Stem Cells 2007; 25(8): 2104-17.
[http://dx.doi.org/10.1634/stemcells.2006-0726] [PMID: 17510219]

[96] Sánchez-Danés A, Consiglio A, Richaud Y, *et al.* Efficient generation of A9 midbrain dopaminergic neurons by lentiviral delivery of LMX1A in human embryonic stem cells and induced pluripotent stem cells. Hum Gene Ther 2012; 23(1): 56-69.
[http://dx.doi.org/10.1089/hum.2011.054] [PMID: 21877920]

[97] Doi D, Samata B, Katsukawa M, *et al.* Isolation of human induced pluripotent stem cell-derived dopaminergic progenitors by cell sorting for successful transplantation. Stem Cell Rep 2014; 2(3): 337-50.
[http://dx.doi.org/10.1016/j.stemcr.2014.01.013] [PMID: 24672756]

[98] Fernando CV, Moses D, Parish CL, Tomas D, Drago J, Horne MK. Creating a ventral midbrain stem cell niche in an animal model for Parkinson's disease. Stem Cells Dev 2010; 19(12): 1995-2007. [http://dx.doi.org/10.1089/scd.2009.0486] [PMID: 20415529]

[99] Cova L, Bossolasco P, Armentero MT, *et al.* Neuroprotective effects of human mesenchymal stem cells on neural cultures exposed to 6-hydroxydopamine: implications for reparative therapy in Parkinson's disease. Apoptosis : an international journal on programmed cell death 2012 Mar.; 17(3): 289-304.

[100] Somoza R, Juri C, Baes M, *et al.* Intranigral transplantation of epigenetically induced BDNF-secreting human mesenchymal stem cells: implications for cell-based therapies in Parkinson's disease. Biology of blood and marrow transplantation : journal of the American Society for Blood and Marrow Transplantation 2010 Nov.; 16(11): 1530-40.

[101] Proschel C, Stripay JL, Shih CH, Munger JC, Noble MD. Delayed transplantation of precursor cell-derived astrocytes provides multiple benefits in a rat model of Parkinsons. EMBO Mol Med 2014; 6(4): 504-18. [http://dx.doi.org/10.1002/emmm.201302878] [PMID: 24477866]

[102] Deng X, Liang Y, Lu H, *et al.* Co-transplantation of GDNF-overexpressing neural stem cells and fetal dopaminergic neurons mitigates motor symptoms in a rat model of Parkinson's disease. PLoS One 2013; 8(12): e80880. [http://dx.doi.org/10.1371/journal.pone.0080880] [PMID: 24312503]

[103] Rutishauser U. Polysialic acid in the plasticity of the developing and adult vertebrate nervous system. Nat Rev Neurosci 2008; 9(1): 26-35. [http://dx.doi.org/10.1038/nrn2285] [PMID: 18059411]

[104] Hallett PJ, Cooper O, Sadi D, Robertson H, Mendez I, Isacson O. Long-term health of dopaminergic neuron transplants in Parkinson's disease patients. Cell Reports 2014; 7(6): 1755-61. [http://dx.doi.org/10.1016/j.celrep.2014.05.027] [PMID: 24910427]

[105] Tuszynski MH, Thal L, Pay M, *et al.* A phase 1 clinical trial of nerve growth factor gene therapy for Alzheimer disease. Nat Med 2005; 11(5): 551-5. [http://dx.doi.org/10.1038/nm1239] [PMID: 15852017]

[106] Popescu IR, Nicaise C, Liu S, *et al.* Neural progenitors derived from human induced pluripotent stem cells survive and differentiate upon transplantation into a rat model of amyotrophic lateral sclerosis. Stem Cells Transl Med 2013; 2(3): 167-74. [http://dx.doi.org/10.5966/sctm.2012-0042] [PMID: 23413376]

[107] Hargus G, Cooper O, Deleidi M, *et al.* Differentiated Parkinson patient-derived induced pluripotent stem cells grow in the adult rodent brain and reduce motor asymmetry in Parkinsonian rats. Proc Natl Acad Sci USA 2010; 107(36): 15921-6. [http://dx.doi.org/10.1073/pnas.1010209107] [PMID: 20798034]

[108] Qiang L, Fujita R, Abeliovich A. Remodeling neurodegeneration: somatic cell reprogramming-based models of adult neurological disorders. Neuron 2013; 78(6): 957-69. [http://dx.doi.org/10.1016/j.neuron.2013.06.002] [PMID: 23791192]

[109] Mohamet L, Miazga NJ, Ward CM. Familial Alzheimer's disease modelling using induced pluripotent

stem cell technology. World J Stem Cells 2014; 6(2): 239-47.
[http://dx.doi.org/10.4252/wjsc.v6.i2.239] [PMID: 24772250]

[110] Kronenberg H. Adult mesenchymal stem cells. StemBook 2009.

[111] Lee JH, Yu HS, Lee GS, *et al.* Collagen gel three-dimensional matrices combined with adhesive proteins stimulate neuronal differentiation of mesenchymal stem cells. J Roy Soc, Interface / Roy Soc 2011 Jul.; 6;8(60): 998-1010.
[http://dx.doi.org/10.1098/rsif.2010.0613]

[112] Egea V, von Baumgarten L, Schichor C, *et al.* TNF-α respecifies human mesenchymal stem cells to a neural fate and promotes migration toward experimental glioma. Cell Death Differ 2011; 18(5): 853-63.
[http://dx.doi.org/10.1038/cdd.2010.154] [PMID: 21127499]

[113] Park HJ, Shin JY, Kim HN, Oh SH, Lee PH. Neuroprotective effects of mesenchymal stem cells through autophagy modulation in a parkinsonian model. Neurobiol Aging 2014; 35(8): 1920-8.
[http://dx.doi.org/10.1016/j.neurobiolaging.2014.01.028] [PMID: 24629674]

[114] Jeong CH, Kim SM, Lim JY, *et al.* Mesenchymal stem cells expressing brain-derived neurotrophic factor enhance endogenous neurogenesis in an ischemic stroke model. BioMed research international 2014. 2014:129145
[http://dx.doi.org/10.1155/2014/129145]

[115] Chan TM, Harn HJ, Lin HP, *et al.* The use of ADSCs as a treatment for chronic stroke. Cell Transplant 2014; 23(4-5): 541-7.
[http://dx.doi.org/10.3727/096368914X678409] [PMID: 24816449]

[116] Ng TK, Fortino VR, Pelaez D, Cheung HS. Progress of mesenchymal stem cell therapy for neural and retinal diseases. World J Stem Cells 2014; 6(2): 111-9.
[http://dx.doi.org/10.4252/wjsc.v6.i2.111] [PMID: 24772238]

[117] Nagahara AH, Merrill DA, Coppola G, *et al.* Neuroprotective effects of brain-derived neurotrophic factor in rodent and primate models of Alzheimer's disease. Nat Med 2009; 15(3): 331-7.
[http://dx.doi.org/10.1038/nm.1912] [PMID: 19198615]

[118] Jönhagen ME. Nerve growth factor treatment in dementia. Alzheimer Dis Assoc Disord 2000; 14 (Suppl. 1): S31-8.
[http://dx.doi.org/10.1097/00002093-200000001-00006] [PMID: 10850728]

[119] Sanchez-Ramos J, Song S, Sava V, *et al.* Granulocyte colony stimulating factor decreases brain amyloid burden and reverses cognitive impairment in Alzheimer's mice. Neuroscience 2009; 163(1): 55-72.
[http://dx.doi.org/10.1016/j.neuroscience.2009.05.071] [PMID: 19500657]

[120] Chong ZZ, Li F, Maiese K. Erythropoietin requires NF-kappaB and its nuclear translocation to prevent early and late apoptotic neuronal injury during beta-amyloid toxicity. Curr Neurovasc Res 2005; 2(5): 387-99.
[http://dx.doi.org/10.2174/156720205774962683] [PMID: 16375720]

[121] Diógenes MJ, Outeiro TF. Neurotrophic factors as a protective strategy in Parkinson's disease. CNS Neurol Disord Drug Targets 2010; 9(6): 754-63.

[http://dx.doi.org/10.2174/187152710793237449] [PMID: 20942787]

[122] Butzkueven H, Emery B, Cipriani T, Marriott MP, Kilpatrick TJ. Endogenous leukemia inhibitory factor production limits autoimmune demyelination and oligodendrocyte loss. Glia 2006; 53(7): 696-703.
[http://dx.doi.org/10.1002/glia.20321] [PMID: 16498619]

[123] Im W, Kim M. Cell Therapy Strategies *vs.* Paracrine Effect in Huntington's Disease. J Mov Disord 2014; 7(1): 1-6.
[http://dx.doi.org/10.14802/jmd.14001] [PMID: 24926403]

[124] Im W, Ban J, Lim J, *et al.* Extracts of adipose derived stem cells slows progression in the R6/2 model of Huntington's disease. PLoS One 2013; 8(4): e59438.
[http://dx.doi.org/10.1371/journal.pone.0059438] [PMID: 23565152]

[125] Jeon D, Chu K, Lee ST, *et al.* Neuroprotective effect of a cell-free extract derived from human adipose stem cells in experimental stroke models. Neurobiol Dis 2013; 54: 414-20.
[http://dx.doi.org/10.1016/j.nbd.2013.01.015] [PMID: 23376682]

CHAPTER 5

Cancer Stem Cells in Brain Tumorigenesis

Maria Linda Cruceru[1,*] and **Adrian Claudiu Popa**[1,2]

[1] *Carol Davila University of Medicine and Pharmacy, Department of Cellular and Molecular Medicine, Bucharest, Romania*

[2] *Army Centre for Medical Research, Bucharest, Romania*

Abstract: The presence of cancer stem cells in brain tumors has been suggested in the recent years, extrapolating from other types of malignancies. The actual debate centers on the possibility that this small group of cells or a single cell can initiate and maintain such an aggressive malignancy status. Cancer stem cells are a major theme for aggressive brain tumors, their extensive study offering potential biomarkers for identification of therapy targets based on abnormal and connected signaling pathways. This chapter presents the phenomenology of cancer stem cells in their microenvironment, in its astonishing complexity. Identification markers, pathologic signaling profiles and cross-talk, microenvironment interactions and novel therapies targeting cancer stem cells are discussed. Continuous studies are researching genetic and epigenetic modifications associated with malignant evolution in glioma cancer stem cells. From the complex entangled signaling pathways presented, one must extract only the important molecules involved in oncogenesis onset and propagation, putting aside the coincidentally modified molecules that can be misleading. After removing the "smoke screen" of oncogenic irrelevant, but modified molecules, there remain the true therapy targets, to address by specific therapies in a combined manner to overcome the adapting processes governed by signaling pathway cross-talk.

Keywords: Anti cancer stem cell therapy, Brain tumors, Cancer stem cells, Signaling pathways.

[*] **Corresponding author Maria Linda Cruceru:** Carol Davila University of Medicine and Pharmacy, Department of Cellular and Molecular Medicine, Bd. Eroii Sanitari 8, Sector 5, Bucharest, Romania; Tel: +40 723169349; Fax: +40 213124885; E-mail: maria_lindabv@yahoo.com.

Cristiana Tanase & Monica Neagu (Eds.)

INTRODUCTION

Potential cancer stem cells have been identified and characterized in several human solid tumors, including aggressive brain tumors [1 - 4].

During the last decade, intensive research is undergoing on various tumors associated with an aggressive course and many studies have been dedicated to the involvement of cancer stem cells in cancer tumorigenesis – between 2000 and 3000 on PubMed. Concerning brain tumors, recent papers have focused on the role of cancer stem cells in tumor initiation, progression and relapse, as well as on their biological and clinical characterization [5 - 9].

The high grade gliomas – anaplastic astrocytoma (WHO grade III) and glioblastoma (WHO grade IV) – have a very poor prognosis and, despite intensive conventional therapy, the median survival rate for patients with glioblastoma (GBM) is approximately 1 year [10].

The most recent discoveries regarding glioma cancer stem cells (GCSCs) have provided a valuable input on the evaluation of aggressiveness, recurrence and treatment resistance [11].

This chapter will highlight the features of cancer stem cells (CSCs) and the challenges that they raise, as well as conventional *versus* specific signaling pathways involved in the initiation, maintenance and propagation of brain tumor malignant phenotype.

Cancer Stem Cells and the Tumor Malignancy Grade

Nowadays, the general consensus is that GCSCs and neural stem cells (NSCs) have some common features, despite their differences: both of them show pluripotency and self-renewal capacity and even the tendency to develop spheres when cultivated *in vitro* [9]. The GCSCs can present *in vitro* or *in vivo* heterogeneity, genetic alterations, abnormal proliferation and differentiation patterns [12 - 15].

Regarding the ability of the GCSCs to form spheres in culture, the more aggressive the tumor is (as shown by injection into immune-compromised mice), the more spheres will appear *in vitro* [16].

In order to study the cancer stem cells in brain tumors, one must follow few generic steps:

1. To isolate the cancer stem cells from tumoral tissue (tissue identified macroscopically by a pathologist), using specific stemness markers;
2. To pursue the development of spheroid cell cultures;
3. To observe the ability of stem cells in the spheroid colonies to differentiate into neurons, astrocytes and oligodendrocytes [16, 17];
4. To verify the malignant character of the cells isolated in the spheroid cultures when injected into mice with compromised immune system.

Another feature observed in GCSCs when compared with NSCs is their resistance to aggressions (chemotherapy drugs). Also, differences may exist regarding proliferation pattern and the markers presented [9, 10]. Other types of tumors derived from nervous tissue, such as medulloblastoma, show high proliferation rates – up to 30% proliferation rate in cells expressing vimentin in the perivascular niche in contrast with 1% proliferation rate of NSCs from the subventricular zone [18].

Compared to NSCs, GCSCs show more resistance to treatment, being susceptible to develop new tumors and to promote recurrence [9].

In order to fully understand the GCSCs we must understand NSCs. Although a great quantity of information is available on NSCs, some aspects remain unclear and are theoreticized based on similarities with the hematopoietic stem cells (HSCs). For example the maintenance of quiescent phenotype of NSCs is supposed to imply epigenetic regulation, similar to HSCs. However, the exact signals are yet to be discovered.

In the same manner, it has been speculated that a cytokine could be the promoter of NSC quiescence, and also stimulate the proliferation of committed progenitors. Such a cytokine could belong to the IL-6 family, since they show strong activation of JAK-STAT signaling, especially of STAT3, similar to the hematopoietic system [10]. Due to genetic modifications in any downstream signaling molecule, the cytokine could become, instead of promoter of quiescence, promoter of self-renewal for the cancer stem cells. The general consensus is that cancer stem cells may acquire genetic mutations that would grant them immortality, increased self-renewal ability and the capacity of blocking the differentiation of their progenitors [9].

The GCSCs could also be much more heterogeneous compared to NSCs, since multiple clones of cancer stem cells can arise from one initial clone, through mechanisms not fully understood [19 - 21].

Regarding the character of tumor inducing cells, some cellular markers were described to help differentiate between high grade and low grade gliomas. It has been found that both high and low grade glioma exhibit cellular markers of glial progenitor cells and only high grade glioma also exhibits markers consistent with potential for neuronal differentiation [22].

Another unanswered question is whether the capacity of multilineage emergence observed *in vivo* for high grade glioma can be achieved also for low grade glioma through progression to high grade. Most of the researchers consider that cells in low grade gliomas are in a more differentiated state. Some cellular markers are identified to associate with benign or malignant intracranial tumors, some of them are characteristic for stem cells.

Comparative studies between high-grade glioma (glioblastoma multiforme), low-grade glioma and benign astrocytoma have discovered genetic modifications that could represent important events for the progression of glioma: from chromosomal gains or losses, to mutations (such as IDH1 mutations found in low-grade astrocytoma), DNA methylation with DNA methylase 1 (DNAMT1) and methyl guanine methyl transferase (MGMT) and even promoters for EGFR [23 - 26].

The GCSCs are considered to be true cancer reservoirs, therefore the therapy strategies for brain cancers are blocked by the resistance to treatment shown by these cells. Even if they remain the only surviving cells after an aggressive treatment that can destroy most of the other tumoral cells, the cancer stem cells are able to regenerate the entire tumor [27, 28]. The same as for glioblastoma, the stem cells isolated from low-grade gliomas [23, 29 - 31] are more resilient to chemotherapy than their more differentiated progenies [32].

Cancer Stem/Tumor-Initiating Cells

The concept of cancer stem cells is not new; Rudolf Wirchow suggested more than 150 years ago that cancer arises from cells similar to those found in embryonic tissue (Virchow, 1858). Years later the concept of cancer stem cells emerged and nowadays there are proofs that cancer cells wrongfully use genetic and epigenetic mechanisms that are present specifically in the embryonic developmental programs.

Two models are proposed for tumor generation: hierarchical and stochastic models [33, 34]. According to the hierarchical model, the malignant transformation of one glial precursor or one stem cell leads to glioma onset and subsequent development. In contrast, the stochastic model sustains that every cell in the tumor has a behavior similar to a stem cell, deriving from committed or differentiated cells that – through genetic mutations – acquire immortal proliferative capacity [33].

CGSCs and NSCs have some common features: multipotency and a great self-renewal capacity [33, 35]. Some of the features of CGSCs are typical to most of cancer stem cells and are comprised in their definition: they are able to initiate cancer upon injection in animals (same species or allogenic, with compromised immune system), they present genetic alterations and aberrant or blocked differentiation.

One should also take into account that the frequency of cancer stem cells in a bulk of tumoral tissue is underestimated due to limitations of procedures used to identify them (e.q. xenograft transplantation, spherical cell cultures).

Latest findings suggest that neural progenitor cells (NPCs) or NSCs are able to give rise to glial tumors through abnormal Notch signaling activation [36]. One hypothesis is that NPC and NSC give rise to high-grade gliomas and the more committed cells (oligodendrocyte lineage-restricted progenitors, astrocyte lineage-restricted progenitors) will generate low-grade gliomas [22].

High-grade glioma was formed from NSCs as substrate cells when epidermal growth factor receptor (EGFR) was introduced and p19ARF and p16INK4a tumor suppressors were knocked out [37].

When the same tumor suppressors are inactivated in a mature astrocyte cell culture, and the cells are maintained in a serum free medium with EGF, cells are prone to transformation and dedifferentiation [37]. So, loss of p16INK4a and p19ARF normal activity, together with a high EGF activation (possible through EGFR constitutive activation) can provoke a high grade glioma, regardless of the cell origin – differentiated astrocyte or NSC.

EGF was found to aid renewal of GCSCs through expression of ID3 (inhibitor of differentiation 3), which induces expressions of IL-8 and IL-6 [30]. In the same time, ID3 inhibits migration of cells through a diminished expression of matrix-metallo-proteinases mediated by inhibition of p27 (KIP1)-RhoA [31]. Therefore, inhibition of EGFR will decrease expression of ID3, which will hinder the differentiation, but will promote migration of cells [23, 38]. Taken together, these studies underline the duality of EGFR modulation and postpone therapy with EGFR targeted agents.

GBMs can be divided into several subtypes based on gene expression patterns for PDGFR alpha, EGFR, neurofibromin 1 (NF1) and isocitrate dehydrogenase 1 (IDH1) [31]. Data from 202 patients identified 1740 genes, which clustered into 4 subtypes with 210 gene signatures: classical, mesenchymal, neural and proneural. An important finding of the same study is that available treatments seem to increase the survival time only for mesenchymal and classical subtypes. There are no clues whether these subtypes originate from the same cell or from different kinds of cells.

Despite various differences in traditional molecular subclasses of high-grade gliomas, further studies are needed in order to divide the GBM subtypes depending on their cell origin or different mutations.

Markers Related with GCSCs

CD133

CD133 (prominin 1) is an intensively studied stem-cell surface marker that is currently used for identification and isolation of cancer stem cells (CSCs) in malignant brain tumors. CD133 is a transmembrane protein of the prominin family, being the first member discovered. It stands as a marker for the self-renewal feature of GCSCs and *in vitro* clonogenic studies [14, 39].

One common hypothesis is that CD133+ CSCs initiate formation of gliomas, since in some tumors up to 50 % of the cells are positive for CD133 (classical percentage 0.1-50%) [17, 40]. In this respect, the percentage of CD133+ cells might serve as a prognostic factor for overall survival rates, independently of tumor grade. It is supposed that increased levels of CD133 expression may confer a high risk of dissemination, although low CD133 expression may not rule out this possibility [41].

Currently, it is well established that CD133-positive GCSCs are highly resistant to conventional therapies [27, 28]. A specific homeobox gene expression was identified in association with CD133, providing the first clinical evidence of a 'glioma stem cell' or 'self-renewal' phenotype involvement in treatment resistance of glioblastoma, thus predicting poor survival [42]. In addition to CD133, combinations of different standard markers have been tested recently. For example, the presence of double CD133+/Ki67+ positive cells may be an indicator of tumor progression and unfavorable prognosis [43].

The tumor microenvironment stimulates the self-renewal ability of CD133-positive human GCSCs [44]. CD133+ cells are characterized by HIF-1alpha up-regulation and expression of CXCR4 chemokine receptor, CD44low and A2B5 surface marker. Due to the fact that CD133-positive glioma stem cells are a potential source for surrounding tissue infiltration, they might represent novel therapeutic targets in brain tumors [5].

Despite the correlation of CD133 expression with poor clinical outcome in brain tumors, the use of CD133 as a single glioma stem cell marker is insufficient to characterize the entire population of self-renewing tumor cells [45].

Association of CD133 with Nestin Expression

Nestin is another protein marker associated with GCSCs, with implication in signaling and metabolism, especially in cell proliferation and migration [46, 47]. Nestin is frequently expressed in high grade gliomas, especially in primary tumors in patients with dissemination, being a predictive marker for poor survival rate [48].

An association between co-expression of nestin/CD133 and clinical outcome predictive power was described in various brain tumors [47]. Patients with tumors positive for nestin and CD133 had the worst prognosis. Positive expression of both CD133 and nestin was detected around blood vessels for CD31 and the association of CD31 with either CD133 or nestin was found in endothelial cells, as well as in cancer stem cells [48].

Generally, positivity for nestin/CD133 was an indicator of poor prognosis and demonstrated a better correlation with clinical outcome and survival rate in comparison with histological grading. Thus, the combination of the two markers might indicate the aggressiveness of gliomas and represents a possible marker of recurrence [49, 50].

Musashi-1

Musashi-1 protein is a regulator of cell fate, displaying anti-differentiation properties. Musashi-1 is expressed in GBM stem cells and in neurospheres derived from brain tumors, being correlated with tumor grade and proliferation rate [51]. Therefore, this molecule could be a marker for proliferating GCSCs. Alongside Musashi-1, maternal embryonic leucine zipper kinase (MELK), nestin, Sox2 and CD133 were also found, being correlated with the ability to re-initiate tumorigenesis [52]. *In vitro* experiments with leukemia inhibitory factor (LIF) and TGF β increased glioma-initiating cell renewal through Musashi-1, Sox2, and nestin expressions [53].

Some markers are co-expressed in GCSCs: nestin, Musashi-1, YB-1 (Y box-protein 1), SOX2, and Bmi-1 [54]. The decrease of Musashi-1 in glioma cells promoted the elongation of the cell cycle, also reducing the signaling through Notch and PI3 kinase-Akt, as a consequence of PTEN (phosphatase and tensin homologue) and Numb up-regulation [55].

In highly malignant tumors, Musashi proteins play an essential role, determining the activation of various signaling pathways, thus blocking differentiation.

Other Markers

SOX2 is one of the important genes implicated in the self-renewal of stem cells, NSCs especially [56]. SOX2 is essential for modulating the activity of the following four transcription factors: Oct4, SOX2, Myc and Klf4 [57]. Recently, it has been demonstrated that induction of SOX2 initiates reprogramming, through a hierarchical program [58]. SOX2 expression in GCSCs [51] was shown to be strongly correlated with stemness maintenance [53], being regulated by TGF β signaling pathway [59]. SOX2 pathway components have been recently demonstrated to function as a therapeutic target, as SOX2 silencing is followed by loss of proliferation and tumorigenicity of GCSCs [60].

MELK, one of the AMP kinases, is overexpressed in NSCs and malignant brain tumors [61]. *In vitro* experiments on GCSC-enriched cell cultures have established their increased expression. *In vivo* studies, performed on transgenic MELK mice, sustain the importance of MELK expression in NSCs. MELK is a prerequisite for self-renewal of NSCs and GCSC proliferation.

MELK is highly expressed in GCSCs. Its level of expression was found to be correlated with the human astricytoma malignancy grade: increased levels of MELK expression were detected in highly invasive GBM in comparison with the benign pilocytic astrocytoma [61]. In addition, an up-regulation of MELK expression was identified in glioblastoma tissue [62]. Consequently, MELK could be targeted in new therapies.

Cancer Stem-Cell Niche

The combination of cell-autonomous properties and extrinsic signals emitted by niche stromal cells promote the self-renewing and differentiation capacity of stem cells [1]. The stem-cell niche has multiple functions: serves as stem cell "anchoring site", which seems to 'bind' them to support cells and to facilitate communication, by tight and gap junctions; it represents a dynamic microenvironment generating extrinsic signals that sustain stemness and regulate stem cell number, division, proliferation and migration in the proximity of blood vessels [63].

Stem-cell niches were discovered in several tissues, including neural structures. They are all characterized by cooperation of stem cells with vascular niches (Fig. **1**). Deregulation of the microenvironment has been proven to cause dysplastic features and malignant transformation in brain tissue [64]. Moreover, a significant population of endothelial cells seemed to have originated from the neoplastic clone in glioblastoma [64], suggesting that tumors are able to generate their own niches. Glioblastoma cell transdifferentiation into endothelial cells was described [65].

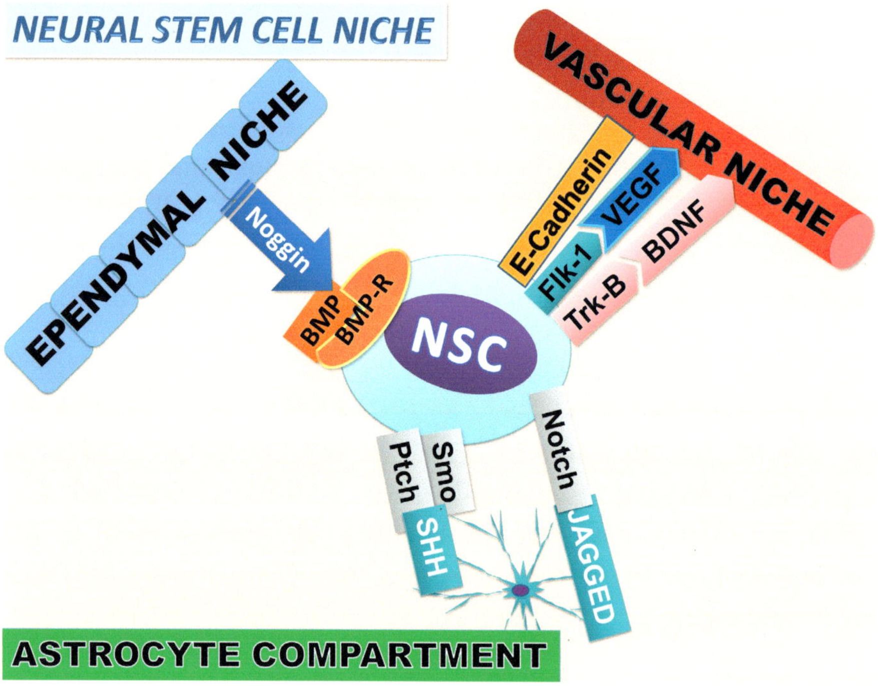

Fig. (1). Neural stem-cell niche – molecular mediators involved in renewal of adult stem cells (adapted from [10]).

It was shown that NSCs are anchored in the vascular microenvironment, and CD133+/nestin+ GCSCs are found in areas adjacent to capillaries in brain tumors [1, 18, 66]. Stem cells interact directly with non-stem cells, defining the niche dynamics.

As it has been shown, stem-cell niches are major players in the regulation of stem cell behavior, highlighting their involvement in controlling cell lineage, proliferation, self-renewal rate; therefore, they can have a contribution in tumorigenesis and in the degree of malignant transformation [67, 68]. Experimental studies demonstrate that, although irradiation induces apoptosis of the main cell population of the brain tumor, perivascular niche tumor cells are relatively radio-resistant, exhibiting a significant activation of the Akt/mTOR pathway [18].

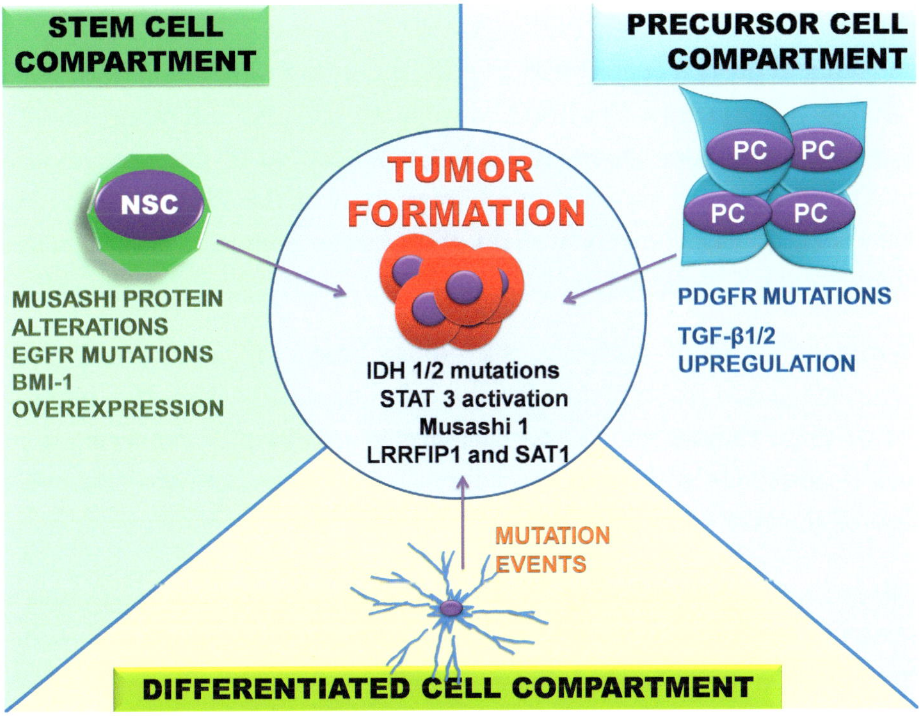

Fig. (2). Events associated with tumorigenesis from stem cells and progenitor cells.

Signaling Pathway Deregulations

GCSCs present common features with NSCs; however, they show subtle genetic and epigenetic alterations that conduct to differentiation blockage, immortal growth and aggressively invasive phenotype. Signaling molecules acting in cancer stem or progenitor cells of the brain tumors can be considered as both potential diagnosis markers and novel therapeutic targets [69]. High-grade gliomas require innovative treatment strategies, probably targeting several signaling pathways simultaneously, since they are very resistant to conventional therapies, probably due to the presence of GCSCs (Fig. **2**).

Tyrosine Kinase Receptors and their Downstream Effectors: PI-3'-Kinase (PI3K) and Akt

Many tyrosine kinase receptors have been extensively studied in gliomas: EGFR, PDGFR, FGFR, Met-receptor [15]. Fibroblast growth factor (FGF) and epidermal growth factor (EGF) stimulate proliferation of both GCSCs and NSCs. Downstream the receptor, the PI3K/Akt/mTOR pathway has been very well characterized and plays a central part in gliomas. In high-grade gliomas, namely GBM, EGFR is frequently amplified, displaying specific activating mutation; PDGFR expression is also increased [31]. The activation of the tyrosine kinase receptor signaling results in constitutive activation of the PI3K pathway, with significant consequences on cell survival, therefore important in several cancers, mostly hematological malignancies [70].

Downstream of the PI3K pathway, Ak mouse breed thymoma (Akt) plays a critical role in the survival and proliferation of glioma cells [15]. The aggressive behavior of these tumors and their resistance to treatment – rendering a poor prognosis – correlates with both the expression of stem-cell markers and the activation of the Akt pathway.

The contribution of Akt signaling to the formation and growth of gliomas has been proven in several *in vitro* studies [9]. A reduced constitutive activation of mitogen activated protein (MAP)-kinase Erk completes the role of Akt, stimulating the proliferative activity of the tumor cells; this establishes both Akt and Erk as important targets for personalized therapy [71]. Notably, in cells where certain EGFR mutants are expressed, Erk1,2 MAP-kinase shows rather low activity, which seems to be due to the induction of MAP-kinase phosphatases

[72], enzymes that hinder MAP-kinase hyperactivity.

The activation of the mTOR (mammalian target of rapamycin) pathway by EGF, *via* protein kinase C alpha (PKCα), seems to be specific for gliomas, since it is using an autonomous circuit different from the standard Akt-mTOR activation [73]. PKCα activation is involved in generation of proliferative signals, by cleavage of the ectodomain in members of EGF family, through a disintegrin and metallo-proteinase (ADAM). As a result, endodomains of transmembrane EGF precursors get phosphorylated at Ser/Thr residues, promoting selection of these receptor precursors as ADAMs substrates [74]. These findings suggest the use of protein kinase C inhibitors in combination therapies.

Studies on other types of cancer suggest that type I pan PI3K inhibitors (including gamma p110, delta, beta and alpha isoforms) are expected to be more effective [70] than inhibitors specific to one isoform.

Apparently, PI3K inhibitors alone are able to diminish proliferation of glioblastoma cells, but not to cause apoptosis [75, 76], therefore there is a need of complex new drug combinations. A promising one is the association of cyclin-dependent kinase (CDK) 1/2 inhibitor roscovitine with the PI3K inhibitor PIK-90, that induces apoptosis of glioma cells only when combined and not in monotherapy – which is defined as synthetic lethality [77].

The most frequent anomaly in high-grade gliomas is EGFR gene amplification; moreover, EGFR is constitutively active in half of the EGFR amplified tumors, determining glioma growth [78]. Glioma cells expressing wild-type EGFR are recruited in accelerated *in vivo* proliferation *via* a paracrine mechanism, involving secretion of IL-6 and LIF, driven by Delta-EGFR [78]. Furthermore, EGFR over-expressing/Delta-EGFR glioblastoma can gain resistance to EGFR inhibitors by phosphorylation of PTEN at tyrosine 240 [79]. As already mentioned, inhibitors of tumor suppressors cooperate with EGF signaling pathway, resulting in inhibition of differentiation and finally inducing inflammatory cytokines like IL-8 and IL-6 [30, 37].

Mutation and/or overexpression of EGFR can be found in various carcinomas: ovarian and prostate, colon, lung, breast, head and neck, excepting the hematological disorders. Nevertheless, some glioblastoma characteristic mutations were identified [80].

Gliomas present plastic and robust oncogenic EGFR signaling networks, which can adjust to single molecule inhibition, and therefore promoting them as targets for combined therapy – with EGFR inhibitors and inhibitors of downstream signaling molecules [32].

Glioma cells reveal multiple genetic alterations, as the PTEN and p53 loss, resulting in significant therapy-resistant tumors. PTEN and p53 are known to inhibit NSC self-renewal; their deficiency determines an important decrease of apoptotic index in neurospheres [9].

The pharmacological inhibition of Akt disrupts formation of neurospheres, stimulates apoptosis, decreases migration and invasion *in vitro* and significantly delays the formation of intracranial tumors [68, 69].

The use of tyrosine kinase inhibitors has, so far, failed to inhibit glioblastoma development in clinical trials (www.clinicaltrials.gov). The reason is speculated to be the possibility that receptor tyrosine Kinase signaling intervenes late in glioma evolution, promoting cell survival and proliferation, while the events that trigger tumor formation remain elusive.

PDGF

All PDGF ligands and their specific receptors, PDGFR α and β, respectively, are expressed in glioblastoma cells. This abundance determines the stimulation of tumor cells by complex PDGF/PDGFR, especially PDGFR α. In human low-grade gliomas PDGFR α is overexpressed, while in high-grade gliomas – GBM – expression is enhanced even further. PDGFR β is shown to induce angiogenesis, which is a characteristic for glioblastoma relapse [81].

Interestingly, intracerebral injection of a PDGF-encoding retrovirus is able to drive brain tumors of either oligo- or monoclonal origin, tumors that reveal a co-expression of PDGF β chain with α receptor (PDGFR α) mRNA [82]. Most of the tumors induced by PDGF exhibit features of glioblastoma multiforme or of a primitive neuro-ectodermal tumor and are nestin positive, suggesting their origin in an early neuroglial progenitor [10, 82]. Moreover, NSCs of the subventricular zone in adult brain express PDGFR α, which is an essential factor for physiological oligodendrocyte formation. PDGF-targeted treatments of NSCs lead to neurogenesis blockage and hyperplasia, an aspect similar to the early phases of tumorigenesis [83].

Notch

Notch signaling inhibits degradation of nestin during differentiation of NSCs, *via* a mechanism possibly involving ubiquitin-proteasome pathway [84].

When occurring in NSCs, constitutive Notch signaling is tumorigenic, promoting astroglial lineage development [85].

Many components of the Notch pathway (ligands, receptors, targets) are detected in GCSCs; they seem to be involved in glioma pathogenesis [22].

Increased Notch signaling promotes tumor development, while blockage of Notch pathway suppresses proliferation and/or survival. Knockdown of Notch1 and its ligands inhibits proliferation and determines apoptosis [86].

In gliomas, Notch pathway can directly determine the up-regulation of nestin expression. Furthermore, constitutive Notch signaling activation in glioma cell lines induces the formation of neurospheres and promotes cell proliferation [84, 87].

Notch signaling seems to be closely linked to MAP-kinase signaling, with demonstrated EGFR and Notch signaling cross-talk and effects on tumor angiogenesis [88].

The expression of Notch receptors correlates with the presence of various receptors that play a role in angiogenesis: PDGFR β, VEGFR2 and VEGFR3 [62]. Components of Notch signaling pathway, especially Delta-like 4, contribute to glioma resistance to anti-VEGF therapy [89, 90]. Overall, inhibition of Notch signaling appears as an attractive therapeutic alternative.

TGF β

Cells that seem to be responsible for glioblastoma initiation present a CD44high/Id1high phenotype [91]. These cells respond to TGF β signals, which contributes to their oncogenicity [59]. Also, patients with glioblastoma reveal high plasma levels of TGF β [92]. TGF β1 and SOX4 are shown to be up-regulated in GBM [91]. GCSC self-renewal can be induced by TGF β, *via* LIF [53]. LIF is a cytokine belonging to the IL-6 family, responsible for self-renewal in embryonic stem cells *via* STAT3 activation. The IL-10 cytokine also correlates with the TGF β expression, both at genetic and protein levels. Both markers are highly expressed in tumor neurospheres (supposed to arise from GCSCs) than in regular cultivated glioma cells, even in case of the same tumor of origin. This overexpression in GCSCs correlates with glioma grading [93]. Furthermore, enhanced TGF β signaling *via* Smad proteins appears to signify unfavorable prognosis in GBM, since it induces proliferation of tumor cells in case the PDGF β gene is induced by demethylation [94].

TGF β receptor inhibitors exert their action on GCSCs in the perivascular niche, which is of great importance from therapeutic point of view [91].

Bone Morphogenetic Proteins

Members of TGF superfamily, bone morphogenetic proteins (BMP) are implicated in regulation of cell maintenance, growth and differentiation for both normal tissue and tumors. BMPs, especially BMP4, negatively regulate GCSCs [95]. Xenotransplantation experiments using BMP4 have shown that tumor expansion is prevented. The mechanism implies BMP receptors and Smad activation, followed by an antiproliferative effect, which acts by inducing neural differentiation [95].

BMPs are vital for cell differentiation in the CNS [96]. Human glioma cells present both types of BMP receptors: type IA (BMPR-1A) as well as type IB (BMPR-1B). Malignant gliomas show a substantial increase in BMPR-1B, in contrast to low-grade astrocytomas [97]. Epigenetic BMPR-1B silencing in a glioma subset resulted in astroglial differentiation blockage [10]. BMP itself can be a valuable target for therapy, as well as its target gene, induced or suppressed by high BMP4 expression.

Sonic Hedgehog and Wnt

Sonic Hedgehog (SHH)-Gli signaling participates in self-renewal regulation and contributes to the tumorigenic abilities of GCSCs; glioblastoma has been associated with active SHH signaling pathway [69, 98]. Self-renewal and proliferation of GCSCs are suppressed by inhibition of SHH signaling, together with apoptosis increase in glioblastoma cells [99, 100]. In PDGF-induced gliomas, the malignancy grade was found to be correlated with Gli activity [53]. Xenotransplantation studies that investigated the functions of Notch, Hedgehog and Wnt signaling found Hedgehog indispensable for tumor initiation [101].

Wnt signaling implies β catenin translocation to the nucleus, where it modulates particular target genes, as transcription factor. It was shown that Wnt/β catenin signaling plays a role in GCSCs, as well as in NSCs [102]. Wnt/β catenin signaling contributes to glioblastoma resistance to radiotherapy, which makes it a suitable therapeutic target for GCSCs [103], since this pathway is involved in the cell migration, leading to glioblastoma invasiveness, also promoting epithelial to mesenchymal transition of tumor cells [104].

STAT3

Tumorigenesis of the brain involves several of the STAT (signal transducer and activator of transcription) family members. There are seven STATs encoded by the human genome (STAT1, 2, 3, 4, 5A, 5B and 6), which become phosphorylated by four Janus kinases (JAKs), in turn connected to over 30 cytokine receptors [105]. Upon binding to the ligand, which requires receptor dimerization, the cytokine receptor suffers conformational changes in the transmembrane domain [106], leading to activation of JAK, pre-bound to the receptor's endodomain [107]. JAKs cross-phosphorylate each other and the receptor endodomains, attracting STAT proteins through their SH2 domains. Upon binding to the receptors, STATs become phosphorylated at tyrosine residues of the transactivation domain, which enables the formation of a STAT dimer that detaches from the receptor and migrates to the nucleus, regulating gene expression [108].

Proofs on STAT3's involvement in glioblastoma biology are accumulating in recent studies [109, 110]. STAT3 activation is also related to mesenchymal transformation in GBM and inversely correlates with the outcome [110, 111]. Using specific STAT3 inhibitors disrupts maintenance and proliferation of GCSCs [112]. Cancer stem cells in gliomas reveal several active signaling circuits upstream of STAT3. Targeting molecules involved in these circuits inhibit the activation of STAT3, leading to self-renewal and growth of GCSCs [113]. Similar to other cancers, survival and proliferation of GBM cells are favored by the constitutive activation of STAT3 [35]. The precise tyrosine kinase accountable for constitutive STAT3 activation in gliomas still remains to be discovered.

Table 1. Key molecules operating in normal neural stem cells and in glioma cancer stem cells (adapted from [10]).

Protein Name	Normal Function	Role in Oncogenesis	Citation
β catenin Catenin β1	Cell adhesion and signal transduction; transcription factor (Wnt pathway) Important for NSC /progenitor cell proliferation.	Mutated in several human glioma, medulloblastomas	[114]
Bmi-1 Polycomb complex protein	Transcription factor (chromatin regulator), gene transcription repressor. Proliferation of neural and hematopoietic stem cells, but not involved in progenitor functionality. Regulated by Shh; regulates p16/INK.	Required for cancer stem-cell function. Overexpressed in brain tumors	[115]
EGFR Epidermal growth factor receptor	Membrane receptor; tyrosine kinase. EGF-dependent NSC proliferation.	Often amplified and mutated in high-grade gliomas	[42, 116]
Gli1, Gli2, Gli3	Components of Shh and Wnt pathway (transcription factors, activators or repressors). Gli1 protein expressed in NSC.	Originally isolated from glioblastoma. GLI expressed in GBM, astrocytoma, oligodendroglioma	[99]

(Table 1) contd.....

Protein Name	Normal Function	Role in Oncogenesis	Citation
IDH1/IDH2 Isocitrate dehydrogenase 1/2	Metabolic enzyme. Catalyses neomorphic formation of 2-hydroxyglutarate.	Frequent in astrocytomas, oligodendrogliomas and glioblastomas; IDH1/2 mutations in acute myeloid leukemia correlate with better outcome	[117, 118]
MELK Maternal embryonic Leucine zipper kinase	Protein kinase. Functional regulation of transcription factor.	Highly expressed in NSC and brain tumors	[61, 119]
MSI1, Musashi, RNA-binding protein Musashi homologue 1	Regulates the expression of target mRNAs at the translation level.	Protein alteration favors tumorigenesis. NSC and HSC display Musashi alteration. Consistently correlated with the tumor proliferation in gliomas	[120]
Nestin	Intermediate filament. Trafficking and distribution of factors during progenitor cell division.	Expression correlated with tumor grade	[47]
N-myc Proto-oncogene protein	Transcription factor. Neural progenitor cell expansion and inhibition of neuronal differentiation.	N-MYC gene is amplified in some human brain tumors. Overexpression correlates with tumor progression	[121]
Notch1 Neurogenic locus notch homologue protein 1	Membrane receptor for Jagged1, Jagged2, Delta1. Involved in the maturation of NSC and NSC function's maintenance.	Collaborates with epigenetic silencers to promote malignant tumors by Rb silencing	[122]
p16-INK4a/CDKN2A Cyclin-dependent kinase inhibitor 2A	p53 activator. Inhibitor of NSC proliferation; overexpressed in Bmi-/- mutant stem cells.	Absent or altered expression in various GBM	[123]
p21 cip1 Cyclin-dependent kinase inhibitor 1	Kinase inhibitor. Quiescence of NSC and HSC. Mediates inhibition of proliferation by p53 pathway.	Down-regulated in tumor transformation	[124]
p53 Cellular tumor antigen	Transcription regulator. Negative regulation of NSC proliferation, tumor suppressor. Synergistic with Ptch, Rb, PARP-1, Ink4c.	Mutated in some brain tumors	[125]
PDGFR α Platelet-derived growth factor receptor α protein	Membrane receptor; tyrosine kinase. Conversion of oligodendrocyte progenitors into neural stem-like cells	Expressed in gliomas. Amplified and mutated in glioblastoma	[83]

(Table 1) contd.....

Protein Name	Normal Function	Role in Oncogenesis	Citation
PROM1/CD133 Prominin-1	Membrane protein / Stem-cell marker. Maintenance of stem-cell properties (differentiation suppressor).	Lost during CSC differentiation, different glycosylation pattern in CSC	[126]
Ptch Protein patched homolog 1	Shh receptor; tumor suppressor function.	Proliferation of NSC and GSC. Ptch highly expressed in astrocytoma, oligodendroglioma, GBM	[99]
PTEN Phosphatase and tensin homologue	Tumor suppressor. Inhibitor of NSC proliferation.	Mutated in human glioblastoma	[5]
Shh Sonic hedgehog protein	Signaling protein. Proliferation of NSC and GCSC.	Activation of Shh pathway in brainstem glioma	[127]
SOX-2	Transcription factor. Involved in proliferation and functionality of NSC and precursors.	Expressed in malignant gliomas (gene expression profiling) and pediatric brain tumors	[56, 128]
TET2 Tet methylcytosine dioxygenase 2	Enzyme. Catalyses the conversion of the modified DNA base methylcytosine into 5-hydroxymethylcytosine.	Frequently mutated in myeloproliferative disorders and brain cancers; absence of TET2 expression is exclusive with IDH1/2 mutations	[129]

Future Therapies in Brain Tumors Targeting Cancer Stem Cells

Lately, many researchers consider that cancer stem cells are linked with the aggressive behavior of glioblastoma and poor survival despite intensive treatment. CSCs are characterized by heterogeneity, fueled by complex interactions with their microenvironment and their niche, and represent the latest research field for emerging therapies in GBM. The CSCs can be addressed for therapies in two ways: direct and indirect [130]. Direct strategies imply targeting signaling pathways in CSCs to overcome resistance to treatment (eq. EGFR/PI-3K/Akt inhibition). Indirect strategies target the CSC microenvironment: hypoxic niche, the perivascular niche, and immune evasion.

Boosting Treatment Sensitivity

GCSCs present drug resistance properties provided by activation of some genes (like ABCG2 and BCRP1), which help the efflux of drugs form cytoplasm, hence increasing their survival. This was noticed frequently in the GBM sphere cell cultures [131]. Cells expressing CD133 – used for identifying GCSCs – present a greater resistance to radiation by comparison with other cells in the tumor, that are CD133 negative [27]. The CD133+ cells can activate DNA repair checkpoints preferentially by phosphorylation and activation of Chk1 and Chk2 [27]. Radio- and chemo-therapy resistance of GCSCs may be promoted by activation of pathways such as: Notch [84, 132 - 134], Sonic Hedgehog (Shh) [99, 100], EGFR [1, 135, 136], PI3K-Akt-PTEN [9, 137] and Bone Morphogenetic Protein (BMP) [95, 138]. Some studies suggest that even endothelial cells can promote the radio-resistance of CD133+ cells [1].

To overcome the complex mechanisms of GCSC resistance to radiation therapy and chemotherapy, a combined strategy is required to eradicate GCSCs (since the use of only one regimen can lead to treatment failure).

Studies on medulloblastoma that express high Notch and CD133 levels discovered that Notch blockade induces decrease of CD133 positivity, with subsequent increase in cell differentiation, reduced proliferation and tumorigenicity [133]. Notch pathway inhibition seems to render CSCs vulnerable and can be a promising therapy target.

Another important aspect would be to target the GCSCs without acting on the normal stem cells. Targeting Hedgehog-Gli signaling may be the answer [99], since the inhibition of this pathway using specific inhibitors leads to major changes in sphere cells such as reduced CD133 expression and decreased tumorigenicity [100]. Moreover, the Hedgehog blockade may increase the radio-sensitivity of GBM cells.

Akt inhibitors can directly target GCSCs, also sensitizing them to radiation therapy. Eradication of brain CSCs can be achieved using an indirect approach such as targeting the vascular niche (similar to the normal stem cells) [1], decreasing the blood supply of the tumor, and disrupting stem cell preservation mechanisms. Thus, using a complex treatment with antiangiogenic drugs and common chemo- and radio-therapy, brain CSCs can be eradicated.

Immunogenicity enhancement is another direction of attack for GBM, which are proved to express low quantities of MHC I (major histocompatibility complex) and natural killer (NK) activating ligands. Interferon gamma (IFN γ) can increase the expression of these molecules and NK lysis [139].

Other research directions connected to GCSCs that can lead to future treatments are nanotechnology and stem cells as drug delivery systems.

Therapy to Induce Differentiation

The first therapy that aimed to induce differentiation is now in clinical trials. It uses all trans retinoic acid (ATRA) [7], that was used previously in promyelocyte leukemia.

ATRA can induce the *in vitro* differentiation of early stage embryonic mouse stem cells into neurons, by unknown mechanisms [140]. ATRA induced differentiation of C6 rat glioma cells into oligodendrocytes [141] and also, in neonatal rats, induced differentiation of striatal neural stem cells into neurons [142].

ATRA induced differentiation of human embryonic neural stem cells into neurons, possibly by Notch1 down-regulation [143]. All these studies promoted ATRA as a possible therapy candidate for brain tumors with GCSCs, by differentiation induction, reaching now the clinical stage. Other studies may soon lead to another candidate for differentiation treatment, BMP4, which promotes differentiation of NSCs in astrocytes. *In vitro* experiments on brain CSCs treated with BMP demonstrated CSC number reduction and *in vivo* murine studies showed that BMP4 reduced the growth of human brain CSCs [95].

Tumor Microenvironment Targeting Therapy

The GCSC function in a microenvironment promotes its sustainability and survival; the vascular niche secretes soluble factors that inhibit differentiation and support self-renewal. Therefore, a modification of this microenvironment can decrease GCSC tumorigenicity. One approach at hand is the use of antiangiogenic drugs that decrease blood flow toward the tumor and will induce local modifications to finally reduce the number of GCSCs or render them sensitive to other therapies [144].

On *in vivo* models, xenografted glioma tumors were depleted of GCSCs and reduced in size after a combined treatment with antiangiogenic drugs and cytotoxic agents [136].

In another animal model study, microenvironment modifications were produced by treatment with polyinosinic-polycytidylic acid stabilized by lysine and carboxymethylcellulose (poly-ICLC), a ligand for toll-like receptor 3 and melanoma differentiation-associated gene 5, combined with vaccinations and epitopes from glioma associated antigens. These treatments induced CXCL10/IP-10 and IFN α, leading to decrease of tumor size and increased survival in the absence of autoimmunity. Using this approach, some groups started phase I/II studies on patients with recurrent malignant glioma [145].

Stem, Angiogenesis and RTK Signaling Cross Talk

Molecular treatment failures that have been observed recently are proposed to be the effect of tumor cell adaptation by receptor mutations or oncogene addition.

Tumoral cells with mutant EFGRvIII are constitutively active and are sensitive to EGFR inhibitors, only if PTEN shows normal expression level and function. In the case of EGFR activation with low PTEN or RTK activation, mTORC1 seems to be the key molecule for treatment resistance, so dual inhibition of EGFR/ mTOR leads to good results [73, 146].

Now the phenomenon of signaling cross-talk is well-known: one active receptor can activate, besides its own signaling pathway, other pathways through loops and shortcuts or by pathway cross-activation in signaling nodes [147].

The case of an activated EGFR pathway responds better to tyrosine kinase inhibitors if Akt levels are low [148]. Akt specific inhibition induces the expression and activation of many RTKs, through combined actions, which are mTORC1 inhibition and receptor expression activation by FOXO (Forkhead box O) [149].

Keeping in mind the importance of cross-talk between Notch, the RAS/MEK/ERK and EGFR downstream PI3K/Akt signaling pathways, their simultaneous inhibition at different levels emerges as a sound treatment with improved results in glioma patients. For example the Notch signaling involvement in angiogenesis through Dll-4 can be a cause for anti-VEGF treatment resistance. After finishing the treatment with bevacizumab, the U87MG glioblastoma cells, positive for Dll-4, maintained the same proliferation rate as the untreated ones [150]. Regardless of bevacizumab sensitivity, all tumors benefitted from Notch inhibition by soluble Dll-4, proving that targeting Notch and VEGF improves the outcome by decreasing cross-talk.

Other synergic treatments targeting the microenvironment of GCSCs (relation with the niche) are represented by concomitant inhibition of EGF and basic fibroblast growth factor (bFGF) [1, 151].

CD133 positive GCSCs are affected by microRNA-124, microRNA-137, microRNA-425, microRNA-451 and microRNA-486. Combination of imatinib mesilate (a c-Kit inhibitor) with microRNA-451 synergically inhibits sphere formation [152].

Targeting the EGFR/PI-3K/Akt Axe

To develop new therapies, the major players in GCSC signaling were addressed; Notch, BMP, Shh and others [95, 99, 100, 133].

Experiments show a depletion of GCSCs and decreased tumor growth after cyclopamine treatment in gliomas [99, 100] or Notch pathway inhibitors in medulloblastoma cell lines [133]. In neurosphere cultures, a low expression of CD133 and nestin was obtained with Notch blocking using gamma-secretase inhibitors. Notch blockade also inhibited the generation of tumors *in vivo* and the formation of *in vitro* colonies [86]. According to Fan X, Notch signaling inhibition increases apoptosis and reduces proliferation in GBM through STAT3 and Akt phosphorylation, suggesting multiple pathway changes through signaling cross-talk.

Monoclonal antibodies that target Notch (Dll-4 region) and inhibit its activation were developed. One of them, OMP-21M18/demcizumab, reached clinical trial phase. Hopes are high that Notch blockade combined with other therapies will improve survival in GBM and other tumors [86].

Stem Cells for Treatment of Malignant Brain Tumors

Mesenchymal stem cells (MSCs) can be used to target glioblastoma by delivering a specific immunotoxin – EphA2 [153]. However, there are concerns regarding MSC therapy and their ability to promote tumor growth, thus protecting GCSCs from the immune anti-tumoral action [154].

Treatment of brain tumors took advantage of stem cell technology due to their ability to infiltrate in the brain tumor region. Thus, NSCs can target malignant brain tumors, as demonstrated in preclinical models. Recently, a solution to train stem cells in an engineered tissue that mimics native tissue milieu was proposed [155].

Personalized Medicine Challenges in Brain Tumor Field

In GBM, as many other cancer types, there is a continuous search to develop personalized treatments based on patient tumoral phenotype [156].

The first step towards personalized medicine is to classify tumors in phenotype clusters, considering the molecular expression of various biomarkers. Then to specific and stratified patient groups specific treatment will be prescribed, to ensure maximal results.

Nowadays, in GBM some molecules were identified as candidate biomarkers and treatment targets, thus the best approach will be to investigate the proteomic profile for every patient and treat accordingly. The future perspectives are to target specific protein interactions and signaling pathway nodes, which are specifically altered in a given patient.

CONCLUSION

Our understanding of malignant gliomas envisions that the future treatment belongs to the personalized therapy, using molecular agents targeting the specific genotype and phenotype of a given patient. In the next years, we will enter in a new era of glioma treatment based on CSC paradigm. The first data from current clinical trials involving small molecule inhibitors and monoclonal antibodies are encouraging, but there is need for more research in the field of glioblastoma signaling machinery. Breakthrough may come from proteomic/biomarker research that will identify new targets for combined therapy regimes. New research areas can lead to therapy agents/targets: microRNA, cancer stem cell signaling pathways and microenvironment: STAT3, BMPs, Notch, Bmi-1, RTKs-Akt, Hedgehog, Wnt/β catenin.

Following the example of blood malignancies, we envision that with the unveiling of the CSC role in brain tumors and appropriate complex treatment based on genotype/phenotype patient grouping, survival and general outcome will improve. Fundamental research will shed light on the links between signaling, genetic and epigenetic changes that drive malignant proliferation. An efficient therapy will be forged after understanding the role of the primordial initiating event, the sustaining signaling constellation (CSCs & their microenvironment) and the acquired modifications, by simultaneously targeting these events. Main directions of therapies involving CSCs in brain tumors are: radio-chemotherapy sensitization, differentiation induction, microenvironment modifications and immune therapy.

New protocols that achieve therapeutic synergy by combining molecular targets/agents, classic radiation/chemotherapy and immunotherapy will be applicable in the near future for brain tumors.

CONFLICT OF INTEREST

The authors confirm that they have no conflict of interest to declare for this publication.

ACKNOWLEDGEMENTS

Declared none.

REFERENCES

[1] Calabrese C, Poppleton H, Kocak M, *et al.* A perivascular niche for brain tumor stem cells. Cancer Cell 2007; 11(1): 69-82. [http://dx.doi.org/10.1016/j.ccr.2006.11.020] [PMID: 17222791]

[2] Singh SK, Hawkins C, Clarke ID, *et al.* Identification of human brain tumour initiating cells. Nature 2004; 432(7015): 396-401. [http://dx.doi.org/10.1038/nature03128] [PMID: 15549107]

[3] Taylor MD, Poppleton H, Fuller C, *et al.* Radial glia cells are candidate stem cells of ependymoma. Cancer Cell 2005; 8(4): 323-35. [http://dx.doi.org/10.1016/j.ccr.2005.09.001] [PMID: 16226707]

[4] Wu A, Oh S, Wiesner SM, *et al.* Persistence of CD133+ cells in human and mouse glioma cell lines: detailed characterization of GL261 glioma cells with cancer stem cell-like properties. Stem Cells Dev 2008; 17(1): 173-84. [http://dx.doi.org/10.1089/scd.2007.0133] [PMID: 18271701]

[5] Cheng JX, Liu BL, Zhang X. How powerful is CD133 as a cancer stem cell marker in brain tumors? Cancer Treat Rev 2009; 35(5): 403-8. [http://dx.doi.org/10.1016/j.ctrv.2009.03.002] [PMID: 19369008]

[6] Gilbert CA, Ross AH. Cancer stem cells: cell culture, markers, and targets for new therapies. J Cell Biochem 2009; 108(5): 1031-8. [http://dx.doi.org/10.1002/jcb.22350] [PMID: 19760641]

[7] Sell S. Cancer stem cells and differentiation therapy. Tumour Biol 2006; 27(2): 59-70. [http://dx.doi.org/10.1159/000092323] [PMID: 16557043]

[8] Yadirgi G, Marino S. Adult neural stem cells and their role in brain pathology. J Pathol 2009; 217(2): 242-53. [http://dx.doi.org/10.1002/path.2480] [PMID: 19040211]

[9] Hambardzumyan D, Squatrito M, Carbajal E, Holland EC. Glioma formation, cancer stem cells, and akt signaling. Stem Cell Rev 2008; 4(3): 203-10. [http://dx.doi.org/10.1007/s12015-008-9021-5] [PMID: 18595010]

[10] Cruceru ML, Neagu M, Demoulin JB, Constantinescu SN. Therapy targets in glioblastoma and cancer stem cells: lessons from haematopoietic neoplasms. J Cell Mol Med 2013; 17(10): 1218-35. [http://dx.doi.org/10.1111/jcmm.12122] [PMID: 23998913]

[11] Sato A, Sakurada K, Kumabe T, *et al.* Association of stem cell marker CD133 expression with dissemination of glioblastomas. Neurosurg Rev 2010; 33(2): 175-83. discussion 183-174. Epub 2010/02/06

[12] Hemmati HD, Nakano I, Lazareff JA, *et al.* Cancerous stem cells can arise from pediatric brain tumors. Proc Natl Acad Sci USA 2003; 100(25): 15178-83. [http://dx.doi.org/10.1073/pnas.2036535100] [PMID: 14645703]

[13] Lee J, Kotliarova S, Kotliarov Y, *et al.* Tumor stem cells derived from glioblastomas cultured in bFGF and EGF more closely mirror the phenotype and genotype of primary tumors than do serum-cultured cell lines. Cancer Cell 2006; 9(5): 391-403.

[http://dx.doi.org/10.1016/j.ccr.2006.03.030] [PMID: 16697959]

[14] Singh SK, Clarke ID, Terasaki M, *et al.* Identification of a cancer stem cell in human brain tumors. Cancer Res 2003; 63(18): 5821-8.
[PMID: 14522905]

[15] Zaidi HA, Kosztowski T, DiMeco F, Quiñones-Hinojosa A. Origins and clinical implications of the brain tumor stem cell hypothesis. J Neurooncol 2009; 93(1): 49-60.
[http://dx.doi.org/10.1007/s11060-009-9856-x] [PMID: 19430882]

[16] Ignatova TN, Kukekov VG, Laywell ED, Suslov ON, Vrionis FD, Steindler DA. Human cortical glial tumors contain neural stem-like cells expressing astroglial and neuronal markers *in vitro*. Glia 2002; 39(3): 193-206.
[http://dx.doi.org/10.1002/glia.10094] [PMID: 12203386]

[17] Galli R, Binda E, Orfanelli U, *et al.* Isolation and characterization of tumorigenic, stem-like neural precursors from human glioblastoma. Cancer Res 2004; 64(19): 7011-21.
[http://dx.doi.org/10.1158/0008-5472.CAN-04-1364] [PMID: 15466194]

[18] Hambardzumyan D, Becher OJ, Rosenblum MK, Pandolfi PP, Manova-Todorova K, Holland EC. PI3K pathway regulates survival of cancer stem cells residing in the perivascular niche following radiation in medulloblastoma *in vivo*. Genes Dev 2008; 22(4): 436-48.
[http://dx.doi.org/10.1101/gad.1627008] [PMID: 18281460]

[19] Crocetti E, Trama A, Stiller C, *et al.* Epidemiology of glial and non-glial brain tumours in Europe. Eur J Cancer 2012; 48(10): 1532-42.
[http://dx.doi.org/10.1016/j.ejca.2011.12.013] [PMID: 22227039]

[20] Wicha MS. Cancer stem cell heterogeneity in hereditary breast cancer. Breast Cancer Res 2008; 10(2): 105.
[http://dx.doi.org/10.1186/bcr1990] [PMID: 18423071]

[21] Wright MH, Calcagno AM, Salcido CD, Carlson MD, Ambudkar SV, Varticovski L. Brca1 breast tumors contain distinct CD44+/CD24- and CD133+ cells with cancer stem cell characteristics. Breast Cancer Res 2008; 10(1): R10.
[http://dx.doi.org/10.1186/bcr1855] [PMID: 18241344]

[22] Rebetz J, Tian D, Persson A, *et al.* Glial progenitor-like phenotype in low-grade glioma and enhanced CD133-expression and neuronal lineage differentiation potential in high-grade glioma. PLoS One 2008; 3(4) e1936. Epub 2008/04/10.

[23] Li Y, Wang D, Wang L, *et al.* Distinct genomic aberrations between low-grade and high-grade gliomas of Chinese patients. PLoS One 2013; 8(2): e57168.
[http://dx.doi.org/10.1371/journal.pone.0057168] [PMID: 23451178]

[24] Gömöri E, Pál J, Kovács B, Dóczi T. Concurrent hypermethylation of DNMT1, MGMT and EGFR genes in progression of gliomas. Diagn Pathol 2012; 7: 8.
[http://dx.doi.org/10.1186/1746-1596-7-8] [PMID: 22264301]

[25] Ahmadi R, Stockhammer F, Becker N, *et al.* No prognostic value of IDH1 mutations in a series of 100 WHO grade II astrocytomas. J Neurooncol 2012; 109(1): 15-22.
[http://dx.doi.org/10.1007/s11060-012-0863-y] [PMID: 22528790]

[26] Juratli TA, Kirsch M, Robel K, *et al.* IDH mutations as an early and consistent marker in low-grade astrocytomas WHO grade II and their consecutive secondary high-grade gliomas. J Neurooncol 2012; 108(3): 403-10.
[http://dx.doi.org/10.1007/s11060-012-0844-1] [PMID: 22410704]

[27] Bao S, Wu Q, McLendon RE, *et al.* Glioma stem cells promote radioresistance by preferential activation of the DNA damage response. Nature 2006; 444(7120): 756-60.
[http://dx.doi.org/10.1038/nature05236] [PMID: 17051156]

[28] Liu G, Yuan X, Zeng Z, *et al.* Analysis of gene expression and chemoresistance of CD133+ cancer stem cells in glioblastoma. Mol Cancer 2006; 5: 67.
[http://dx.doi.org/10.1186/1476-4598-5-67] [PMID: 17140455]

[29] Dalerba P, Kalisky T, Sahoo D, *et al.* Single-cell dissection of transcriptional heterogeneity in human colon tumors. Nat Biotechnol 2011; 29(12): 1120-7.
[http://dx.doi.org/10.1038/nbt.2038] [PMID: 22081019]

[30] Jin X, Yin J, Kim SH, *et al.* EGFR-AKT-Smad signaling promotes formation of glioma stem-like cells and tumor angiogenesis by ID3-driven cytokine induction. Cancer Res 2011; 71(22): 7125-34.
[http://dx.doi.org/10.1158/0008-5472.CAN-11-1330] [PMID: 21975932]

[31] Verhaak RG, Hoadley KA, Purdom E, *et al.* Integrated genomic analysis identifies clinically relevant subtypes of glioblastoma characterized by abnormalities in PDGFRA, IDH1, EGFR, and NF1. Cancer Cell 2010; 17(1): 98-110.
[http://dx.doi.org/10.1016/j.ccr.2009.12.020] [PMID: 20129251]

[32] Huang PH, Xu AM, White FM. Oncogenic EGFR signaling networks in glioma. Sci Signal 2009; 2(87): re6.
[http://dx.doi.org/10.1126/scisignal.287re6] [PMID: 19738203]

[33] Bengzon JE, Salford L, *et al.* Glioma stem cells in the context of oncogenesis. In: Bagley R, Teicher B, Eds. Stem cells and cancer. Framingham, MA: Humana Press 2009; pp. 115-26.
[http://dx.doi.org/10.1007/978-1-60327-933-8_9]

[34] Nguyen LV, Vanner R, Dirks P, Eaves CJ. Cancer stem cells: an evolving concept. Nat Rev Cancer 2012; 12(2): 133-43.
[PMID: 22237392]

[35] Altaner C. Glioblastoma and stem cells. Neoplasma 2008; 55(5): 369-74.
[PMID: 18665745]

[36] Morrison SJ, Perez SE, Qiao Z, *et al.* Transient Notch activation initiates an irreversible switch from neurogenesis to gliogenesis by neural crest stem cells. Cell 2000; 101(5): 499-510.
[http://dx.doi.org/10.1016/S0092-8674(00)80860-0] [PMID: 10850492]

[37] Bachoo RM, Maher EA, Ligon KL, *et al.* Epidermal growth factor receptor and Ink4a/Arf: convergent mechanisms governing terminal differentiation and transformation along the neural stem cell to astrocyte axis. Cancer Cell 2002; 1(3): 269-77.
[http://dx.doi.org/10.1016/S1535-6108(02)00046-6] [PMID: 12086863]

[38] Jin X, Jin X, Sohn YW, *et al.* Blockade of EGFR signaling promotes glioma stem-like cell invasiveness by abolishing ID3-mediated inhibition of p27(KIP1) and MMP3 expression. Cancer Lett

2013; 328(2): 235-42.
[http://dx.doi.org/10.1016/j.canlet.2012.09.005] [PMID: 23022473]

[39] McCord AM, Jamal M, Williams ES, Camphausen K, Tofilon PJ. CD133+ glioblastoma stem-like cells are radiosensitive with a defective DNA damage response compared with established cell lines. Clin Cancer Res 2009; 15(16): 5145-53.
[http://dx.doi.org/10.1158/1078-0432.CCR-09-0263] [PMID: 19671863]

[40] Yuan X, Curtin J, Xiong Y, *et al.* Isolation of cancer stem cells from adult glioblastoma multiforme. Oncogene 2004; 23(58): 9392-400.
[http://dx.doi.org/10.1038/sj.onc.1208311] [PMID: 15558011]

[41] Zeppernick F, Ahmadi R, Campos B, *et al.* Stem cell marker CD133 affects clinical outcome in glioma patients. Clin Cancer Res 2008; 14(1): 123-9.
[http://dx.doi.org/10.1158/1078-0432.CCR-07-0932] [PMID: 18172261]

[42] Murat A, Migliavacca E, Gorlia T, *et al.* Stem cell-related "self-renewal" signature and high epidermal growth factor receptor expression associated with resistance to concomitant chemoradiotherapy in glioblastoma. J Clin Oncol 2008; 26(18): 3015-24.
[http://dx.doi.org/10.1200/JCO.2007.15.7164] [PMID: 18565887]

[43] Pallini R, Ricci-Vitiani L, Banna GL, *et al.* Cancer stem cell analysis and clinical outcome in patients with glioblastoma multiforme. Clin Cancer Res 2008; 14(24): 8205-12.
[http://dx.doi.org/10.1158/1078-0432.CCR-08-0644] [PMID: 19088037]

[44] Soeda A, Park M, Lee D, *et al.* Hypoxia promotes expansion of the CD133-positive glioma stem cells through activation of HIF-1alpha. Oncogene 2009; 28(45): 3949-59.
[http://dx.doi.org/10.1038/onc.2009.252] [PMID: 19718046]

[45] Clément V, Dutoit V, Marino D, Dietrich PY, Radovanovic I. Limits of CD133 as a marker of glioma self-renewing cells. Int J Cancer 2009; 125(1): 244-8.
[http://dx.doi.org/10.1002/ijc.24352] [PMID: 19350631]

[46] Joo KM, Kim SY, Jin X, *et al.* Clinical and biological implications of CD133-positive and CD133-negative cells in glioblastomas. Lab Invest 2008; 88(8): 808-15.
[http://dx.doi.org/10.1038/labinvest.2008.57] [PMID: 18560366]

[47] Zhang M, Song T, Yang L, *et al.* Nestin and CD133: valuable stem cell-specific markers for determining clinical outcome of glioma patients. J Exp Clin Cancer Res 2008; 27: 85.
[http://dx.doi.org/10.1186/1756-9966-27-85] [PMID: 19108713]

[48] He H, Niu CS, Li MW. Correlation between glioblastoma stem-like cells and tumor vascularization. Oncol Rep 2012; 27(1): 45-50.
[PMID: 21971709]

[49] Strojnik T, Rosland GV, Sakariassen PO, *et al.* Neural stem cell markers, nestin and musashi proteins, in the progression of human glioma: correlation of nestin with prognosis of patient survival. Surg Neurol 2007; 68(2): 133-43. discussion 143-134. Epub 2007/06/01.
[http://dx.doi.org/10.1016/j.surneu.2006.10.050]

[50] Dell'Albani P. Stem cell markers in gliomas. Neurochem Res 2008; 33(12): 2407-15.
[http://dx.doi.org/10.1007/s11064-008-9723-8] [PMID: 18493853]

[51] deCarvalho AC, Nelson K, Lemke N, *et al.* Gliosarcoma stem cells undergo glial and mesenchymal differentiation *in vivo*. Stem Cells 2010; 28(2): 181-90.
[PMID: 19937755]

[52] Singh SK, Clarke ID, Hide T, Dirks PB. Cancer stem cells in nervous system tumors. Oncogene 2004; 23(43): 7267-73.
[http://dx.doi.org/10.1038/sj.onc.1207946] [PMID: 15378086]

[53] Peñuelas S, Anido J, Prieto-Sánchez RM, *et al.* TGF-beta increases glioma-initiating cell self-renewal through the induction of LIF in human glioblastoma. Cancer Cell 2009; 15(4): 315-27.
[http://dx.doi.org/10.1016/j.ccr.2009.02.011] [PMID: 19345330]

[54] Fotovati A, Abu-Ali S, Wang PS, *et al.* YB-1 bridges neural stem cells and brain tumor-initiating cells *via* its roles in differentiation and cell growth. Cancer Res 2011; 71(16): 5569-78.
[http://dx.doi.org/10.1158/0008-5472.CAN-10-2805] [PMID: 21730024]

[55] Muto J, Imai T, Ogawa D, *et al.* RNA-binding protein Musashi1 modulates glioma cell growth through the post-transcriptional regulation of Notch and PI3 kinase/Akt signaling pathways. PLoS One 2012; 7(3): e33431.
[http://dx.doi.org/10.1371/journal.pone.0033431] [PMID: 22428049]

[56] Guo Y, Liu S, Wang P, *et al.* Expression profile of embryonic stem cell-associated genes Oct4, Sox2 and Nanog in human gliomas. Histopathology 2011; 59(4): 763-75.
[http://dx.doi.org/10.1111/j.1365-2559.2011.03993.x] [PMID: 22014056]

[57] Takahashi K, Yamanaka S. Induction of pluripotent stem cells from mouse embryonic and adult fibroblast cultures by defined factors. Cell 2006; 126(4): 663-76.
[http://dx.doi.org/10.1016/j.cell.2006.07.024] [PMID: 16904174]

[58] Buganim Y, Faddah DA, Cheng AW, *et al.* Single-cell expression analyses during cellular reprogramming reveal an early stochastic and a late hierarchic phase. Cell 2012; 150(6): 1209-22.
[http://dx.doi.org/10.1016/j.cell.2012.08.023] [PMID: 22980981]

[59] Ikushima H, Todo T, Ino Y, Takahashi M, Miyazawa K, Miyazono K. Autocrine TGF-beta signaling maintains tumorigenicity of glioma-initiating cells through Sry-related HMG-box factors. Cell Stem Cell 2009; 5(5): 504-14.
[http://dx.doi.org/10.1016/j.stem.2009.08.018] [PMID: 19896441]

[60] Gangemi RM, Griffero F, Marubbi D, *et al.* SOX2 silencing in glioblastoma tumor-initiating cells causes stop of proliferation and loss of tumorigenicity. Stem Cells 2009; 27(1): 40-8.
[http://dx.doi.org/10.1634/stemcells.2008-0493] [PMID. 18948646]

[61] Marie SK, Okamoto OK, Uno M, *et al.* Maternal embryonic leucine zipper kinase transcript abundance correlates with malignancy grade in human astrocytomas. Int J Cancer 2008; 122(4): 807-15.
[http://dx.doi.org/10.1002/ijc.23189] [PMID: 17960622]

[62] Yoshimoto K, Ma X, Guan Y, *et al.* Expression of stem cell marker and receptor kinase genes in glioblastoma tissue quantified by real-time RT-PCR. Brain Tumor Pathol 2011; 28(4): 291-6.
[http://dx.doi.org/10.1007/s10014-011-0046-0] [PMID: 21691733]

[63] Abkowitz JL, Chen J. Studies of c-Mpl function distinguish the replication of hematopoietic stem cells

from the expansion of differentiating clones. Blood 2007; 109(12): 5186-90.
[http://dx.doi.org/10.1182/blood-2006-08-044503] [PMID: 17347409]

[64] Ricci-Vitiani L, Pallini R, Biffoni M, *et al.* Tumour vascularization *via* endothelial differentiation of glioblastoma stem-like cells. Nature 2010; 468(7325): 824-8.
[http://dx.doi.org/10.1038/nature09557] [PMID: 21102434]

[65] Soda Y, Marumoto T, Friedmann-Morvinski D, *et al.* Transdifferentiation of glioblastoma cells into vascular endothelial cells. Proc Natl Acad Sci USA 2011; 108(11): 4274-80.
[http://dx.doi.org/10.1073/pnas.1016030108] [PMID: 21262804]

[66] Naveiras O, Daley GQ. Stem cells and their niche: a matter of fate. Cell Mol Life Sci 2006; 63(7-8): 760-6.
[http://dx.doi.org/10.1007/s00018-005-5469-5] [PMID: 16568245]

[67] Pérez Castillo A, Aguilar-Morante D, Morales-García JA, Dorado J. Cancer stem cells and brain tumors. Clin Transl Oncol 2008; 10(5): 262-7.
[http://dx.doi.org/10.1007/s12094-008-0195-8] [PMID: 18490242]

[68] Croker AK, Allan AL. Cancer stem cells: implications for the progression and treatment of metastatic disease. J Cell Mol Med 2008; 12(2): 374-90.
[http://dx.doi.org/10.1111/j.1582-4934.2007.00211.x] [PMID: 18182063]

[69] Cheng L, Bao S, Rich JN. Potential therapeutic implications of cancer stem cells in glioblastoma. Biochem Pharmacol 2010; 80(5): 654-65.
[http://dx.doi.org/10.1016/j.bcp.2010.04.035] [PMID: 20457135]

[70] Vanhaesebroeck B, Stephens L, Hawkins P. PI3K signalling: the path to discovery and understanding. Nat Rev Mol Cell Biol 2012; 13(3): 195-203.
[http://dx.doi.org/10.1038/nrm3290] [PMID: 22358332]

[71] Molina JR, Hayashi Y, Stephens C, Georgescu MM. Invasive glioblastoma cells acquire stemness and increased Akt activation. Neoplasia 2010; 12(6): 453-63.
[http://dx.doi.org/10.1593/neo.10126] [PMID: 20563248]

[72] Amit I, Wides R, Yarden Y. Evolvable signaling networks of receptor tyrosine kinases: relevance of robustness to malignancy and to cancer therapy. Mol Syst Biol 2007; 3: 151.
[http://dx.doi.org/10.1038/msb4100195] [PMID: 18059446]

[73] Fan QW, Cheng C, Knight ZA, *et al.* EGFR signals to mTOR through PKC and independently of Akt in glioma. Sci Signal 2009; 2(55): ra4.
[http://dx.doi.org/10.1126/scisignal.2000014] [PMID: 19176518]

[74] Dang M, Armbruster N, Miller MA, *et al.* Regulated ADAM17-dependent EGF family ligand release by substrate-selecting signaling pathways. Proc Natl Acad Sci USA 2013; 110(24): 9776-81.
[http://dx.doi.org/10.1073/pnas.1307478110] [PMID: 23720309]

[75] Wen PY, Lee EQ, Reardon DA, Ligon KL, Alfred Yung WK. Current clinical development of PI3K pathway inhibitors in glioblastoma. Neuro-oncol 2012; 14(7): 819-29.
[http://dx.doi.org/10.1093/neuonc/nos117] [PMID: 22619466]

[76] Cruceru ML, Enciu AM, Popa AC, *et al.* Signal transduction molecule patterns indicating potential glioblastoma therapy approaches. Onco Targets Ther 2013; 6: 1737-49.

[http://dx.doi.org/10.2147/OTT.S52365] [PMID: 24348050]

[77] Cheng CK, Gustafson WC, Charron E, *et al.* Dual blockade of lipid and cyclin-dependent kinases induces synthetic lethality in malignant glioma. Proc Natl Acad Sci USA 2012; 109(31): 12722-7. [http://dx.doi.org/10.1073/pnas.1202492109] [PMID: 22802621]

[78] Inda MM, Bonavia R, Mukasa A, *et al.* Tumor heterogeneity is an active process maintained by a mutant EGFR-induced cytokine circuit in glioblastoma. Genes Dev 2010; 24(16): 1731-45. [http://dx.doi.org/10.1101/gad.1890510] [PMID: 20713517]

[79] Fenton TR, Nathanson D, Ponte de Albuquerque C, *et al.* Resistance to EGF receptor inhibitors in glioblastoma mediated by phosphorylation of the PTEN tumor suppressor at tyrosine 240. Proc Natl Acad Sci USA 2012; 109(35): 14164-9. [http://dx.doi.org/10.1073/pnas.1211962109] [PMID: 22891331]

[80] Hatanpaa KJ, Burma S, Zhao D, Habib AA. Epidermal growth factor receptor in glioma: signal transduction, neuropathology, imaging, and radioresistance. Neoplasia 2010; 12(9): 675-84. [http://dx.doi.org/10.1593/neo.10688] [PMID: 20824044]

[81] Nazarenko I, Hede SM, He X, *et al.* PDGF and PDGF receptors in glioma. Ups J Med Sci 2012; 117(2): 99-112. [http://dx.doi.org/10.3109/03009734.2012.665097] [PMID: 22509804]

[82] Uhrbom L, Hesselager G, Nistér M, Westermark B. Induction of brain tumors in mice using a recombinant platelet-derived growth factor B-chain retrovirus. Cancer Res 1998; 58(23): 5275-9. [PMID: 9850047]

[83] Jackson EL, Garcia-Verdugo JM, Gil-Perotin S, *et al.* PDGFR alpha-positive B cells are neural stem cells in the adult SVZ that form glioma-like growths in response to increased PDGF signaling. Neuron 2006; 51(2): 187-99. [http://dx.doi.org/10.1016/j.neuron.2006.06.012] [PMID: 16846854]

[84] Shih AH, Holland EC. Notch signaling enhances nestin expression in gliomas. Neoplasia 2006; 8(12): 1072-82. [http://dx.doi.org/10.1593/neo.06526] [PMID: 17217625]

[85] Tchorz JS, Tome M, Cloëtta D, *et al.* Constitutive Notch2 signaling in neural stem cells promotes tumorigenic features and astroglial lineage entry. Cell Death Dis 2012; 3: e325. [http://dx.doi.org/10.1038/cddis.2012.65] [PMID: 22717580]

[86] Fan X, Khaki L, Zhu TS, *et al.* NOTCH pathway blockade depletes CD133-positive glioblastoma cells and inhibits growth of tumor neurospheres and xenografts. Stem Cells 2010; 28(1): 5-16. [PMID: 19904829]

[87] Ying M, Wang S, Sang Y, *et al.* Regulation of glioblastoma stem cells by retinoic acid: role for Notch pathway inhibition. Oncogene 2011; 30(31): 3454-67. [http://dx.doi.org/10.1038/onc.2011.58] [PMID: 21383690]

[88] Zeng Q, Li S, Chepeha DB, *et al.* Crosstalk between tumor and endothelial cells promotes tumor angiogenesis by MAPK activation of Notch signaling. Cancer Cell 2005; 8(1): 13-23. [http://dx.doi.org/10.1016/j.ccr.2005.06.004] [PMID: 16023595]

[89] Yan M, Plowman GD. Delta-like 4/Notch signaling and its therapeutic implications. Clin Cancer Res

2007; 13(24): 7243-6.
[http://dx.doi.org/10.1158/1078-0432.CCR-07-1393] [PMID: 18094402]

[90] Li JL, Sainson RC, Shi W, *et al.* Delta-like 4 Notch ligand regulates tumor angiogenesis, improves tumor vascular function, and promotes tumor growth *in vivo.* Cancer Res 2007; 67(23): 11244-53.
[http://dx.doi.org/10.1158/0008-5472.CAN-07-0969] [PMID: 18056450]

[91] Anido J, Sáez-Borderías A, Gonzàlez-Juncà A, *et al.* TGF β receptor inhibitors target the CD44(high)/Id1(high) glioma-initiating cell population in human glioblastoma. Cancer Cell 2010; 18(6): 655-68.
[http://dx.doi.org/10.1016/j.ccr.2010.10.023] [PMID: 21156287]

[92] Schneider T, Sailer M, Ansorge S, Firsching R, Reinhold D. Increased concentrations of transforming growth factor beta1 and beta2 in the plasma of patients with glioblastoma. J Neurooncol 2006; 79(1): 61-5.
[http://dx.doi.org/10.1007/s11060-005-9116-7] [PMID: 16614941]

[93] Qiu B, Zhang D, Wang C, *et al.* IL-10 and TGF β2 are overexpressed in tumor spheres cultured from human gliomas. Mol Biol Rep 2011; 38(5): 3585-91.
[http://dx.doi.org/10.1007/s11033-010-0469-4] [PMID: 21088899]

[94] Bruna A, Darken RS, Rojo F, *et al.* High TGFbeta-Smad activity confers poor prognosis in glioma patients and promotes cell proliferation depending on the methylation of the PDGF-B gene. Cancer Cell 2007; 11(2): 147-60.
[http://dx.doi.org/10.1016/j.ccr.2006.11.023] [PMID: 17292826]

[95] Piccirillo SG, Reynolds BA, Zanetti N, *et al.* Bone morphogenetic proteins inhibit the tumorigenic potential of human brain tumour-initiating cells. Nature 2006; 444(7120): 761-5.
[http://dx.doi.org/10.1038/nature05349] [PMID: 17151667]

[96] Mehler MF, Mabie PC, Zhu G, Gokhan S, Kessler JA. Developmental changes in progenitor cell responsiveness to bone morphogenetic proteins differentially modulate progressive CNS lineage fate. Dev Neurosci 2000; 22(1-2): 74-85.
[http://dx.doi.org/10.1159/000017429] [PMID: 10657700]

[97] Yamada N, Kato M, ten Dijke P, *et al.* Bone morphogenetic protein type IB receptor is progressively expressed in malignant glioma tumours. Br J Cancer 1996; 73(5): 624-9.
[http://dx.doi.org/10.1038/bjc.1996.108] [PMID: 8605097]

[98] Xu Q, Yuan X, Liu G, Black KL, Yu JS. Hedgehog signaling regulates brain tumor-initiating cell proliferation and portends shorter survival for patients with PTEN-coexpressing glioblastomas. Stem Cells 2008; 26(12): 3018-26.
[http://dx.doi.org/10.1634/stemcells.2008-0459] [PMID: 18787206]

[99] Clement V, Sanchez P, de Tribolet N, Radovanovic I, Ruiz i Altaba A. HEDGEHOG-GLI1 signaling regulates human glioma growth, cancer stem cell self-renewal, and tumorigenicity. Curr Biol 2007; 17(2): 165-72.
[http://dx.doi.org/10.1016/j.cub.2006.11.033] [PMID: 17196391]

[100] Bar EE, Chaudhry A, Lin A, *et al.* Cyclopamine-mediated hedgehog pathway inhibition depletes stem-like cancer cells in glioblastoma. Stem Cells 2007; 25(10): 2524-33.
[http://dx.doi.org/10.1634/stemcells.2007-0166] [PMID: 17628016]

[101] Takezaki T, Hide T, Takanaga H, Nakamura H, Kuratsu J, Kondo T. Essential role of the Hedgehog signaling pathway in human glioma-initiating cells. Cancer Sci 2011; 102(7): 1306-12. [http://dx.doi.org/10.1111/j.1349-7006.2011.01943.x] [PMID: 21453386]

[102] Fodde R, Brabletz T. Wnt/beta-catenin signaling in cancer stemness and malignant behavior. Curr Opin Cell Biol 2007; 19(2): 150-8. [http://dx.doi.org/10.1016/j.ceb.2007.02.007] [PMID: 17306971]

[103] Kim Y, Kim KH, Lee J, *et al.* Wnt activation is implicated in glioblastoma radioresistance. Lab Invest 2012; 92(3): 466-73. [http://dx.doi.org/10.1038/labinvest.2011.161] [PMID: 22083670]

[104] Kahlert UD, Maciaczyk D, Doostkam S, *et al.* Activation of canonical WNT/β-catenin signaling enhances *in vitro* motility of glioblastoma cells by activation of ZEB1 and other activators of epithelial-to-mesenchymal transition. Cancer Lett 2012; 325(1): 42-53. [http://dx.doi.org/10.1016/j.canlet.2012.05.024] [PMID: 22652173]

[105] Vainchenker W, Constantinescu SN. JAK/STAT signaling in hematological malignancies. Oncogene 2013; 32(21): 2601-13. [http://dx.doi.org/10.1038/onc.2012.347] [PMID: 22869151]

[106] Seubert N, Royer Y, Staerk J, *et al.* Active and inactive orientations of the transmembrane and cytosolic domains of the erythropoietin receptor dimer. Mol Cell 2003; 12(5): 1239-50. [http://dx.doi.org/10.1016/S1097-2765(03)00389-7] [PMID: 14636581]

[107] Constantinescu SN, Ghaffari S, Lodish HF. The erythropoietin receptor: structure, activation and intracellular signal transduction. Trends Endocrinol Metab 1999; 10(1): 18-23. [http://dx.doi.org/10.1016/S1043-2760(98)00101-5] [PMID: 10322390]

[108] Darnell JE Jr, Kerr IM, Stark GR. Jak-STAT pathways and transcriptional activation in response to IFNs and other extracellular signaling proteins. Science 1994; 264(5164): 1415-21. [http://dx.doi.org/10.1126/science.8197455] [PMID: 8197455]

[109] de la Iglesia N, Puram SV, Bonni A. STAT3 regulation of glioblastoma pathogenesis. Curr Mol Med 2009; 9(5): 580-90. [http://dx.doi.org/10.2174/156652409788488739] [PMID: 19601808]

[110] Sherry MM, Reeves A, Wu JK, Cochran BH. STAT3 is required for proliferation and maintenance of multipotency in glioblastoma stem cells. Stem Cells 2009; 27(10): 2383-92. [http://dx.doi.org/10.1002/stem.185] [PMID: 19658181]

[111] Cooper LA, Gutman DA, Chisolm C, *et al.* The tumor microenvironment strongly impacts master transcriptional regulators and gene expression class of glioblastoma. Am J Pathol 2012; 180(5): 2108-19. [http://dx.doi.org/10.1016/j.ajpath.2012.01.040] [PMID: 22440258]

[112] Sai K, Wang S, Balasubramaniyan V, *et al.* Induction of cell-cycle arrest and apoptosis in glioblastoma stem-like cells by WP1193, a novel small molecule inhibitor of the JAK2/STAT3 pathway. J Neurooncol 2012; 107(3): 487-501. [http://dx.doi.org/10.1007/s11060-011-0786-z] [PMID: 22249692]

[113] Nilsson CL, Dillon R, Devakumar A, *et al.* Quantitative phosphoproteomic analysis of the STAT3/IL-

6/HIF1alpha signaling network: an initial study in GSC11 glioblastoma stem cells. J Proteome Res 2010; 9(1): 430-43.
[http://dx.doi.org/10.1021/pr9007927] [PMID: 19899826]

[114] Liu X, Wang L, Zhao S, Ji X, Luo Y, Ling F. β-Catenin overexpression in malignant glioma and its role in proliferation and apoptosis in glioblastoma cells. Med Oncol 2011; 28(2): 608-14.
[http://dx.doi.org/10.1007/s12032-010-9476-5] [PMID: 20300972]

[115] Häyry V, Tynninen O, Haapasalo HK, *et al.* Stem cell protein BMI-1 is an independent marker for poor prognosis in oligodendroglial tumours. Neuropathol Appl Neurobiol 2008; 34(5): 555-63.
[http://dx.doi.org/10.1111/j.1365-2990.2008.00949.x] [PMID: 18346113]

[116] Larysz D, Kula D, Kowal M, *et al.* Epidermal growth factor receptor gene expression in high grade gliomas. Folia Neuropathol 2011; 49(1): 28-38.
[PMID: 21455841]

[117] Yan H, Parsons DW, Jin G, *et al.* IDH1 and IDH2 mutations in gliomas. N Engl J Med 2009; 360(8): 765-73.
[http://dx.doi.org/10.1056/NEJMoa0808710] [PMID: 19228619]

[118] Patel JP, Gönen M, Figueroa ME, *et al.* Prognostic relevance of integrated genetic profiling in acute myeloid leukemia. N Engl J Med 2012; 366(12): 1079-89.
[http://dx.doi.org/10.1056/NEJMoa1112304] [PMID: 22417203]

[119] Nakano I, Kornblum HI. Methods for analysis of brain tumor stem cell and neural stem cell self-renewal. Methods Mol Biol 2009; 568: 37-56.
[http://dx.doi.org/10.1007/978-1-59745-280-9_4] [PMID: 19582420]

[120] Toda M, Iizuka Y, Yu W, *et al.* Expression of the neural RNA-binding protein Musashi1 in human gliomas. Glia 2001; 34(1): 1-7.
[http://dx.doi.org/10.1002/glia.1034] [PMID: 11284014]

[121] Rao G, Pedone CA, Coffin CM, Holland EC, Fults DW. c-Myc enhances sonic hedgehog-induced medulloblastoma formation from nestin-expressing neural progenitors in mice. Neoplasia 2003; 5(3): 198-204.
[http://dx.doi.org/10.1016/S1476-5586(03)80052-0] [PMID: 12869303]

[122] Lino MM, Merlo A, Boulay JL. Notch signaling in glioblastoma: a developmental drug target? BMC Med 2010; 8: 72.
[http://dx.doi.org/10.1186/1741-7015-8-72] [PMID: 21078177]

[123] Molofsky AV, Pardal R, Iwashita T, Park IK, Clarke MF, Morrison SJ. Bmi-1 dependence distinguishes neural stem cell self-renewal from progenitor proliferation. Nature 2003; 425(6961): 962-7.
[http://dx.doi.org/10.1038/nature02060] [PMID: 14574365]

[124] Jung GA, Yoon JY, Moon BS, *et al.* Valproic acid induces differentiation and inhibition of proliferation in neural progenitor cells *via* the beta-catenin-Ras-ERK-p21Cip/WAF1 pathway. BMC Cell Biol 2008; 9: 66.
[http://dx.doi.org/10.1186/1471-2121-9-66] [PMID: 19068119]

[125] Meletis K, Wirta V, Hede SM, Nistér M, Lundeberg J, Frisén J. p53 suppresses the self-renewal of

adult neural stem cells. Development 2006; 133(2): 363-9.
[http://dx.doi.org/10.1242/dev.02208] [PMID: 16368933]

[126] Kato H, Fujimura M, Kumabe T, Ishioka C, Kanamaru R, Yoshimoto T. PTEN gene mutation and high MIB-1 labeling index may contribute to dissemination in patients with glioblastoma. J Clin Neurosci 2004; 11(1): 37-41.
[http://dx.doi.org/10.1016/j.jocn.2002.09.001] [PMID: 14642363]

[127] Xin Y, Hao SY, Tian YJ, *et al.* Expression and significance of sonic hedgehog signaling pathway-related components in brainstem and supratentorial astrocytomas. Chin Med J (Engl) 2011; 124(21): 3515-20.
[PMID: 22340170]

[128] Schmitz M, Temme A, Senner V, *et al.* Identification of SOX2 as a novel glioma-associated antigen and potential target for T cell-based immunotherapy. Br J Cancer 2007; 96(8): 1293-301.
[http://dx.doi.org/10.1038/sj.bjc.6603696] [PMID: 17375044]

[129] Jin SG, Jiang Y, Qiu R, *et al.* 5-Hydroxymethylcytosine is strongly depleted in human cancers but its levels do not correlate with IDH1 mutations. Cancer Res 2011; 71(24): 7360-5.
[http://dx.doi.org/10.1158/0008-5472.CAN-11-2023] [PMID: 22052461]

[130] Binello E, Germano IM. Targeting glioma stem cells: a novel framework for brain tumors. Cancer Sci 2011; 102(11): 1958-66.
[http://dx.doi.org/10.1111/j.1349-7006.2011.02064.x] [PMID: 21848914]

[131] Salmaggi A, Boiardi A, Gelati M, *et al.* Glioblastoma-derived tumorospheres identify a population of tumor stem-like cells with angiogenic potential and enhanced multidrug resistance phenotype. Glia 2006; 54(8): 850-60.
[http://dx.doi.org/10.1002/glia.20414] [PMID: 16981197]

[132] Clark PA, Treisman DM, Ebben J, Kuo JS. Developmental signaling pathways in brain tumor-derived stem-like cells. Dev Dyn 2007; 236(12): 3297-308.
[http://dx.doi.org/10.1002/dvdy.21381] [PMID: 18000980]

[133] Fan X, Matsui W, Khaki L, *et al.* Notch pathway inhibition depletes stem-like cells and blocks engraftment in embryonal brain tumors. Cancer Res 2006; 66(15): 7445-52.
[http://dx.doi.org/10.1158/0008-5472.CAN-06-0858] [PMID: 16885340]

[134] Zhang XP, Zheng G, Zou L, *et al.* Notch activation promotes cell proliferation and the formation of neural stem cell-like colonies in human glioma cells. Mol Cell Biochem 2008; 307(1-2): 101-8.
[http://dx.doi.org/10.1007/s11010-007-9589-0] [PMID: 17849174]

[135] Bao S, Wu Q, Sathornsumetee S, *et al.* Stem cell-like glioma cells promote tumor angiogenesis through vascular endothelial growth factor. Cancer Res 2006; 66(16): 7843-8.
[http://dx.doi.org/10.1158/0008-5472.CAN-06-1010] [PMID: 16912155]

[136] Folkins C, Man S, Xu P, Shaked Y, Hicklin DJ, Kerbel RS. Anticancer therapies combining antiangiogenic and tumor cell cytotoxic effects reduce the tumor stem-like cell fraction in glioma xenograft tumors. Cancer Res 2007; 67(8): 3560-4.
[http://dx.doi.org/10.1158/0008-5472.CAN-06-4238] [PMID: 17440065]

[137] Eyler CE, Foo WC, LaFiura KM, McLendon RE, Hjelmeland AB, Rich JN. Brain cancer stem cells

display preferential sensitivity to Akt inhibition. Stem Cells 2008; 26(12): 3027-36. [http://dx.doi.org/10.1634/stemcells.2007-1073] [PMID: 18802038]

[138] Lee J, Son MJ, Woolard K, *et al.* Epigenetic-mediated dysfunction of the bone morphogenetic protein pathway inhibits differentiation of glioblastoma-initiating cells. Cancer Cell 2008; 13(1): 69-80. [http://dx.doi.org/10.1016/j.ccr.2007.12.005] [PMID: 18167341]

[139] Liu Y, Wu C, Lyu Q, *et al.* Germline stem cells and neo-oogenesis in the adult human ovary. Dev Biol 2007; 306(1): 112-20. [http://dx.doi.org/10.1016/j.ydbio.2007.03.006] [PMID: 17428461]

[140] Zhong S, Salomoni P, Pandolfi PP. The transcriptional role of PML and the nuclear body. Nat Cell Biol 2000; 2(5): E85-90. [http://dx.doi.org/10.1038/35010583] [PMID: 10806494]

[141] Bianchi MG, Gazzola GC, Tognazzi L, Bussolati O. C6 glioma cells differentiated by retinoic acid overexpress the glutamate transporter excitatory amino acid carrier 1 (EAAC1). Neuroscience 2008; 151(4): 1042-52. [http://dx.doi.org/10.1016/j.neuroscience.2007.11.055] [PMID: 18207650]

[142] Deng H, Zou F, Luo HJ. J First Mil Med Univ 2005; 25(11): 1357-1360, 1374. [Differentiation of neonatal rat striatal neural stem cells induced by all-trans retinoic acid].

[143] Wang F, Li ST, Huang Q, Lan Q. [Expression of Notch1 gene in the differentiation of the human embryonic neural stem cells to neurons]. Xibao Yu Fenzi Mianyixue Zazhi 2004; 20(6): 769-72. [Expression of Notch1 gene in the differentiation of the human embryonic neural stem cells to neurons]. [PMID: 15555458]

[144] Shen Q, Goderie SK, Jin L, *et al.* Endothelial cells stimulate self-renewal and expand neurogenesis of neural stem cells. Science 2004; 304(5675): 1338-40. [http://dx.doi.org/10.1126/science.1095505] [PMID: 15060285]

[145] Okada H. Brain tumor immunotherapy with type-1 polarizing strategies. Ann N Y Acad Sci 2009; 1174: 18-23. [http://dx.doi.org/10.1111/j.1749-6632.2009.04932.x] [PMID: 19769732]

[146] Wang MY, Lu KV, Zhu S, *et al.* Mammalian target of rapamycin inhibition promotes response to epidermal growth factor receptor kinase inhibitors in PTEN-deficient and PTEN-intact glioblastoma cells. Cancer Res 2006; 66(16): 7864-9. [http://dx.doi.org/10.1158/0008-5472.CAN-04-4392] [PMID: 16912159]

[147] Stommel JM, Kimmelman AC, Ying H, *et al.* Coactivation of receptor tyrosine kinases affects the response of tumor cells to targeted therapies. Science 2007; 318(5848): 287-90. [http://dx.doi.org/10.1126/science.1142946] [PMID: 17872411]

[148] Haas-Kogan DA, Prados MD, Tihan T, *et al.* Epidermal growth factor receptor, protein kinase B/Akt, and glioma response to erlotinib. J Natl Cancer Inst 2005; 97(12): 880-7. [http://dx.doi.org/10.1093/jnci/dji161] [PMID: 15956649]

[149] Chandarlapaty S, Sawai A, Scaltriti M, *et al.* AKT inhibition relieves feedback suppression of receptor tyrosine kinase expression and activity. Cancer Cell 2011; 19(1): 58-71.

[http://dx.doi.org/10.1016/j.ccr.2010.10.031] [PMID: 21215704]

[150] Perri F, Cotugno R, Piepoli A, *et al.* Aberrant DNA methylation in non-neoplastic gastric mucosa of H. Pylori infected patients and effect of eradication. Am J Gastroenterol 2007; 102(7): 1361-71. [http://dx.doi.org/10.1111/j.1572-0241.2007.01284.x] [PMID: 17509026]

[151] Jin L, Hope KJ, Zhai Q, Smadja-Joffe F, Dick JE. Targeting of CD44 eradicates human acute myeloid leukemic stem cells. Nat Med 2006; 12(10): 1167-74. [http://dx.doi.org/10.1038/nm1483] [PMID: 16998484]

[152] Gal H, Pandi G, Kanner AA, *et al.* MIR-451 and Imatinib mesylate inhibit tumor growth of Glioblastoma stem cells. Biochem Biophys Res Commun 2008; 376(1): 86-90. [http://dx.doi.org/10.1016/j.bbrc.2008.08.107] [PMID: 18765229]

[153] Sun XL, Xu ZM, Ke YQ, *et al.* Molecular targeting of malignant glioma cells with an EphA2-specific immunotoxin delivered by human bone marrow-derived mesenchymal stem cells. Cancer Lett 2011; 312(2): 168-77. [http://dx.doi.org/10.1016/j.canlet.2011.07.035] [PMID: 21924825]

[154] Rameshwar P. Would cancer stem cells affect the future investment in stem cell therapy. World J Exp Med 2012; 2(2): 26-9. [http://dx.doi.org/10.5493/wjem.v2.i2.26] [PMID: 24520530]

[155] Li SC, Kabeer MH, Vu LT, *et al.* Training stem cells for treatment of malignant brain tumors. World J Stem Cells 2014; 6(4): 432-40. [http://dx.doi.org/10.4252/wjsc.v6.i4.432] [PMID: 25258664]

[156] Tanase CP, Enciu AM, Mihai S, Neagu AI, Calenic B, Cruceru ML. Anti-cancer Therapies in High Grade Gliomas. Curr Proteomics 2013; 10(3): 246-60. [http://dx.doi.org/10.2174/1570164611310030007] [PMID: 24228024]

CHAPTER 6

Adult Pituitary Stem Cells

Ancuta Augustina Gheorghisan-Galateanu[1,2,*]

[1] *Department of Cellular and Molecular Biology and Histology, "Carol Davila" University of Medicine and Pharmacy, 8 Eroii Sanitari Blvd., 050474 Bucharest, Romania*

[2] *"C.I.Parhon" National Institute of Endocrinology, 34-36 Aviatorilor Blvd., 011853 Bucharest, Romania*

Abstract: Stem cells were first identified in adult organs with high regenerative capacity including skin, liver, intestine and bone marrow. The pituitary gland is an organ with low cell turnover, and while differentiated cells can re-enter the cell cycle, most hormone producing cells are not dividing. The adult pituitary gland has some capacity to regenerate after tissue injury. Recent studies have reported potential populations of stem cells in the pituitary. Stem cells population belongs to the chromophobe/folliculo-stellate population of adenohypophysis. Studies to date suggest that at least a part of the endocrine cells originates from marginal layer adjacent to Rathke's cleft. Different groups, using diverse approaches, have demonstrated the presence in the pituitary of cells with progenitor or stem cell capacities and have characterized the pituitary stem cells, progenitors, and transit amplifying cells. A large variety of markers used to identify pituitary progenitors and stem cells make the integrated view difficult over the results obtained. Until now the nature of pituitary stem cells remains a matter of debate, and additional studies are needed to define the progenitors and stem cells in adenohypophysis.

Keywords: Adenohypophysis, Folliculo-stellate cells, Markers of stem cells, Progenitor/stem.

INTRODUCTION

The anterior pituitary (adenohypophysis) is composed of the anterior, intermediate

* **Corresponding author Ancuta Augustina Gheorghisan-Galateanu:** : "Carol Davila" University of Medicine and Pharmacy, 8 Eroii Sanitari Blvd., 050474 Bucharest, Romania ; Tel: +40 213 180 862; Fax: +40 21318 0730; E-mail: agheorghisan.a@gmail.com

Cristiana Tanase & Monica Neagu (Eds.)

and tuberal lobes and develops from the oral ectoderm by invagination in the early stage of the embryo. During embryonic organogenesis, five types of polypeptide hormone-producing cells (granular cells) are differentiated. These cells include: somatotrophs that produce GH, lactotrophs producing prolactin, gonadotrophs that secrete both LH and FSH, thyrotrophs producing TSH, and corticotrophs that express proopiomelanocortin (POMC) and cleave it to produce ACTH. These pituitary hormones steer essential physiological processes such as reproduction, lactation, growth, metabolism and immune system activity.

In addition to hormone-producing cells, the anterior pituitary gland also comprises of non-endocrine cells (agranular cells), such as folliculo-stellate cells. In hematoxylin and eosin-stained sections, three principal cell types are identified in the normal adenohypophysis: acidophils, basophils, and chromophobes. Basophilic cells can be strongly stained *via* PAS method and are adrenocorticotropin (ACTH) and glycoprotein hormones (LH, FSH, TSH) producing. Acidophilic cells are engaged in growth hormone (GH) or in prolactin (PRL) production. The chromophobe cells were identified as folliculo-stellate cells.

THE FOLLICULO-STELLATE CELLS

The folliculo-stellate cells (FS cells) were first identified in 1953, by Rinehart JF and Farquhar MG, using electron microscopy. FS cells are located in the center of the anterior lobe and constitute about 5–10% of the whole cells population of the anterior pituitary lobe [1]. They are star-shaped with long cytoplasmic projections that envelop hormone- producing cells and extending to the blood capillaries [2]. FS cells are located in the parenchymal tissue mainly around the lumen of large follicles scattered throughout the lobe [3]. Occasionally, the follicles contain colloid, an electron dense material. Clusterin, the major protein of colloid, which is not present in the cytoplasm of FS cells, is related to the phagocytic activity of FS cells. FS cells are recognized by the lack of cytoplasmic granules, many microvilli on the apical pole facing the follicular lumen, numerous fibrous structures in their cytoplasmic processes, and primary cilia (central cilia) which face the follicular lumen [4].

The impact of aging on pituitary FS cells is not well-known. It seems that only moderate changes in the endoplasmic reticulum were observed in old and senescent animals [5], and generally they increase with age [6, 7].

FS cells currently are considered as functionally and phenotypically heterogeneous and seem to have multifunctional properties. They are connected by gap junctions and form an extensive and complex three dimensional network. FS cells form numerous interconnections with neighboring FS cells in the presence of laminin, an extracellular matrix component of the basal lamina [8]. It was hypothesized that FS cells transmit signals *via* gap junctions so as to regulate hormone release from hormone-producing cells, in addition to the hypophyseal-portal system [9]. They also act as scavenger cells with phagocytic activity [1]. FS cells seems to be important in the paracrine regulation of anterior pituitary (particularly gonadotroph activity) by producing radical nitric oxide [10], various growth factors and peptides such as basic fibroblast growth factor (bFGF) [11], vascular endothelial cell growth factor (VEGF) and follistatin [12], as well as cytokines, such as interleukin-6 (IL-6) [13], leukemia inhibitory factor (LIF) [14], macrophage migration inhibitory factor (MIF) [15], and macrophage colony-stimulating factor (M-CSF) [16]. In addition, FS cells respond to β– adrenergic stimuli and enhance cAMP-accumulation upon β-adrenergic stimulation [17].

Over the past years, several studies have suggested the progenitor potential of a subpopulation of FS cells, on the basis of their ability to form colonies *in vitro* and to differentiate *in vivo*. Their role is supposed to be as the stem cell in the adult gland, involved in the normal cell turnover. A subset of the FS cells may be a source of pituitary stem cells, and another subset may be involved in creating a niche [18]. The possibility that FS cell be a type of stem cell with the potential to differentiate in endocrine cells or at least to be involved in their differentiation has been suggested by Yoshimura *et al.* [19].

Indirect evidence suggests that there is the possibility of retro-differentiation of endocrine cells into folliculo-stellate cells [20], or expansion of the folliculo-stellate compartment in parallel with gonadotrophs cells after castration [21].

It has been shown that FS cells can be detected by S100 calcium binding protein B (S100β) [22, 23], glial fibrillary acidic protein (GFAP) [24], and nestin. In addition, S100β- positive FS cells could be classified into subtypes depending on the marker protein, such as GFAP and keratin [25], or GFAP and vimentin [26]. However, according to some authors, it seems that not all FS cells express S100β and use it when folliculo-stellate marker is uncertain.

Thanks to their pleiotropic features, S100β-positive cells are assumed to be of a heterogeneous or even a non-pituitary origin. The observation of various markers has allowed these cells to be classified into populations such as stem/progenitor cells, epithelial cells, astrocytes and dendritic cells [27]. A recent study has found S100β to co-localize with SOX2 in a proportion of about 89 % [28].

If the anterior pituitary FS cell population contains more subpopulations with different phenotypes that can represent different cell types, or different stages of development/ activation of the same cell type, remains a question that is still looking for an answer [29].

STEM CELLS OF ADULT PITUITARY GLAND

Nowadays, there are more data showing the presence of stem cells in the adult pituitary which are involved in tissue turnover, plasticity and repair. In basal conditions, a continuous proliferation of stem cells generating transit-amplifying cells was reported [30].

Evidence of stem cells in adult pituitary was suggested by several conditions. The number of gonadotropins increases substantially during puberty [31]. The high prolactin levels identified during pregnancy and lactation were initially thought to be obtained through hypertrophy, but an increase in lactotroph cell number and transdifferentiation of somatotroph cells, may occur [32, 33]. The excessive expansion of TSH producing cells occurs in acquired hypothyroidism [34]. The pituitary tissue may regenerate after damages from immune aggressions or surgery [35]. Transgenic mouse studies demonstrate regeneration of GH cells after massive cell loss [36].

The existence of two populations of stem cells, one involved in embryogenesis and one involved in maintenance functions after birth, is still a matter of debate [37].

Several studies assume that populations of new hormonal cells can generate from choromophobes/FS cells.

It has been suggested that in adult pituitary a group of chromophobes cells could be progenitors or stem cells and as a result they could evolve and differentiate further in response to hypothalamic signaling hormones [38]. Chromophobes transplanted into the hypothalamus of hypophysectomized rats underwent proliferation and differentiation into mature basophils (thyrotrophs, gonadotrophs and corticotrophs) and acidophils (somatotrophs and lactotrophs) [39].

Considering the large spectrum of stem cells (totipotent, pluripotent, multipotent, oligopotent or unipotent), it is believed that a small proportion of multipotent or oligopotent stem cells can be identified in the adult pituitary. These cells can generate pituitary cell lineages and at the same time may have an important role in pituitary plasticity.

Different groups, using diverse approaches, characterized the possible candidates as pituitary stem cells, progenitors, and transit-amplifying cells.

In the last years studies using innovative approaches based on potential markers have identified new possible multipotent pituitary progenitor cells. The use of new information about stem cells and progenitor markers can offer large opportunities in understanding the response of pituitary in different situations and in the same time in elucidating the role of transdifferentiation, hypertrophy and implications of new cells generation in this response. Studies to date suggest the origin of some pituitary endocrine cells in marginal zone cells. Marginal zone cells localized between the anterior and the intermediate lobes show weak secretory organelles, absence of secretory granules, excessive free ribosomes and polysomes, aspects which can be considered specific for stem cells properties [40]. They are immunoreactive for nestin and express SOX2, suggesting a stem/progenitor cell role [41].

Identification of the stem cell niche is very helpful in understanding the process of organ development. The niche could be associated with Rathke's cleft in humans pituitary considered to be the origin of all cell lineages responsible for pituitary hormone secretion [42]. The marginal cell layer facing the residual lumen has been indicated as a stem/progenitor cell niche of the pituitary. Cell signaling between progenitors and other cells in the niche probably regulates the decision for stem cell maintenance *vs.* division to produce transit-amplifying cells. Identifying these pathways and defining the microenvironment for stem cell survival are critically important for establishing regenerative therapies [18].

One characteristic of progenitors and stem cells is the ability to form colonies *in vitro*. Lepore DA *et al.* were the first to demonstrate this for the pituitary. In the anterior pituitary, the stem cells having the ability to form colonies represent 0.2% from the whole cell population. These cells import fluorescent dipeptide β-Al--Lys- Ne -AMCA (Y-amino-4-methylcoumarin-3-acetic acid) and express S100β and GFAP. Their ability to express both S-100β and GFAP is an argument that these cells belong to FS cell compartment [43, 44].

Some studies were focused on different markers of stem cells to identify pituitary stem cells and progenitors, including primarily SOX family, GFRa2, nestin, PRRX1/ PRRX2 and SCA1.

Members of the **SOX family** of high-mobility group (HMG) box transcription factors are likely to be involved in the earliest steps of pituitary stem cell and the earliest transition to differentiation.

SOX2+ cells have the capability of self-renewal and capacity to differentiate in anterior pituitary cell lineages and do not express markers for pituitary hormones. During development, SOX2 is expressed in all cells of Rathke's pouch [45], however, in adults, these cells, regarded as different from those in embryo and neonate [46], become progressively restricted and represent approximately 3% of adult pituitary cells [18]. SOX2+ cells were identified mainly in the cleft between the anterior and intermediate lobes but also in small clusters in the anterior lobe [47]. It is not clear if the scattered SOX2+ cells generate a niche-independent status or serve as mini-niches. About 50% of the SOX2+ cells population express epithelial markers like E-cadherin, a nonspecific marker of marginal zone cells, suggesting the existence of a epithelial –mesenchymal transition process before the cells become differentiating cells [48]. **SOX9**, a marker for stem cells in pancreas, retina, and central nervous system [49 - 51], which is expressed in the embryo apparently later than SOX2, show a similar pattern expression to SOX2 in the mature pituitary gland [40].

In a recent study, Rizzoti *et al.* have demonstrated for the first time that SOX2+, and SOX9+ cells are the pituisphere-forming cells and give rise to pituitary endocrine cells, proving that they are progenitors [52].

In adult pituitaries, the majority of SOX2+ cells also express SOX9, whereas only rare cells are SOX2+, SOX9−.

Some reports suggest that the SOX2+/SOX9+ cells divide rapidly and represent transit-amplifying cells, while, by their ability to divide slowly, an attribute of stem cells, the SOX2+/SOX9-could act as multipotent cells involved in organ maintenance [48].

In specific culture conditions, a small (0.03%) population of progenitors in the adult pituitary gland, that are SOX2+/ SOX9−, can form pituispheres *in vitro*, which can self-renew, giving rise to secondary spheres, and they can differentiate into all the five endocrine cells of the anterior pituitary, as well as FS cells [53]. The labeling of stem cells requires at least five passages to demonstrate self-renewal clearly. It is possible that SOX2+/SOX9− cells to have the capacity for multiple passages of self-renewal if they are cultured in a milieu that better mimics the niche [18]. Vankelecom identified the coexpression of SOX2 and pituitary hormone markers in some adult pituitary cells and consider that during the differentiation stages SOX2 might be localized in the cytoplasm while during the stem cell state it might be localized in the nucleus [54].

Another marker expressed in adult pituitary is **Glial cell line-derived neurotrophic factor (GDNF) receptor α2 (GFRa2)**. It belongs to a family of glycosyl-phosphatidyl receptors that modulate signaling pathways induced by their ligands, the main one being GDNF and is expressed in putative stem cells of the testis and ovaries [55]. In the adult, the GFRa2+ cells are less than 1% from the anterior pituitary cell population. Most GFRa2+ cells are located in the marginal zone of the pituitary where they form a single layer, and a few cells are scattered in the anterior pituitary parenchyma [56]. More than 90% of these cells expressed SOX2 and SOX9 and about 50% of these cells expressed S100β. In cells expressing markers for pituitary hormones GFRa2 is not detectable. Also, GFRa2+ cells expressed PROP1, a pituitary transcription factor responsible for the POU1F1 lineage including somatolactotroph and thyrotroph cells types [45, 57].

The GFRa2+/PROP1+ cells showed the ability to form primary and secondary pituispheres and to differentiate in the main pituitary cell lineages. The fact that only a limited number of passages of GFRa2+ cells are possible, similar to SOX2+ cells, suggests the attribute of multipotent progenitors/transit-amplifying cells, with a transient role during the differentiation process, but not stem cells [53].

An important element in the consideration of GFRa2+ as progenitor cells rather than stem cells is the size of the pituispheres. The pituispheres derived from stem cells are larger than those derived from progenitors which have a more limited potential [40].

Nestin, an intermediate filament protein mostly expressed in nerve cells, is a marker of stem cells in several types of tissues, including adult and embryonic neural stem cells [41].

Using a genetic approach, transgenic mice expressing green fluorescent protein (GFP) driven by regulatory elements of the nestin gene, an adult pituitary stem cell population expressing nestin, was identified [46].

The number of GFP+ cells does not change during embryogenesis but increases after birth when the demand for new progenitor cells is high. This fact suggests that these cells did not play a significant role in organogenesis, but undergo postnatal expansion [58].

GFP+ cells represent only 2% of the anterior pituitary cells immediately after birth and in the adult pituitary, but only very few cells showed transgenic activity. The transgene-expressing cells seem to give birth to all five pituitary lineages. A small fraction of the GFP+ cells expressed POU1F1 (PIT1), which is essential for establishing the fate of somatolactotroph and thyrotroph lineages [56].

Fauquier *et al.* reported nestin expression in pituispheres after SOX2, suggesting that nestin play a role in stem/progenitor cell development [53].

Recently, it has been shown that paired-related transcription factors, **PRRX1** and **PRRX2**, which are present in mesenchymal tissues and participate in mesenchymal cell differentiation, were found in the stem/progenitor cells of the pituitary gland. Immunohistochemistry confirmed that PRRX1-positive cells appeared inside the pituitary during embryonic development, whereas significant expression of PRRX2 was found only in the postnatal pituitary. PRRX2 emerged in cells positive for SOX2, but negative for PROP1. Thus, PRRX2 might participate distinctly in the postnatal cell-supply system that occurs in the pituitary gland [59].

Side population cells (SPC), identified by flow cytometry, are related to their capacity of excluding a fluorescent dye, Hoechst 33342 which binds DNA, a technique developed by Goodell in 1996 for hematopoietic stem [60], and first time applied to pituitary gland by Vankelecom *et al.* [61].

Castinetti F *et al.* have shown that side population cells represent 1.7% of the whole pituitary cell population from the anterior pituitary of 3- to 8-wk-old mice [40].

It seems that SPC are a population with stem cell like activity, but highly heterogeneous. These cells express **SCA1 (Stem cell antigen 1)**, a putative stem cell marker [62]. The side population is composed of cells expressing high and low levels of the stem cell marker SCA1, representing 60% and 40% of the population, respectively [40].

The SCA1low cells express transcription factors, including Hesx1, Prop1, Pax6, and Lhx4, do not express pituitary hormone markers, and can form spheres that can give rise to all endocrine cell types of the anterior pituitary [61].

The SCA1high cells express higher levels of nestin, BMI-1, Notch1, and HES1. Some of the SCA1high cells also expressed S100β, suggesting that the SCA1high cell population might contain folliculo- stellate cells [54, 61].

Since SOX2 and SOX9 were expressed in about 50% of SCA low cells and not expressed in SCA1high cells, it has been suggested that pituitary progenitor cells may found low fraction in the SCA1.

Because SCA1 is involved in endothelial cell development [63], some authors considered that SCA1high cells are rather endothelial progenitors than pituitary progenitors.

It is important to point out that the Notch signaling system is active in the side population cells of the postnatal pituitary [61]. Activation of Notch signaling increased the number of side population cells. SOX9-deficient progenitors have reduced the expression of the Notch target HES1, suggesting possible interactions between SOX9 and the Notch signaling pathway in stem cell maintenance [49].

CONCLUSION

The pituitary progenitors and stem cells were identified by using different approaches. Even this fact makes difficult an integrated view over the results obtained and future studies are necessary. None of the cells studied until now has shown unequivocal characteristics of stem cells.

The nature of stem cells in adult pituitary and the identification of cells having the ability of self-renewal and to generate transit-amplifying cells in adult pituitary still remain a challenge.

CONFLICT OF INTEREST

The author confirms that author has no conflict of interest to declare for this publication.

ACKNOWLEDGEMENTS

Declared none.

REFERENCES

[1] Devnath S, Inoue K. An insight to pituitary folliculo-stellate cells. J Neuroendocrinol 2008; 20(6): 687-91.
[http://dx.doi.org/10.1111/j.1365-2826.2008.01716.x] [PMID: 18601690]

[2] Kagayama M. The follicular cell in the pars distalis of the dog pituitary gland: an electron microscope study. Endocrinology 1965; 77(6): 1053-60.
[http://dx.doi.org/10.1210/endo-77-6-1053] [PMID: 5849540]

[3] Vila-Porcile E. [The network of the folliculo-stellate cells and the follicles of the adenohypophysis in the rat (pars distalis)]. Z Zellforsch Mikrosk Anat 1972; 129(3): 328-69.
[http://dx.doi.org/10.1007/BF00307293] [PMID: 5042774]

[4] Harrisson F. Primary cilia associated with striated rootlets in granulated and folliculo-stellate cells of the avian adenohypophysis. Anat Embryol (Berl) 1989; 180(6): 543-7.
[http://dx.doi.org/10.1007/BF00300551] [PMID: 2610386]

[5] Cónsole GM, Jurado SB, Riccillo FL, Gómez Dumm CL. Immunohistochemical and ultrastructural study of pituitary folliculostellate cells during aging in rats. Cells Tissues Organs (Print) 2000; 167(1): 25-32.
[http://dx.doi.org/10.1159/000016763] [PMID: 10899713]

[6] Ciocca DR, Puy LA, Stati AO. Constitution and behavior of follicular structures in the human anterior pituitary gland. Am J Pathol 1984; 115(2): 165-74.
[PMID: 6326578]

[7] Pavlović M, Jovanović I, Ugrenović S, *et al.* Morphometric analysis of the human anterior pituitary's folliculostellate cells during the aging process. Ann Anat 2013; 195(3): 231-7.
[http://dx.doi.org/10.1016/j.aanat.2012.11.002] [PMID: 23295121]

[8] Horiguchi K, Kouki T, Fujiwara K, *et al.* Expression of the proteoglycan syndecan-4 and the mechanism by which it mediates stress fiber formation in folliculostellate cells in the rat anterior pituitary gland. J Endocrinol 2012; 214(2): 199-206.
[http://dx.doi.org/10.1530/JOE-12-0156] [PMID: 22645300]

[9] Shirasawa N, Mabuchi Y, Sakuma E, *et al.* Intercellular communication within the rat anterior pituitary gland: X. Immunohistocytochemistry of S-100 and connexin 43 of folliculo-stellate cells in the rat anterior pituitary gland. Anat Rec A Discov Mol Cell Evol Biol 2004; 278(1): 462-73.
[http://dx.doi.org/10.1002/ar.a.20040] [PMID: 15103742]

[10] Ceccatelli S, Hulting AL, Zhang X, Gustafsson L, Villar M, Hökfelt T. Nitric oxide synthase in the rat anterior pituitary gland and the role of nitric oxide in regulation of luteinizing hormone secretion. Proc Natl Acad Sci USA 1993; 90(23): 11292-6.
[http://dx.doi.org/10.1073/pnas.90.23.11292] [PMID: 7504302]

[11] Ferrara N, Schweigerer L, Neufeld G, Mitchell R, Gospodarowicz D. Pituitary follicular cells produce basic fibroblast growth factor. Proc Natl Acad Sci USA 1987; 84(16): 5773-7.
[http://dx.doi.org/10.1073/pnas.84.16.5773] [PMID: 2441393]

[12] Gospodarowicz D, Lau K. Pituitary follicular cells secrete both vascular endothelial growth factor and follistatin. Biochem Biophys Res Commun 1989; 165(1): 292-8.
[http://dx.doi.org/10.1016/0006-291X(89)91068-1] [PMID: 2590228]

[13] Vankelecom H, Carmeliet P, Van Damme J, Billiau A, Denef C. Production of interleukin-6 by folliculo-stellate cells of the anterior pituitary gland in a histotypic cell aggregate culture system. Neuroendocrinology 1989; 49(1): 102-6.
[http://dx.doi.org/10.1159/000125097] [PMID: 2785659]

[14] Ferrara N, Winer J, Henzel WJ. Pituitary follicular cells secrete an inhibitor of aortic endothelial cell growth: identification as leukemia inhibitory factor. Proc Natl Acad Sci USA 1992; 89(2): 698-702.

[http://dx.doi.org/10.1073/pnas.89.2.698] [PMID: 1370585]

[15] Tierney T, Patel R, Stead CA, Leng L, Bucala R, Buckingham JC. Macrophage migration inhibitory factor is released from pituitary folliculo-stellate-like cells by endotoxin and dexamethasone and attenuates the steroid-induced inhibition of interleukin 6 release. Endocrinology 2005; 146(1): 35-43. [http://dx.doi.org/10.1210/en.2004-0946] [PMID: 15388650]

[16] Hoshino S, Kurotani R, Miyano Y, *et al.* Macrophage colony-stimulating factor induces prolactin expression in rat pituitary gland. Zoolog Sci 2014; 31(6): 390-7. [http://dx.doi.org/10.2108/zs130226] [PMID: 24882100]

[17] Swennen L, Baes M, Schramme C, Denef C. Beta-adrenergic stimulation of adenosine-3′,5--monophosphate (c-AMP) in primary cultures of rat anterior pituitary cell populations separated by unit gravity sedimentation. Relationship to growth hormone and prolactin release and to nonsecreting cells. Neuroendocrinology 1985; 40(1): 78-83. [http://dx.doi.org/10.1159/000124055] [PMID: 2982112]

[18] Davis SW, Ellsworth BS, Peréz Millan MI, *et al.* Pituitary gland development and disease: from stem cell to hormone production. Curr Top Dev Biol 2013; 106: 1-47. [http://dx.doi.org/10.1016/B978-0-12-416021-7.00001-8] [PMID: 24290346]

[19] Yoshimura F, Harumiya K. Electron microscopy of the anterior lobe of pituitary in normal and castrated rats. Endocrinol Jpn 1965; 12(2): 119-52. [http://dx.doi.org/10.1507/endocrj1954.12.119] [PMID: 5897652]

[20] Horvath E, Kovacs K. Folliculo-stellate cells of the human pituitary: a type of adult stem cell? Ultrastruct Pathol 2002; 26(4): 219-28. [http://dx.doi.org/10.1080/01913120290104476] [PMID: 12227947]

[21] Shirasawa N, Kihara H, Yamaguchi S, Yoshimura F. Pituitary folliculo-stellate cells immunostained with S-100 protein antiserum in postnatal, castrated and thyroidectomized rats. Cell Tissue Res 1983; 231(2): 235-49. [http://dx.doi.org/10.1007/BF00222177] [PMID: 6850801]

[22] Ishikawa H, Nogami H, Shirasawa N. Novel clonal strains from adult rat anterior pituitary producing S-100 protein. Nature 1983; 303(5919): 711-3. [http://dx.doi.org/10.1038/303711a0] [PMID: 6406910]

[23] Cocchia D, Miani N. Immunocytochemical localization of the brain-specific S-100 protein in the pituitary gland of adult rat. J Neurocytol 1980; 9(6): 771-82. [http://dx.doi.org/10.1007/BF01205018] [PMID: 7009796]

[24] Velasco ME, Roessmann U, Gambetti P. The presence of glial fibrillary acidic protein in the human pituitary gland. J Neuropathol Exp Neurol 1982; 41(2): 150-63. [http://dx.doi.org/10.1097/00005072-198203000-00005] [PMID: 7062085]

[25] Höfler H, Denk H, Walter GF. Immunohistochemical demonstration of cytokeratins in endocrine cells of the human pituitary gland and in pituitary adenomas. Virchows Arch A Pathol Anat Histopathol 1984; 404(4): 359-68. [http://dx.doi.org/10.1007/BF00695220] [PMID: 6208678]

[26] Tachibana O, Yamashima T. Immunohistochemical study of folliculo-stellate cells in human pituitary

adenomas. Acta Neuropathol 1988; 76(5): 458-64.
[http://dx.doi.org/10.1007/BF00686384] [PMID: 2847475]

[27] Horiguchi K, Fujiwara K, Yoshida S, *et al.* Isolation of dendritic-cell-like S100β-positive cells in rat anterior pituitary gland. Cell Tissue Res 2014; 357(1): 301-8.
[http://dx.doi.org/10.1007/s00441-014-1817-9] [PMID: 24737488]

[28] Higuchi M, Kanno N, Yoshida S, *et al.* GFP-expressing S100β-positive cells of the rat anterior pituitary differentiate into hormone-producing cells. Cell Tissue Res 2014; 357(3): 767-79.
[http://dx.doi.org/10.1007/s00441-014-1890-0] [PMID: 24842050]

[29] Allaerts W, Vankelecom H. History and perspectives of pituitary folliculo-stellate cell research. Eur J Endocrinol 2005; 153(1): 1-12.
[http://dx.doi.org/10.1530/eje.1.01949] [PMID: 15994739]

[30] Vankelecom H. Pituitary stem cells drop their mask. Curr Stem Cell Res Ther 2012; 7(1): 36-71.
[http://dx.doi.org/10.2174/157488812798483467] [PMID: 22023621]

[31] Ebling FJ. The neuroendocrine timing of puberty. Reproduction 2005; 129(6): 675-83.
[http://dx.doi.org/10.1530/rep.1.00367] [PMID: 15923383]

[32] Castrique E, Fernandez-Fuente M, Le Tissier P, Herman A, Levy A. Use of a prolactin-Cre/ROS--YFP transgenic mouse provides no evidence for lactotroph transdifferentiation after weaning, or increase in lactotroph/somatotroph proportion in lactation. J Endocrinol 2010; 205(1): 49-60.
[http://dx.doi.org/10.1677/JOE-09-0414] [PMID: 20139144]

[33] Fu Q, Vankelecom H. Regenerative capacity of the adult pituitary: multiple mechanisms of lactotrope restoration after transgenic ablation. Stem Cells Dev 2012; 21(18): 3245-57.
[http://dx.doi.org/10.1089/scd.2012.0290] [PMID: 22970885]

[34] Alkhani AM, Cusimano M, Kovacs K, Bilbao JM, Horvath E, Singer W. Cytology of pituitary thyrotroph hyperplasia in protracted primary hypothyroidism. Pituitary 1999; 1(3-4): 291-5.
[http://dx.doi.org/10.1023/A:1009966812195] [PMID: 11081211]

[35] Thodou E, Asa SL, Kontogeorgos G, Kovacs K, Horvath E, Ezzat S. Clinical case seminar: lymphocytic hypophysitis: clinicopathological findings. J Clin Endocrinol Metab 1995; 80(8): 2302-11.
[PMID: 7629223]

[36] Borrelli E, Heyman RA, Arias C, Sawchenko PE, Evans RM. Transgenic mice with inducible dwarfism. Nature 1989; 339(6225): 538-41.
[http://dx.doi.org/10.1038/339538a0] [PMID: 2733785]

[37] Nassiri F, Cusimano M, Zuccato JA, *et al.* Pituitary stem cells: candidates and implications. Pituitary 2013; 16(3): 413-8.
[http://dx.doi.org/10.1007/s11102-013-0470-8] [PMID: 23423660]

[38] Chen J, Hersmus N, Van Duppen V, Caesens P, Denef C, Vankelecom H. The adult pituitary contains a cell population displaying stem/progenitor cell and early embryonic characteristics. Endocrinology 2005; 146(9): 3985-98.
[http://dx.doi.org/10.1210/en.2005-0185] [PMID: 15932930]

[39] Yoshimura F, Ishikawa H. Identification of the thyrotrophs with the gonadotrophs in the anterior

pituitaries of thyroidectomized rats. Endocrinol Jpn 1969; 16(1): 69-85.
[http://dx.doi.org/10.1507/endocrj1954.16.69] [PMID: 5394334]

[40] Castinetti F, Davis SW, Brue T, Camper SA. Pituitary stem cell update and potential implications for treating hypopituitarism. Endocr Rev 2011; 32(4): 453-71.
[http://dx.doi.org/10.1210/er.2010-0011] [PMID: 21493869]

[41] Gilyarov AV. Nestin in central nervous system cells. Neurosci Behav Physiol 2008; 38(2): 165-9.
[http://dx.doi.org/10.1007/s11055-008-0025-z] [PMID: 18197384]

[42] Krylyshkina O, Chen J, Mebis L, Denef C, Vankelecom H. Nestin-immunoreactive cells in rat pituitary are neither hormonal nor typical folliculo-stellate cells. Endocrinology 2005; 146(5): 2376-87.
[http://dx.doi.org/10.1210/en.2004-1209] [PMID: 15677762]

[43] Lepore DA, Roeszler K, Wagner J, Ross SA, Bauer K, Thomas PQ. Identification and enrichment of colony-forming cells from the adult murine pituitary. Exp Cell Res 2005; 308(1): 166-76.
[http://dx.doi.org/10.1016/j.yexcr.2005.04.023] [PMID: 15916758]

[44] Nolan LA, Levy A. A population of non-luteinising hormone/non-adrenocorticotrophic hormone-positive cells in the male rat anterior pituitary responds mitotically to both gonadectomy and adrenalectomy. J Neuroendocrinol 2006; 18(9): 655-61.
[http://dx.doi.org/10.1111/j.1365-2826.2006.01459.x] [PMID: 16879164]

[45] Kelberman D, Rizzoti K, Lovell-Badge R, Robinson IC, Dattani MT. Genetic regulation of pituitary gland development in human and mouse. Endocr Rev 2009; 30(7): 790-829.
[http://dx.doi.org/10.1210/er.2009-0008] [PMID: 19837867]

[46] Rizzoti K, Michurina T, Encinas JM, *et al.* Adult pituitary progenitors/stem cells: from *in vitro* characterization to *in vivo* function. Eur J Neurosci 2010; 32(12): 2053-62.
[http://dx.doi.org/10.1111/j.1460-9568.2010.07524.x] [PMID: 21143660]

[48] Lloyd RV, Hardin H, Montemayor-Garcia C, *et al.* Stem cells and cancer stem-like cells in endocrine tissues. Endocr Pathol 2013; 24(1): 1-10.
[http://dx.doi.org/10.1007/s12022-013-9235-1] [PMID: 23435637]

[49] Seymour PA, Freude KK, Tran MN, *et al.* SOX9 is required for maintenance of the pancreatic progenitor cell pool. Proc Natl Acad Sci USA 2007; 104(6): 1865-70.
[http://dx.doi.org/10.1073/pnas.0609217104] [PMID: 17267606]

[50] Poché RA, Furuta Y, Chaboissier MC, Schedl A, Behringer RR. Sox9 is expressed in mouse multipotent retinal progenitor cells and functions in Müller glial cell development. J Comp Neurol 2008; 510(3): 237-50.
[http://dx.doi.org/10.1002/cne.21746] [PMID: 18626943]

[51] Scott CE, Wynn SL, Sesay A, *et al.* SOX9 induces and maintains neural stem cells. Nat Neurosci 2010; 13(10): 1181-9.
[http://dx.doi.org/10.1038/nn.2646] [PMID: 20871603]

[52] Rizzoti K, Akiyama H, Lovell-Badge R. Mobilized adult pituitary stem cells contribute to endocrine regeneration in response to physiological demand. Cell Stem Cell 2013; 13(4): 419-32.
[http://dx.doi.org/10.1016/j.stem.2013.07.006] [PMID: 24094323]

[53] Fauquier T, Rizzoti K, Dattani M, Lovell-Badge R, Robinson IC. SOX2-expressing progenitor cells generate all of the major cell types in the adult mouse pituitary gland. Proc Natl Acad Sci USA 2008; 105(8): 2907-12.
[http://dx.doi.org/10.1073/pnas.0707886105] [PMID: 18287078]

[54] Vankelecom H. Stem cells in the postnatal pituitary? Neuroendocrinology 2007; 85(2): 110-30.
[http://dx.doi.org/10.1159/000100278] [PMID: 17337880]

[55] Hofmann MC. Gdnf signaling pathways within the mammalian spermatogonial stem cell niche. Mol Cell Endocrinol 2008; 288(1-2): 95-103.
[http://dx.doi.org/10.1016/j.mce.2008.04.012] [PMID: 18485583]

[56] Garcia-Lavandeira M, Quereda V, Flores I, *et al.* A GRFa2/Prop1/stem (GPS) cell niche in the pituitary. PLoS One 2009; 4(3): e4815.
[http://dx.doi.org/10.1371/journal.pone.0004815] [PMID: 19283075]

[57] Himes AD, Raetzman LT. Premature differentiation and aberrant movement of pituitary cells lacking both Hes1 and Prop1. Dev Biol 2009; 325(1): 151-61.
[http://dx.doi.org/10.1016/j.ydbio.2008.10.010] [PMID: 18996108]

[58] Galichet C, Lovell-Badge R, Rizzoti K. Nestin-Cre mice are affected by hypopituitarism, which is not due to significant activity of the transgene in the pituitary gland. PLoS One 2010; 5(7): e11443.
[http://dx.doi.org/10.1371/journal.pone.0011443] [PMID: 20625432]

[59] Higuchi M, Yoshida S, Ueharu H, Chen M, Kato T, Kato Y. PRRX1 and PRRX2 distinctively participate in pituitary organogenesis and a cell-supply system. Cell Tissue Res 2014; 357(1): 323-35.
[http://dx.doi.org/10.1007/s00441-014-1861-5] [PMID: 24770895]

[60] Goodell MA, Brose K, Paradis G, Conner AS, Mulligan RC. Isolation and functional properties of murine hematopoietic stem cells that are replicating *in vivo.* J Exp Med 1996; 183(4): 1797-806.
[http://dx.doi.org/10.1084/jem.183.4.1797] [PMID: 8666936]

[61] Chen J, Gremeaux L, Fu Q, Liekens D, Van Laere S, Vankelecom H. Pituitary progenitor cells tracked down by side population dissection. Stem Cells 2009; 27(5): 1182-95.
[http://dx.doi.org/10.1002/stem.51] [PMID: 19418455]

[62] Asakura A, Seale P, Girgis-Gabardo A, Rudnicki MA. Myogenic specification of side population cells in skeletal muscle. J Cell Biol 2002; 159(1): 123-34.
[http://dx.doi.org/10.1083/jcb.200202092] [PMID: 12379804]

[63] Cherqui S, Kurian SM, Schussler O, Hewel JA, Yates JR III, Salomon DR. Isolation and angiogenesis by endothelial progenitors in the fetal liver. Stem Cells 2006; 24(1): 44-54.
[http://dx.doi.org/10.1634/stemcells.2005-0070] [PMID: 16099996]

CHAPTER 7

Cancer Stem Cells in Pancreatic and Hepatocellular Carcinoma: Similarities and Differences

Simona Olimpia Dima[1,*], Dana Cucu[1,2], Nicolae Bacalbasa[3], Valeria Tica[1] and Irinel Popescu[1]

[1] *"Dan Setlacec" Center of General Surgery and Liver Transplantation, Fundeni Clinical Institute, Bucharest, Romania*

[2] *The Department of Anatomy, Physiology and Biophysics, Faculty of Biology, University of Bucharest, Bucharest, Romania*

[3] *"Carol Davila" University of Medicine and Pharmacy, Bucharest, Romania*

Abstract: The objective of the present chapter is to discuss the identification of cancer stem cells (CSC) in pancreatic ductal adenocarcinomas (PDAs) and hepatocellular carcinoma (HCC) and to briefly overview the current therapies targeting this cell population. The reason to restrict the discussion to these particular cancer types is that both are described by a poor prognosis and resistance to conventional therapies. Therefore, CSC may be an important hint in improving the therapeutic approach in both instances. Due to the capacity of CSC to escape chemo and radiotherapy, novel means of approaching these cells are necessary in order to improve outcome therapeutically.

Keywords: Cancer stem cell, Cell surface marker, Drug resistance, Hepatic cellular carcinoma, Liver cancer, Metastases, Pancreatic cancer.

INTRODUCTION

In the last decades, tremendous progress has been made in "the war on cancer".

* **Corresponding author Simona Olimpia Dima:** Fundeni Clinical Institute, Fundeni Ave, no 258, 022328 Bucharest Romania; Tel: 0040 21 318 04 17; Fax: +0040 21 318 04 17; Email: jarmilavojtkova@gmail.com.

Since 2007, when the National Cancer Act was signed, early detection and new therapies brought hope to patients whose life span had considerably improved. Nowadays, 5-years survival rate for prostate cancer is almost 100%, in Hodgkin's lymphoma and breast cancer about 80%; and for colon cancer and non-Hodgkin's lymphoma more than 60%. However, in strong contrast with these results, for two types of cancers -hepatocellular carcinoma and pancreatic adenocarcinoma-the improvements are still below expectations. For pancreatic adenocarcinoma (PA), 5 years survival rate is about 6% and around 16% for hepatocellular carcinoma (HCC) (https://www.baylorhealth.com/Site CollectionDocuments/BUMC%20Cancer%20Care/Spring_2013_Cancer-Update.pdf). The reason why these types of cancer are still undertreated is because patients are usually asymptomatic and the diagnosis is late. Therefore, metastatic dissemination and resistance to chemotherapy became the determinant factors in PA and HCC prognosis. Unfortunately, new cancer therapies are facing the same fate: surgical removal of tumours and cytotoxic drugs cannot completely eradicate the malignancy. This situation led to a new paradigm that sustains the idea of the heterogeneity of cancer cells designed in a specific microenvironment which makes the treatment extremely difficult.

In the last years, many studies focus on the description of cancer –stroma interactions, which imply a dynamic network between inflammation, epithelial to mesenchymal transition (EMT), specific mutations and cancer stem cells. In this chapter, we will bring an integrative view of the molecular, cellular and extracellular milieu of cancer stem cells in pancreatic and hepatocellular tumours, and discuss thoroughly in the perspective of personalized therapeutic approaches.

Cancer Stem Cells: Definition and Origin

The cancer stem cells (CSC) concept indicates specific cell populations with particular characteristics, which may represent a promising tool in cancer therapy. In order to assign CSC as potential druggable targets, it is important to have a clear definition, that specific markers and unambiguous description of the processes that maintain a distinct lineage within tumours.

It has been recognized for a long time that tumours are formed by a heterogeneous mass of cells in different stages of development, but the first experiments showing that cells with specific growth and division characteristics coexist with other tumour cells as described in the studies of acute myelogenous leukemia (AML) [1]. Using xenotransplatation assays, these experiments provided the first evidence of a distinct immature cell population able to produce postmytotic progeny. Furthermore, there are defined as "cells with the capacity of cell renewal and differentiation", this rather small pool of cells was identified and described in many other cancers. Cancer stem cells (CSC) share some common characteristics with normal stem cells in terms of controlling their self-renewal and proliferation, and residence in a physiological niche which manages these processes [2]. However, this definition created some uncertainties, when CSC are defined as cells derived from normal stem cells. Although, in many cancers CSC are derived from normal stem cells that bared several mutations and transform them into cancerous cells, in other cancers a progenitor acquires "stemness" properties by later mutations. It is therefore feasible that more differentiated cells that supported several mutagenic events may acquire self-renewal capacity and personify the CSC.

An even more interesting feature, similar to normal stem cells is the dormancy pattern of CSC. In tumours, these cells are mainly quiescent, preventing them to be targeted by chemotherapeutic drugs. The quiescent state, which is in fact a reversible cell cycle arrest maintained by many regulatory mechanisms, provides CSC with the fuel for triggering metastasis when the microenvironment is suited for it [3]. A similar model was described for hematopoietic stem cells that have stopped the entire DNA replication and moved to an almost eternal G_0 phase which allows them to preserve the energy for stress situation such as injury or toxic insult. Moreover, an intriguing fact is that tumour cells may, in certain situations, spontaneously convert to CSCs. This result is in comparison with a clear model of specific CSC and sustains the idea that in a tumour population any cell may acquire stemness properties at a certain moment [4].

The theory of CSC became very appealing because of the easy intuitive picture that only a part of tumour cells possesses the metastatic phenotype and these cells are resistant to chemotherapy. This would explain why removing the tumour mass maintains the capacity of proliferation and metastasis. With a slow rate of growing CSCs are less sensitive to treatments. An even more attractive theory was presented by Yachida *et al.* who gave a timing genetic evolution of PDA, indicating that a single clonal population of "parental" cells support mutations and that this event appears one decade after these cells are born [5].

The origin of CSC in both PDA and HCC is still under debate. In adult pancreas, only the cells of the Pdx1-positive lineage are susceptible to acquiring mutations. However, at least theoretically, a population of cells comprising progenitor cells or transdifferentiated bone marrow cells able to regenerate the organ would be prone to malignant transformation. However, there is a poor description of the exact nature of these cellular populations. In the liver, progenitor adult cells residing in the biliary duct that are activated upon liver injury have been described. Due to the capacity of CSC to escape chemo- and radiotherapy, novel means of approaching these cells are necessary in order to improve therapeutically outcome.

Specific Markers for CSC in PA and HCC

If the definition is under debate, the identification of CSC becomes even more challenging. In the last years considerable efforts were put in the characterisation of CSC markers and a plethora of molecules were proposed in order to clearly delineate the small population of CSC (comprising at most 6% of the whole tumour cell population) from the other cancer cells. However, few of them may be designated as specific markers for a certain cancer type and even less were confirmed by independent publications. In PDA and HCC several proteins often used in combinations, are employed to identify CSC.

In early studies, in pancreatic cancer CSC were identified by the expression of the surface antigen CD133, which is also found in normal or tumoral tissues of neural, hematopoietic, epithelial, or endothelial origin [6]. Later on, simultaneous expression of the cell surface markers CD44, CD24, and epithelial-specific antigen (ESA) define as the specific phenotype CD44+CD24+ESA+ was proved to possess a 100-fold increased tumorigenic potential and stem cell properties [7]. Some studies also connect aldehyde dehydrogenase (ALDH) or c-Met with increased capacity of tumour formation [8, 9], while more recent studies indicate Dclk1 (Doublecortin-like kinase 1, formerly known as DCAMKL-1 or doublecortin and CaM kinase-like 1) as an important CSC makers in pancreatic cancer. Dclk1 regulates several oncogenes such as c-MYC, KRAS, NOTCH1 and epithelial to mezemchimal transition (EMT) and is up-regulated in PDA [10]. CSC in HCC can be identified by surface cell markers described as well in PDA such as CD133, EpCAM (epithelial adhesion molecule designated also as ESA), and CD44 but also by other surface proteins CD90 or CD13 [6, 7].

During embryonic development liver precursors are bi-potent stem cells with the ability to develop towards two cell lineages hepatocytes and cholangiocytes, which are in normal conditions dormant cells with a low proliferation rate. Upon challenging with a harmful stimulus, they will change towards a highly dedifferentiation state in order to compensate the injury [8]. A recent theory sustains that human liver stem cells (LSC) are likely to be at the origin of HCC. This hypothesis is based on studies showing that HCC is formed by a heterogeneous cell population from which ~50% express markers of progenitors [9]. The scientific question arises:what molecular pathways determine the transformation of liver stem cells in cancer stem cells?

Previous studies showed that several signalling pathways such as Notch, Wnt, Hedgehog, transforming growth factor (TGF) participate to the malignant transformation of LSC in CSC. For instance, activation of Wnt/β-catenin pathways stimulates self-renewal and tumorigenesis contributing to the initiation of CSC [10]. Moreover, a discrete balance between Wnt and Notch signalling pathways, the latest highly conserved in mammalian cells, stays at the background of oncogenic transformation of LCS in CSC [11]. These studies assume that the biliary specification of LCS cells may be avoided when Wnt pathway is activated, which in turn inhibits Notch signalling. In contrast, Notch activation leads to biliary specification. Another signalling pathway conserved in the development of liver cells is the Hedgehog signalling pathway. Previous reports show that a specific homolog, the sonic Hedgehog, is activated in CD133+ CSC cells [12] whereas others emphasize the inhibition of Hedgehog invert hepatocarcinogenesis in HCC [13].

The TGF-β signalling pathway participates in cell growth, development differentiation and apoptosis of cells as a tumour suppressor. During carcinogenesis, TGF-β promotes cell motility and metastasis. In liver cancer TGF-β expression is decreased in early stages but increased in later carcinogenesis phases. The mechanisms behind this complex regulation are not yet revealed, but the results may be the effect of TGF-β on tumour microenvironment.

Apart from all these proteins, a new view focus on the glycans bound to protein or lipids, which would make the stem cells easily identifiable. The hypothesis is feasible, because during the malignant transformation from a normal progenitor cell to a stem cell the glycosylation process would be perturbed [14]. However, there are only few studies describing glycans as specific markers for CSC often associated with glioblastoma [15]. Recently, using lectin microarray analysis on CSC-like fractions of human pancreatic cancer cell lines, a glycosylation marker was found for three CSC fraction: gemcitabine-resistant, sphere-forming cells, and $CD24^{high}/CD44^{high}$ [16]. These findings may develop new avenues in the effort to target specifically CSC, because disruption of glycosylation is potentially linked to therapeutic resistance.

CSC Contribution to Epithelial to Mesenchymal Transition in PDA and HCC

As described above, to maintain the specific pattern of self-renewal and pluripotency, CSCs have to sustain a plethora of signalling pathways. Wnt/β-catenin pathway, the Sonic Hedgehog pathway, the Notch pathway and many others as being involved in the regulation and preservation of stemness and in the tumorigenesis process.

Besides, a special attention is dedicated to the process of epithelial to mesenchymal transition (EMT) that occurs at the invasive border of cancers. In PDA as well as in HCC, EMT is an indicator of poor prognosis. In physiological conditions, EMT is an essential factor in the gastrulation phase of the embryonic development. It consists in the transition of the epithelial to mesenchymal cells and down-regulation of proteins participating to cell adhesion (such as E-cadherins), while stimulating mesenchymal molecules N-cadherin, vimentin, fibronectin. In adult cells, repression of adhesion molecules is associated with cell migration, infiltration and metastasis, conditions particularly described for PDA [17], but also for HCC [18]. The EMT process is regulated by many signalling pathways, with Notch, Wnt, TGF-β being the most mentioned. The Hedgehog pathway is a significant embryonic pathway connected to PDA.

During invasion, cancer cells stop the proliferation, probably consecutive to changes in the cytoskeleton. Moreover, the process of EMT impedes the division of cells. If CSCs are considered cells with slow proliferation this could be related to EMT. This assumption is in conflict with the theory that states CSC as highly proliferative, which is supported by many recent studies [19]. Nevertheless, in pancreatic cancer, based on overlapped signalling pathways in both processes, new evidence suggests that EMT promotes CSC [20]. The regulatory crosstalk is mediated *via* Hedgehog pathway and enhances the metastatic ability.

In HCC, recently it was suggested that CSC are promoted by the oncogenic hepatitis virus B (HBV) and C (HCV). The viruses have very different life cycles, but they produce persistent infection responsible for cystic fibrosis and finally HCC. Replication of HCV showed induction of CSC markers such as Dclk1 and CD133 together with liver progenitor cell makers such as α-fetoprotein (AFP). The abundance of Dclk1 positive CSC increased with the disease stage [21]. In the same line of evidence, previous studies showed that the expression of HBx protein of HBV in hepatoma cells increased the abundance of the markers EpCAM and β-catenin specific for CSC, but also for EMT promotion. Moreover, the expression of pluripotency transcription factors Nanog, Klf-4, Oct-4 was markedly higher [22]. Based on these results it became increasingly clear that HBV and HCV infection uphold the EMT and stem-like characteristics. This feature is essential for the description of CSC microenvironment, which will make them targetable for new therapies for HCC.

CSC Targeted Therapies in PDA

Despite great efforts, pancreatic ductal adenocarcinomas (PDA), are characterized by extremely poor prognosis, with a 5-year mortality of 97-98% [8]. Unfortunately, metastasis remains responsible for the vast majority of the cancer-related mortality and morbidity in pancreatic cancer. Also, the aggressiveness of pancreatic cancer can be attributed to heterogeneity of the tumours, but also to the chemotherapy-resistant cancer cells.

Even though CSCs represent only between 1 and 6% of the bulk tumours, they may participate to therapeutic resistance. Recently, some studies revealed that CSCs contribute to the resistance of PDA at chemotherapy because they have a higher expression of Notch-1 and the ABC -binding cassette transporter ABCG2 [23].

Moreover, it was recently reported that gemcitabine-resistant pancreatic cancer cells have several characteristics specific to CSC [24]. Firstly, these cells showed particular morphological changes such as spindle-shape, reduction of adherent properties or development of pseudopodia. Secondly, they were functionally described with increased migratory properties and augmented invasiveness with a higher expression of vimentin and E-cadherin expression. Lee *et al.* observed that treatment of primary human PDA xenografts with gembcitabine and ionizing radiation results in the enrichment of the $CD44^{+}CD24^{+}ESA^{+}$ population [25]. Other studies report that some specific signalling pathways such as Wnt/β-catenin or Notch are altered in chemo-resistance and radio-resistance processes, although the exact mechanisms are not very clear so far. Altogether, these studies imply that successful therapeutic strategies able to overcome multiple drug resistance mechanisms should use approaches that allow the inhibition of CSC with those that target the differentiated, nontumorigenic bulk of pancreatic cancer cell. Forthcoming clinical trials will reveal the feasibility of this approach.

Until now, preclinical and clinical studies showed that the inhibition of Hedgehog pathway in animal models reduces tumorigenesis consecutive to the inhibition of CSC ability to proliferate [26]. Hedgehog inhibitors such as vismodegib, IPI-26909, GDC-0449, and cyclopamine are under investigation and are promising therapeutic agents. Using IPI-26909 *in vitro* studies showed decreased cell migration, colony formation, and expression of ALDH, whereas in an orthotropic xenograft model, the drug blocked metastases [27]. An additional potential CSC target in PDA was c-Met. Inhibition of c-Met with XL184 decreased tumour sphere formation in CD44+CD24+ESA+ cells. Moreover, a combination of Xl184- and gemcitabine reduced tumour growth [28]. Cabozantinib, another c-Met inhibitor, decreased sphere formation and increased apoptosis of PDA cells [29]. Other stem cell targeted therapies include inhibition of Notch and TGF-β signalling pathways. AMD3100, an inhibitor of a cell adhesion molecule decreased migration of CSC and the metastatic process in a mouse model [30].

CSC Targeted Therapies in HCC

Hepatocellular carcinoma (HCC) is the fifth most frequent of the solid tumours worldwide and is the third leading cause of cancer mortality, having a death rate estimated at 11,000 cases per year [31]. Currently, liver resection (LR), liver transplantation (LT), and local ablative therapy or transarterial chemoembolization (TACE) are the main treatment options for HCC. Although LR and LT are considered as curative treatment options for HCC, intrahepatic *de novo* recurrence or metastasis occurs in a significant number of patients [32]; recurrence rate at 5-year after LR for HCC is 69-75% [33]. For patients with a high potential of recurrence, postoperative adjuvant therapies have to be added. Existing loco-regional and systemic chemotherapeutic treatment have a low efficiency largely due to the resistance of cancer cells and their ability to stimulate neoangiogenesis [34]. Moreover, systemic chemotherapy in advanced HCC did not show any significant improvement in survival [35]. The only first-line drug available for advanced (non-resectable) HCC is based on the tyrosine kinase inhibitor, sorafenib recommended as standard of care for the treatment of advanced HCC worldwide [36]. However, some studies have shown the mechanism of acquired resistance to Sorafenib is multifactorial and indicate CSC as one of the important factors [37]. Thus, it is important to identify different cell types involved in HCC and to develop new therapeutic agents. Accumulating evidence suggest that specific agents targeting the CSC population may overcome chemotherapy resistance. The main therapeutic approaches proposed to target CSC in HCC were the signalling pathways participating in CSC regulation and specific surface markers of HCC stem cells.

As previously discussed, an important regulatory cascade is the Hedgehog pathway, which is deregulated in hepatocarcinogenesis and is important stem cell biology. Interestingly, inhibition of Hedgehog not only decreased HCC cell proliferation but also increased sensitization of cells to 5-fluorouracil and stimulated the process of apoptosis [38]. Many other compounds have been used to inhibit Hedgehog pathway. As in PDA, vismodegib has been successfully used for the inhibition of CSC and it is now in Phase I trials for HCC (http://www.roche-trials.com/resultsByDiseaseCategoryGet.action?diseaseCategoryId=282).

The Wnt pathway, which participates also to the process of EMT transition, was evaluated as a possible target for HCC. The downstream gene in Wnt pathway is EpCAM that supports CSC proliferation and plays a role in EMT. Previous results demonstrated that inhibition of EpCAM with the specific inhibitor VB4-845, suppressed sphere formation and decreased the abundance of CD 133^{+} cell population in subcutaneous and orthotropic liver xenograft models [39].

The transforming growth factor beta (TGF-β) ligands were studied especially in relation to immune system and tumour progression, being part of a signalling cascade related especially to inflammatory conditions in different cancer types [40]. Most of HCC cases are preceded by fibrosis, which is a consequence of an inflammatory process; therefore TGF-β signalling is a mediator of the immune and inflammatory responses that increase pluripotency and tumorigenesis. Moreover, it promotes resistance to therapies of liver CSC. The TGF-β inhibitor LY2157299 monohydrate (LY) is currently under phase II study for patients with HCC (http://meetinglibrary.asco.org/content/111562-132). Other studies showed that, $CD133^{+}$ cells are also more resistant to conventional chemotherapeutic drugs such as doxorubicin and 5-fluorouracil, the underlying mechanism being the preferential activation of the Akt/protein kinase B and Bcl-2 survival pathways [41].

Many surface markers of HCC were proposed as targets for therapeutic strategies in HCC. For instance, the antibody against CD44 delivered as liposomal nanoparticles produces regression of HCC cells [42]. An agent that downregulated the expression of both CD133 and CD 44 8-bromo-7-methoxychrysin was described as an inhibitor of CSC proliferation and EMT transition in relation to β-catenin suppression [43].

From all these results and many other studies, it is clear, that a tremendous effort is focused on the field of "stemness inhibitor" (Boston Biomedical's) and currently clinical candidates are on the pipeline of different pharmaceutical companies (*e.g.* www.bostonbiomedical.com/bbi-product-pipeline.htm).

PERSPECTIVES

A substantial number of studies sustain the hypothesis that a small population of cell is connected to the process of tumorigenesis characterized by increased proliferation, migration, invasion, metastatic capacity, therapies resistance, and recurrence, all of which eventually contribute to poor prognosis. The effort to better characterise these cells brought new specific markers, isolation and description of CSC, helping to decipher the complex nature of CSC and their implication in tumour microenvironment and metastasis. These new insights have the ultimate role to improve the therapy in the field of cancer by targeting and eliminating CSC. Additional clinical and preclinical studies are required, but the progress made so far, brings hope for the development of CSC-targeting drugs and the management of PDA and HCC.

CONFLICT OF INTEREST

The authors confirm that they have no conflict of interest to declare for this publication.

ACKNOWLEDGEMENTS

This study was financially supported by the project: EEA-JRP-RO-NO no. 4SEE/2014.

REFERENCES

[1] Lapidot T, Sirard C, Vormoor J, *et al.* A cell initiating human acute myeloid leukaemia after transplantation into SCID mice. Nature 1994; 367(6464): 645-8. [http://dx.doi.org/10.1038/367645a0] [PMID: 7509044]

[2] Li HC, Stoicov C, Rogers AB, Houghton J. Stem cells and cancer: evidence for bone marrow stem cells in epithelial cancers. World J Gastroenterol 2006; 12(3): 363-71. [http://dx.doi.org/10.3748/wjg.v12.i3.363] [PMID: 16489634]

[3] Giancotti FG. Mechanisms governing metastatic dormancy and reactivation. Cell 2013; 155(4): 750-64. [http://dx.doi.org/10.1016/j.cell.2013.10.029] [PMID: 24209616]

[4] Chaffer CL, Brueckmann I, Scheel C, *et al.* Normal and neoplastic nonstem cells can spontaneously convert to a stem-like state. Proc Natl Acad Sci USA 2011; 108(19): 7950-5. [http://dx.doi.org/10.1073/pnas.1102454108] [PMID: 21498687]

[5] Yachida S, Jones S, Bozic I, *et al.* Distant metastasis occurs late during the genetic evolution of pancreatic cancer. Nature 2010; 467(7319): 1114-7. [http://dx.doi.org/10.1038/nature09515] [PMID: 20981102]

[6] Mishra L, Banker T, Murray J, *et al.* Liver stem cells and hepatocellular carcinoma. Hepatology 2009;

49(1): 318-29.
[http://dx.doi.org/10.1002/hep.22704] [PMID: 19111019]

[7] Oishi N, Yamashita T, Kaneko S. Molecular biology of liver cancer stem cells. Liver Cancer 2014; 3(2): 71-84.
[http://dx.doi.org/10.1159/000343863] [PMID: 24944998]

[8] Gaudio E, Carpino G, Cardinale V, Franchitto A, Onori P, Alvaro D. New insights into liver stem cells. Dig Liver Dis 2009; 41(7): 455-62.
[http://dx.doi.org/10.1016/j.dld.2009.03.009] [PMID: 19403350]

[9] Durnez A, Verslype C, Nevens F, *et al.* The clinicopathological and prognostic relevance of cytokeratin 7 and 19 expression in hepatocellular carcinoma. A possible progenitor cell origin. Histopathology 2006; 49(2): 138-51.
[http://dx.doi.org/10.1111/j.1365-2559.2006.02468.x] [PMID: 16879391]

[10] Xu LB, Liu C. Role of liver stem cells in hepatocarcinogenesis. World J Stem Cells 2014; 6(5): 579-90.
[http://dx.doi.org/10.4252/wjsc.v6.i5.579] [PMID: 25426254]

[11] Strazzabosco M, Fabris L. The balance between Notch/Wnt signaling regulates progenitor cells' commitment during liver repair: mystery solved? J Hepatol 2013; 58(1): 181-3.
[http://dx.doi.org/10.1016/j.jhep.2012.08.006] [PMID: 22902547]

[12] El Khatib M, Kalnytska A, Palagani V, *et al.* Inhibition of hedgehog signaling attenuates carcinogenesis *in vitro* and increases necrosis of cholangiocellular carcinoma. Hepatology 2013; 57(3): 1035-45.
[http://dx.doi.org/10.1002/hep.26147] [PMID: 23172661]

[13] Philips GM, Chan IS, Swiderska M, *et al.* Hedgehog signaling antagonist promotes regression of both liver fibrosis and hepatocellular carcinoma in a murine model of primary liver cancer. PLoS One 2011; 6(9): e23943.
[http://dx.doi.org/10.1371/journal.pone.0023943] [PMID: 21912653]

[14] Karsten U, Goletz S. What makes cancer stem cell markers different? Springerplus 2013; 2(1): 301.
[http://dx.doi.org/10.1186/2193-1801-2-301] [PMID: 23888272]

[15] Tucker-Burden C, Chappa P, Krishnamoorthy M, *et al.* Lectins identify glycan biomarkers on glioblastoma-derived cancer stem cells. Stem Cells Dev 2012; 21(13): 2374-86.
[http://dx.doi.org/10.1089/scd.2011.0369] [PMID: 22435486]

[16] Terao N, Takamatsu S, Minehira T, *et al.* Fucosylation is a common glycosylation type in pancreatic cancer stem cell-like phenotypes. World J Gastroenterol 2015; 21(13): 3876-87.
[http://dx.doi.org/10.3748/wjg.v21.i13.3876] [PMID: 25852272]

[17] Krantz SB, Shields MA, Dangi-Garimella S, Munshi HG, Bentrem DJ. Contribution of epithelial-t--mesenchymal transition and cancer stem cells to pancreatic cancer progression. J Surg Res 2012; 173(1): 105-12.
[http://dx.doi.org/10.1016/j.jss.2011.09.020] [PMID: 22099597]

[18] Yopp AC, Singal AG. Epithelial to mesenchymal transition expression profiles as predictive biomarkers of recurrence following resection of HCC: implications for current clinical use and future

stratification for systemic therapy. Ann Surg Oncol 2014; 21(12): 3723-4. [http://dx.doi.org/10.1245/s10434-014-3790-7] [PMID: 25074660]

[19] Floor S, van Staveren WC, Larsimont D, Dumont JE, Maenhaut C. Cancer cells in epithelial-t--mesenchymal transition and tumor-propagating-cancer stem cells: distinct, overlapping or same populations. Oncogene 2011; 30(46): 4609-21. [http://dx.doi.org/10.1038/onc.2011.184] [PMID: 21643013]

[20] Castellanos JA, Merchant NB, Nagathihalli NS. Emerging targets in pancreatic cancer: epithelial-mesenchymal transition and cancer stem cells. Onco Targets Ther 2013; 6: 1261-7. [PMID: 24049451]

[21] Ali N, Allam H, May R, *et al.* Hepatitis C virus-induced cancer stem cell-like signatures in cell culture and murine tumor xenografts. J Virol 2011; 85(23): 12292-303. [http://dx.doi.org/10.1128/JVI.05920-11] [PMID: 21937640]

[22] Arzumanyan A, Friedman T, Ng IO, Clayton MM, Lian Z, Feitelson MA. Does the hepatitis B antigen HBx promote the appearance of liver cancer stem cells? Cancer Res 2011; 71(10): 3701-8. [http://dx.doi.org/10.1158/0008-5472.CAN-10-3951] [PMID: 21464043]

[23] Wang YH, Li F, Luo B, *et al.* A side population of cells from a human pancreatic carcinoma cell line harbors cancer stem cell characteristics. Neoplasma 2009; 56(5): 371-8. [http://dx.doi.org/10.4149/neo_2009_05_371] [PMID: 19580337]

[24] Shah AN, Summy JM, Zhang J, Park SI, Parikh NU, Gallick GE. Development and characterization of gemcitabine-resistant pancreatic tumor cells. Ann Surg Oncol 2007; 14(12): 3629-37. [http://dx.doi.org/10.1245/s10434-007-9583-5] [PMID: 17909916]

[25] Lee CJ, Li C, Simeone DM. Human pancreatic cancer stem cells: implications for how we treat pancreatic cancer. Transl Oncol 2008; 1(1): 14-8. [http://dx.doi.org/10.1593/tlo.08013] [PMID: 18607507]

[26] Rasheed ZA, Matsui W. Biological and clinical relevance of stem cells in pancreatic adenocarcinoma. J Gastroenterol Hepatol 2012; 27 (Suppl. 2): 15-8. [http://dx.doi.org/10.1111/j.1440-1746.2011.07015.x] [PMID: 22320910]

[27] Feldmann G, Fendrich V, McGovern K, *et al.* An orally bioavailable small-molecule inhibitor of Hedgehog signaling inhibits tumor initiation and metastasis in pancreatic cancer. Mol Cancer Ther 2008; 7(9): 2725-35. [http://dx.doi.org/10.1158/1535-7163.MCT-08-0573] [PMID: 18790753]

[28] Li C, Wu J J, Hynes M, *et al.* c-Met is a marker of pancreatic cancer stem cells and therapeutic target. Gastroenterology 2011 Dec; 141: 2218-27. e5 [http://dx.doi.org/10.1053/j.gastro.2011.08.009]

[29] Hage C, Rausch V, Giese N, *et al.* The novel c-Met inhibitor cabozantinib overcomes gemcitabine resistance and stem cell signaling in pancreatic cancer. Cell Death Dis 2013; 4: e627. [http://dx.doi.org/10.1038/cddis.2013.158] [PMID: 23661005]

[30] Hermann PC, Huber SL, Herrler T, *et al.* Distinct populations of cancer stem cells determine tumor growth and metastatic activity in human pancreatic cancer. Cell Stem Cell 2007; 1(3): 313-23. [http://dx.doi.org/10.1016/j.stem.2007.06.002] [PMID: 18371365]

[31] Song GW, Hwang S, Lee SG. [Liver transplantation in patients with hepatocellular carcinoma]. Korean J Gastroenterol 2010; 55(6): 350-60.
[http://dx.doi.org/10.4166/kjg.2010.55.6.350] [PMID: 20571302]

[32] Poon RT, Fan ST, Lo CM, Liu CL, Wong J. Long-term survival and pattern of recurrence after resection of small hepatocellular carcinoma in patients with preserved liver function: implications for a strategy of salvage transplantation. Ann Surg 2002; 235(3): 373-82.
[http://dx.doi.org/10.1097/00000658-200203000-00009] [PMID: 11882759]

[33] King AJ, Breen DJ. Understanding the current status of image-guided ablation for metastatic colorectal disease. Abdom Imaging 2013; 38(6): 1234-44.
[http://dx.doi.org/10.1007/s00261-013-0020-x] [PMID: 23764908]

[34] Zhang X, Yang XR, Huang XW, *et al.* Sorafenib in treatment of patients with advanced hepatocellular carcinoma: a systematic review. HBPD INT 2012; 11(5): 458-66.
[http://dx.doi.org/10.1016/S1499-3872(12)60209-4] [PMID: 23060390]

[35] Boucher E, Corbinais S, Brissot P, Boudjema K, Raoul JL. Treatment of hepatocellular carcinoma (HCC) with systemic chemotherapy combining epirubicin, cisplatinum and infusional 5-fluorouracil (ECF regimen). Cancer Chemother Pharmacol 2002; 50(4): 305-8.
[http://dx.doi.org/10.1007/s00280-002-0503-x] [PMID: 12357305]

[36] Federico A, Orditura M, Cotticelli G, *et al.* Safety and efficacy of sorafenib in patients with advanced hepatocellular carcinoma and Child-Pugh A or B cirrhosis. Oncol Lett 2015; 9(4): 1628-32.
[PMID: 25789012]

[37] Zhai B, Sun XY. Mechanisms of resistance to sorafenib and the corresponding strategies in hepatocellular carcinoma. World J Hepatol 2013; 5(7): 345-52.
[http://dx.doi.org/10.4254/wjh.v5.i7.345] [PMID: 23898367]

[38] Wang Q, Huang S, Yang L, *et al.* Down-regulation of Sonic hedgehog signaling pathway activity is involved in 5-fluorouracil-induced apoptosis and motility inhibition in Hep3B cells. Acta Biochim Biophys Sin (Shanghai) 2008; 40(9): 819-29.
[http://dx.doi.org/10.1093/abbs/40.9.819] [PMID: 18776995]

[39] Ogawa K, Tanaka S, Matsumura S, *et al.* EpCAM-targeted therapy for human hepatocellular carcinoma. Ann Surg Oncol 2014; 21(4): 1314-22.
[http://dx.doi.org/10.1245/s10434-013-3430-7] [PMID: 24370904]

[40] Dima S, Cucu D, Bacalbasa N, Popescu I. Chronic Pancreatitis as an Inductor of Pancreatic Cancer—Correlations With Inflammatory Pathways. InTech 2015.

[41] Ma S, Lee TK, Zheng BJ, Chan KW, Guan XY. CD133+ HCC cancer stem cells confer chemoresistance by preferential expression of the Akt/PKB survival pathway. Oncogene 2008; 27(12): 1749-58.
[http://dx.doi.org/10.1038/sj.onc.1210811] [PMID: 17891174]

[42] Wang L, Su W, Liu Z, *et al.* CD44 antibody-targeted liposomal nanoparticles for molecular imaging and therapy of hepatocellular carcinoma. Biomaterials 2012; 33(20): 5107-14.
[http://dx.doi.org/10.1016/j.biomaterials.2012.03.067] [PMID: 22494888]

[43] Ren KQ, Cao XZ, Liu ZH, *et al.* 8-bromo-5-hydroxy-7-methoxychrysin targeting for inhibition of the

properties of liver cancer stem cells by modulation of Twist signaling. Int J Oncol 2013; 43(5): 1719-29.
[PMID: 23970349]

CHAPTER 8

Immunogenicity of Stem Cell in Tumorigenesis *Versus* Regeneration

Monica Neagu[1,2,*] and **Carolina Constantin**[1]

[1] *Immunobiology Laboratory, "Victor Babeş" National Institute of Pathology, Bucharest, Romania*

[2] *Faculty of Biology, University of Bucharest, Romania*

Abstract: This chapter will focus on the main immunological issues in stem cell approaches. The scientific world faces an important immunological dilemma when investigating stem cells immunogenicity. One is the need to have low stem cell's immunogenicity, property that provides modest inflammatory reaction microenvironment and hence lack of rejection of the transplanted stem cells. On the other side any neoplastic transformation can increase the natural immune evasive properties of stem cells linking immune escape and tumorigenicity. In this light oncogenes expression can directly orchestrate inflammation and immune escape to drive the multistep process of cancer progression, independent of any immuno-editing in the tumor microenvironment. The chapter starts with characterizing the immunogenicity of stem cells, where major histocompatibility expression is the immune mold that can drive toward generation or tumorigenesis. The chapter continues with the processes that are involved in stem cells modulating the immune system elements whether in *Regeneration or Tumorigenesis*. The chapter ends with the main immune hurdles in the most recent clinical trials using stem cells.

Keywords: Adaptive, Immunity, Innate, Stem cells, Regeneration, Tumorigenesis.

[*] **Corresponding author Monica Neagu:** Immunobiology Laboratory, "Victor Babeş" National Institute of Pathology, Bucharest, Romania, Faculty of Biology, University of Bucharest; Tel/Fax: +40-213194528; E-mails: neagu.monica@gmail.com, monica.neagu@ivb.ro.

Cristiana Tanase & Monica Neagu (Eds.)

INTRODUCTION

As we are facing an escalating number of clinical trials using stem cell therapies, the immunological issues that emerged are complex. Therefore, the immunological properties of the administered stem cell can easily direct a perfectly designed therapy toward a clinical failure. Then, the relative immune tolerance of the site where stem cells are administered can favour the clinical outcome or not. Last, but not the least, the immune status of the patient can be the main issue of transforming a regenerative therapy into an unwanted tumorigenesis process [1].

Stem cell-based therapy stands on the exploitation of pluripotent stem cells (PSCs) that are manipulated to repair and/or replace tissues or organs. The main immunological advantage in this field is that functional PSC are not tumorigenic and moreover they do not trigger any immune response that can induce host *versus* graft rejection (GVHD). GVHD is an inflammatory syndrome initiated by alloreactive donor T cells [2].

In transplant and tissue regeneration immune rejection is a crucial issue, thus induced PSC-based therapy, due to its decreased immunogenicity can represent the future solution for an unmet medical problem. Characterizing the phenotype of stem cells furnishes the investigator with the boundaries in which immune response triggering can be kept under control. Thus only in the last years, PSCs entered the cell-therapy armamentarium, therapies applied for patient-tailored advanced therapy. The main issue arisen within the domain was the fact that during the reprogramming methodology, several genetic and epigenetic aberrations can appear. Reprogramming process represents the removal and the remodelling of epigenetic markers, such as DNA methylation, a process that takes place during mammalian development [3]. If the medical community is planning the large clinical utilization of these cells, it is essential to comprehend the effects of these abnormalities inflicted upon the stem cells' immunogenicity, further ensuring the survival of PSC grafts [4].

There are several types of stem cells foreseen in this domain and an overview of these cell populations along with their phenotype is *depicted below.*

Embryonic Stem Cells. Although considered great promising tool for cell-based therapies, embryonic stem cells (ESCs) could be eliminated by the immune effector cells of an immunocompetent allogeneic host [5, 6]. The major drawback resides in the fact that ESCs can trigger an immune response in recipient even though they have reduced immunomodulatory potential. This potential resides in the low expression of MHC I and II molecules (major histocompatibility complex), the absence of co-stimulatory molecules such as CD80 and CD86 and in the TGF (transforming growth factor) expression that can inhibit T cell proliferation [7].

Induced Pluripotent Stem Cells. Keeping the typical stem cells abilities related to indefinitely proliferation and differentiation, induced pluripotent stem cells (iPSC) are very promising for therapeutical approaches. There are four transcription factors overexpressed in ESCs which turned on to iPSC phenotype when they are retrovirally induced. Thus, octamer-binding transcription factor 4 (Oct4), sex determining region Y-box 2 (Sox2), Krueppel-like factor 4 (Klf4) and c-myelocytomatosis viral oncogene homolog (c-Myc) are the main responsible for iPSCs derived from ESC [8].

Established as an alternative for stem cells therapeutics, iPSC are designed to solve the ethical issues associated with exploitation of human ESCs. Balancing similarities (morphology, proliferation capacity or cell surface antigens) with differences (global gene expression, genomic imprinting and somatic mutations) between ESC and iPSC, there is a continuous search for the best clinically used formula of iPSCs. Very recently a study with iPSCs derived neurons *via* fibroblast isolated from familial and sporadic Alzheimer's disease patients [9] allowed expression levels evaluation for some key pathological biomarkers such as Aβ, phospho-tau and active glycogen synthase kinase-3β (GSK-3β). This study opened the opportunity for related studies in generating neural progenitor like cells for neurodegenerative diseases treatment [10].

The iPSC regenerative potential could be turned unfortunately to a tumorigenic one by genomic alterations onset during differentiation step. Thus, in a study of Mullally *et al.*, it was emphasized that aneuploidy and hence the tumorigenic potential of iPSC is triggered in some circumstances, by accumulating cell cycle related genes upon duplications on chromosome 12 [11, 12]. Although regarded with cautious and scepticism, these reported models could advance our knowledge regarding the biomedical potential of iPSCs.

Tissue/Adult Stem Cells. Resident stem cells are found in adult organisms during the entire life and are ensuring the necessary cell number throughout life and tissue repair in regenerative processes [13]. The following types of tissue stem cells are widely accepted: hematopoietic stem cells (HSCs), mesenchymal stem cells (MSCs), neural (NSCs) and skeletal stem cells [14].

HSCs have a robust transplantation potential that resides in their ability to evade immune rejection *in vivo*, due to surface expression of some "protective" immune molecules that interact with the immune system by a fine balance of "out" and "in" signals. These arrays of signals modulate their status and position in stem cell niche in relation with the immune system as challenger [15]. The biomolecules that protect HSCs from the immune attack are acting both on innate and adaptive immunity as "out" signals. For instance CD47 protects HSCs from macrophages phagocytosis [16] whereas the immune adaptive attack is counteracted by CD274 expression [17]. Inflammatory signals are "in" type of signals inducing HSCs' differentiation [15].

Recent studies claim that HSCs could be genetically engineered by introducing gene encoding artificial T cell receptors directed to tumor associated antigens in cancer immunotherapy, especially for leukaemia [18]. In haematological malignant therapies a model of HSC and MSC co-infusion was also proposed to avoid adverse reaction and even rejection. But this model is intended to minimize the possible side effects, aiding bone marrow recovery after chemotherapy and/or radiation procedure [19].

MSCs are able to differentiate in osteocytes, chondrocytes, adipocytes, neurons and stromal cells. MSCs express on their surface immune-modulatory molecules with both immuno-suppressive and immuno-regulatory effects on innate and specific immune cells, including antigen presenting cells, T lymphocytes and natural killer cells [20]. In addition, MSCs can elude immune recognition and inhibit *in vivo* several immune responses to maintain body homeostasis in individual areas such as bone marrow or mother - foetus interface [21]. MSCs are also age-and disease-related sensitive having a decreased growth rate, undergoing senescence or altering their morphology in relation to ageing. Abnormal MSCs can be involved in auto-immune diseases pathogenesis, for instance in systemic lupus erythematosus [22].

According to the *International Society of Cellular Therapy*, the foremost surface markers that characterize MSCs are CD13, CD29, CD44, CD54, CD73, CD90, CD105, CD106 and Stro-1 [20, 23, 24], these markers having variable expression dependent on *in vitro* or *in vivo* conditions. Commonly, MSCs display *in vitro* low levels of MHC I, but lack MHC II and a series of co-stimulatory molecules (*e.g.* FasL, B7-1 or B7-2), whereas *in vivo* MSCs have these markers [25].

In addition to the mentioned surface markers repertoire, MSCs can release certain soluble biomolecules with regulatory effects on immune cells. For instance IDO (indoleamine 2,3-dioxygenase), PGE2 (prostaglandin E2), NO (nitric oxide) or IL-10 creates an immunosuppressive milieu for immune cells, hence inhibiting their proliferation, cell activation and/or apoptosis [20, 26].

MSCs fulfill an integrated immuno-regulatory background amongst innate and adaptive immunity cells, equally controlled by their surface markers and soluble released cytokines, growth factors or hormones (Table **1**).

Regarding the *immunosuppressive* properties of MSC there are still controversies. There are studies that claim that constitutive expression of HLA-G and B7-H4 is directly linked to T lymphocytes suppression, hindering proliferation and cellular mediated cytotoxicity [26, 27].

Table 1. Immunoregulatory networks of mesenchymal stem cell population.

Immune Cells	MSCs Mechanism of Action	Soluble Factors
Adaptive immunity cells		
T lymphocytes	• Inhibition of proliferation induced by mitogens or alloantigens with no MHC restriction • T cells blocked in G0/G1 cell cycle phase	IDO, PGE2, NO, TGFβ, IL-10
Regulatory T cells Tregs	• *De novo* induction and expansion of Tregs	PGE-2, TGFβ

(Table 1) contd.....

Immune Cells	MSCs Mechanism of Action	Soluble Factors
B cells	• Proliferation inhibition • Antibody production hampered • Inhibition of chemokine secretion for B cell migration	IFNγ
Dendritic cells DCs	• *In vitro* interfering with DC differentiation from CD14+ and CD34+progenitors • Down-regulation of CD40, CD80, CD83, CD86 and also of MHC II antigens on DCs surface	PGE-2 mainly
Innate immunity cells		
PMN granulocytes	• Enhances anti-microbial functions/bacterial uptake by neutrophils-LPS stimulated • Recruits and activates neutrophils	IL-6, IL-8; MIF
NK cells	• Inhibition of IL-2 mediated proliferation, cytotoxicity to virus-infected cells • Activated NK cells can kill MSCs	TGFβ, HLA-G, PGE-2; PGE-2, IDO

NSCs are expressing a myriad of immune and adhesion molecules suggesting their immunomodulatory competence [27]. In undifferentiated state, NSCs exhibit *in vitro* low levels of MHC molecules, and some degree of CD80 and CD86 co-stimulatory molecules expression. Progenitor stem cells display various integrins with clear role in inflammation (α2, α6 and β1 integrins), CD44 and chemokine receptors (CCR3, CCR6, CCR7, CCR9, and CXCR3).

NSCs exert modulatory effects on adaptive immunity cells by suppressing T cell proliferation or interfering with activation and antigen presenting performed by myeloid dendritic cells [5]. The regenerative potential of NSCs shows an interesting clinical opportunity in neuro diseases but still a difficult to achieve procedure [28 - 30].

From this comprehensive family of stem cells, it seems that hematopoietic stem cells cover the most numerous differentiated cells because this type has to provide for the extended necessities of immune system.

IMMUNOGENICITY OF STEM CELLS – CHARACTERIZING MHC EXPRESSION

Major histocompatibility complex (MHC) in murine strains or human leukocyte antigens (HLAs) in humans represents a family of molecules that sustains the immunological self-non-self recognition [31]. MHC is a set of cellular membrane molecules encoded by a large gene family having complexing peptide fragments as main function derived from any pathogenic structure and further displaying them on the cell surface for appropriate T-cells recognition [32]. There are two major molecules, MHC I and II. As MHC class II can be expressed by various cell types upon specific conditions, it normally occurs only on professional antigen-presenting cells (APCs) such as: macrophages, B cells, and dendritic cells (DCs). An APC takes up the entire antigen molecule, processes it, and expresses only a peptide on its surface complexed with MHC II molecule. This complex is presented to the naive helper T cell (Th) that gets primed/activated. After activation, an immune response is generated whether through a memory or an effector Th cell (Th1, Th2, Th17), or a regulatory/suppressor T cell (Treg). MHC class I occurs on all cells, with the only exception being red blood cells. MHC I presents processed antigens to killer T cells (CTLs). MHC class I mediates cellular immunity, and is involved in elimination of intracellular pathogens, (*e.g.* viruses, bacteria) and in eliminating neoplastic transformed cells.

In normal physiology, it is important that the immune system triggers an efficient immune response against non-self. When transplantation is intended this immune response is the main clinical draw-back. Due to MHC's high genetic polymorphism, transplantation tissue harvested from donor can be subjected to graft rejection by the recipient's immune system [33]. In transplantation, the adaptive immunity is the major mechanism that rejects the transplanted tissue, where the principal targets are donor's MHC molecules [34].

Stem cells with their low MHC expression are attractive as therapeutic approaches in tissue regeneration because the graft rejection is, at least in theory, reduced. Undifferentiated PSCs can display immune privilege characteristics.

According to classical immune privilege notion certain tissues and organs do not comply with strict transplantation immunology mechanisms. Actually, these experimentally proven mechanisms show that heterologous tissue grafts introduced in "immune privileged sites" (eye, brain, testis, pregnant uterus) benefit from commonly indefinite survival. Meanwhile the same grafts when introduced in skin, beneath kidney capsule and so on, are rapidly rejected. Grafts from immune privileged tissues reside un-damaged when implanted in conventional body sites. Probably *immune privilege* is a network of complex mechanisms that endows vulnerable tissues, less capable of regeneration, with an immune tolerance that preserves therefore major physiological functions, like vision, reproduction, and brain activity [35].

Although human ESCs (hESCs) have low MHC class I and II expression, when start to differentiate and/or during experimental *in vitro* interferon-γ (IFNγ) stimulation, MHC increases and therefore the cells intended for regeneration become visible for the immune system and immune rejection can be initiated [36]. There are differences even at the origin of ESC, thus it was shown by several reports that ESCs-derived from cells producing insulin have an increased MHC expression. In the same line, undifferentiated ESCs, if subjected to IFNγ have a higher expression of MHC [37 - 39]. These observations trigger an important alarm regarding ESC transplanted in sites that favor MHC expression augmentation, thus an eventual failure of the engraftment due to immunological recognition could emerge [8].

In a perfect *scenario*, autologous iPSCs are not expressing any alloantigens that could initiate a graft rejection. In a recent study analyzing the immunological phenotype of hESCs and iPSCs it was reported that MHC class I diminution is related to absence of two proteases: transporter associated with antigen processing 1 (TAP-1) and tapasin (TPN). These molecules are known to be associated with antigenic peptides transport mechanisms. Another major molecule in the MHC scaffold is beta2-microglobulin (beta2m) light chain. The lack of beta2m hinders the trimeric structure of MHC class I. Studying the genes involved in the expression of this important immune-related molecules it was highlighted that there is an epigenetic control of MHC in both hESCs and iPSCs. Recently published epigenetic analysis has shown that H3K9me3, found in constitutively repressed genes, repressed the gene encoding for TPN in undifferentiated cells. When embryoid bodies (EBs) start to form in the PSC's differentiation process, H3K4me3 is modified and there is an increase in HLA-B and beta2m [40]. H3K9me3 (trimethylation on Lys 9 of histone 3) is a histone that represses genic and nongenic regions [41].

Recent overviews have shown that the absence or low expressions of HLA-DR and HLA-G are under DNA methylation control. Epigenetic mechanisms control an array of molecules expressions and networks: expression of MHC class I, II, antigen-processing mechanism, NKG2D ligands (MICA, MICB). These mechanisms take place in hESCs and iPSCs and, probably, in all stem cell populations. HLA-G is found increased in embryonic stem cells and in MSCs, although earlier studies describing over-expression of MHC I on stem cells did not differentiate between MHC class Ia (HLA-A,-B,-C) that *induces* an immune response and MHC class Ib (HLA-E,-F,-G) that *hinders* the immune response [26, 42, 43].

All these findings converge to the conclusion that MHC expression on PSCs can be induced and controlled. Probably, the expression of MHC molecule is the clue for controlling the immunogenic potential of both stem cells and their progenitors [4].

Developing cell-based therapy viewed through the immunogenicity of iPSCs, transplant of syngeneic cells would lack immune rejection. However, recent studies have drawn some concerns upon this issues [44]. The concerns that were highlighted regarding immunogenicity, the specific clones that should be identified and the outcome of reprogramming mechanisms were flagged by recent studies regarding immunogenicity of syngeneic iPSCs. The methods for immunogenicity assessment comprised the frequency of teratoma formation when skin and bone marrow stem cells were transplanted in experimental models using ESCs and iPSCs. In C57BL/6 mouse experimental model, no differences were reported when iPSCs or ES originating from skin and bone marrow were used. T-lymphocyte infiltration in the transplanted tissues generated from iPSCs or ES cells was low, with no increase in Zg16 and Hormad1 genes involved in immunogenicity-related processes [41]. ESCs were reported as displaying MHC expression variability after differentiation [38, 45].

A seminal study published in 2007 has shown that iPSCs possess germ line transmission capacity, by transduction of identified transcription factors, Oct3/4, Sox2, Klf4, and c-Myc. In this study, iPSCs were generated from adult human dermal fibroblasts and were similar to ESCcells in all aspects including surface antigens [46]. Few years later, it was shown that the immune system can distinguish pluripotency antigen Oct4, this recognition being sustained by specific memory T lymphocytes [47]. Cells that did not revert to the undifferentiated state from the pool of cells that will be subjected to transplantation could generate an immune response toward Oct4 and/or could generate teratoma. Thus it should be a reasonable ground to eliminate them before transplantation.

ESCs having very low or even absent MHC classI / classII, are clearly ineffective as antigen presenting cells to T lymphocytes. ESCs have low expression of ligands for NKp44 (NK cell receptor involved in activation) and no ligands for other activating receptors (NKp30, NKp46, CD16) [37]. Co-stimulatory molecules such as CD40, CD80, CD86 (Fig. **1**), involved in the T lymphocyte interactions are not expressed by MSCs [48, 49]. MSCs secrete prostaglandin E2 (PGE2), a factor known to hinder T proliferation and macrophage activation [50].

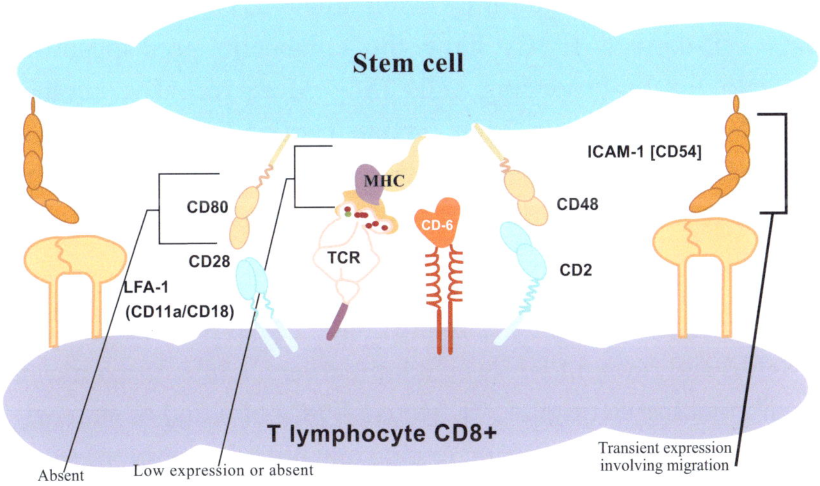

Fig. (1). Immunogenicity of stem cells-MHC and co-stimulatory expression are low, absent or are transiently expressed in processes like cell migration.

The scientific world is still apprehended over the genetic pattern of re-programmed cells, over their genetic instability and over still unknown mechanisms that can amplify oncogene expression [51]. Clones of iPSCs that have slightly different genetic patterns, *e.g.* only partial reprogramming, differential developmental stages, different antigenicity, can generate an immune-related process and hence immune-damaging mechanisms during transplantation [52].

Expression of MHC class I is low in MSC, but is upregulated during hESC differentiation, cell maturation and in an inflammatory milieu [53]. All these conditions can increase immune rejection through recipient's T lymphocyte allorecognition [54]. If MHC is pharmaceutically down regulated, the risks that these cells will be killed by host's natural killer (NK) cells increases.

Immune-Related Epigenetic Mechanisms Sustaining Re-Programming Pathways

Epigenetic memory represents the background to induce iPSCs to a reverted state that is analogous to the cellular state of ESCs. When the clinical intention is to develop a special regenerative cell type, it is highly probable that iPSC lines derived from the intended lineage would differentiate in the needed cell type. For example, beta cell-derived iPSCs have the epigenetic predisposition to be differentiated into insulin-producing cells [55]. More recently, another study comparing iPSC-cardiomyocytes from human cardiac progenitor cells *versus* autologous fibroblasts demonstrated that epigenetic memory is important regarding *in vitro* differentiation efficiency of cardiac *versus* non-cardiac somatic cell, with identical *in vivo* functionality [56].

Moreover, there it is a clear "genetic imprinting", *e.g.* iPSC of musculoskeletal origin has CXCR4 and Integrin B1 as cellular markers, while lysozyme and Gr-1 define iPSCs of granulocyte origin [57]. Mouse iPSCs of blood or skin origin will generate / regenerate elements of blood or skin and not of other types of cells [58, 59]. Keratin-14 gene was found expressed in an increased level in iPSC of skin origin when compared to cells of other origin [58].

Inherent low immunogenicity cells are the best candidates to be transformed into iPSC. For example, umbilical cord mesenchymal cells, have lower immunogenic phenotype than other somatic fibroblasts, hence neural progenitor cells generated from these types of cells have lower immunogenic phenotype than other somatic fibroblasts [60]. Furthermore, different findings reflect the presence of epigenetic memory and sustain the already presented low immunogenicity characteristics.

INNATE IMMUNE RESPONSES TOWARD PLURIPOTENT STEM CELLS

Innate immunity is one of the immune systems arm, consisting of monocytes, granulocytes and NK cells; cell populations that sustain the inflammatory processes associated to transplantation [61, 62]. First immune reactions are rapid and efficient, but non-specific; the first immune mechanisms trigger the activation of adaptive immunity. This later immune mechanism generates slower reactions, is specific and antigen-dependent and induces long-term immune memory. The major link between innate and adaptive immunity is sustained by MHC molecules, the major barrier that sustains the discriminatory self. From the immune point of view, ESCs and iPSCs, have low MHC expression, are less "visible" to the immune recognition and thus can evade T lymphocyte mediated immune responses [35].

Cells with low self-immune recognition (the same as tumor cells), have high probability to be attacked by NK cells. Hence, ESC-derived tissue (low MHC class I expression) and grafts could be damaged by the action of NK cells. In 2007, one of the first studies focusing on iPSC grafts showed that there is an immunologic barrier composed of NK cells that contribute to the graft rejection [63]. Moreover, later it was demonstrated that un-differentiated ESCs and ESC-derived progenitor cells are expressing ligands that upon linking with NK can induce their activation. Natural Killer group 2D (NKG2D) expressed on NK cells is involved in cell activation. A number of NKG2D target ligands have been identified [64]. One of the most intriguing ligands is the minor histocompatibility antigen, H-60. Another related molecule identified by expression cloning is an NKG2D ligand. Early studies in mouse models have shown that an H-60 peptide was presented by MHC class I molecule, H-2K^b [65]. In 2009, the group of Zavazava showed that increased H-60 expression was involved in the destruction of ESC-derived HPCs, process that was mediated by NK cells and by their secretion of perforin-granzyme equipment. Moreover, the group showed that, if enhancing MHC class I expression, after IFNγ stimulation, HPCs are less destroyed by the NK cell attack [66]. After one more year, another study showed that IFNγ up-regulates MHC class I and this is a key point in syngeneic transplant positive outcome of ESC-derived vascular progenitor cells, process that overcomes in a significant degree NK cell mediated lysis [46].

However, contradiction exists in this field because reports have shown hESC-derived mesenchymal progenitor cells (MPCs) resistant to NK cell attack [67] or lack of NKG2D ligands on hESC-derived progenitors, thus resistant to NK lysis [68].

Probably un-identified molecular peculiarities of PSC and/or the existence of NK cell subtypes with different potencies stand on the basis of these contradictory reports. Shedding new light on NK subpopulations and upon their involvement in MHC expression will bring some answers in PSCs' escape from NK cell mediated cytotoxicity.

In a *nutshell*, innate immune cells (granulocytes, macrophages and neutrophils), induce the inflammatory processes immediately after transplantation. When ESCs are introduced, neutrophils and macrophages are recruited and start to infiltrate the implant by secreting inflammatory cytokines and chemokines [37, 43]. It is still a matter of debate if these early events triggered by the innate immunity activate the adaptive immune cells that will continue the graft rejection and if these therapeutically controlled early events can have a more positive clinical outcome on long-term engraftment.

ADAPTIVE IMMUNE RESPONSES TOWARD PLURIPOTENT STEM CELLS

The adaptive immunity triggers antigen specific responses. If immune-suppression is not therapeutically induced, allogeneic transplantation fails due to CD4 and/or CD8 T lymphocytes [69]. APCs activate specific recipient Th, whether the APCs are of donor or recipient origin. Because PSCs lack MHC class II and co-stimulatory molecules it is highly probable that they will not function as APCs.

But, if epithelial cells, can become APCs, and hence present an antigen to T lymphocytes, a graft failure can be experienced owing to the allogeneic T cells activation and their damaging attack [70]. In stem cell-based transplantation, CD4+ T helper cells coordinate an array of immune processes where donor's B and cytotoxic T lymphocytes (CTL) damage the graft by directly lysing cells that express allogeneic MHC I and/or by secreting auto-antibodies that aim these targets [58]. Another immunological issue regarding differentiated progeny that has an increased MHC class I expression retains its capacity to be immunologically "invisible" and hence upon differentiation eluding the attack by the damaging immune response. Cells that have originated from ESCs or iPSCs can/cannot be immunologically identical. T cell anergy, as already known in immunology, is induced by low expression of both MHC and co-stimulatory molecules. Recently, studying this difference it was shown that iPSC-derived HPCs are prone to induce anergy of allogeneic CTL [71].

If IFN-γ stimulation induces increased MHC class I expression, this action does not enhance concomitantly co-stimulatory molecules. Several years ago it was reported in a prospective, randomized, controlled clinical trial that co-transplantation of hematopoietic stem cell with renal allograft increases allograft function, without graft-*versus*-host disease [72]. This is another clinical driven application aiming to reduce immune-related rejection.

As already stated in this chapter, hindering the host's immune response can be the major mechanism for alleviating stem cell transplantation rejection. Thus, in suppressing the anti-PSCs immune response, several mechanisms were reported. Arginase I expressed on the ESCs is reported as inhibiting T lymphocyte proliferation and activation; therefore the locally L-arginine depletion would hinder T activation. It was shown that this process can be manipulated by simple addition of L-arginine in the microenvironment [73]. This study had optimistic results, but a recently published report confirmed this *clear-cut* mechanism. Thus, Wang's team showed that high L-arginine concentration did not modulate T lymphocyte proliferation. This later recent study opens the possibility to search for the possible involvement of other additional mechanisms [74, 75].

Publications regarding the soluble factors secreted by ESC do not abound, but Wang's group found that human ESCs cellular protein extracts can suppress T lymphocyte proliferation and hinder DC maturation. Moreover, the study showed that soluble factors secreted by ESC reduce Th1 cytokine production probably by the inhibition of intracellular signalling through PKC-teta phosphorylation [61,

62].

Treg cells, intensively blamed for the immunosuppression induced in cancer, are another immune player in the stem cell transplantation domain. Induction or enrolment of Treg into the microenvironment of the graft would support the local immune-suppression, establishing a local tolerance and would contribute to the good clinical outcome of the graft. Regarding Tregs, there are several reports that show ESCs as inducing local Tregs polarization towards immune privilege gain [76 - 78]. Stem cells' immune privilege was proven 15 years ago in an experimental model, when EB graft in allogeneic mice was accepted only if anti-CD8 antibody was introduced along with the graft [79].

In 2014, another experimental report showed that, in a porcine model, freshly isolated autologous intra-myocardial Tregs when co-administered with MSCs increased significantly MSC's survival rate, proliferation, and induced angiogenesis. After 6 weeks upon co-injection organized clusters of MSCs were found. In these clusters, CD25+ T lymphocytes were identified between CD90+ MSC. This report showed clear evidence that Tregs sustained the local immune-suppression in order to organize MSCs in cellular clusters. Besides MSC's proliferation, cells that have a factor VIII+ phenotype were present in the co-injection site suggesting an active angiogenesis. This recent report flags a new future procedure for clinical application of MSC-driven therapy [80].

To shed light on the conundrum of stem cells clinical application, recent reports underline the involvement of infiltrating Tregs and the need to hinder CTL-mediated killing. This molecular approach can sustain stem cells efficiently grafting and further lead to differentiated progenitors and tissue regeneration.

UNFOLDING IMMUNE-RELATED MECHANISMS FOR DRIVING PLURIPOTENT STEM CELLS TOWARDS TUMORIGENESIS OR REGENERATION

iPSCs are in the spot light of regeneration cell-therapy-based technology, but, as this book is trying to convey that, there is a very delicate balance between immunogenicity and tumorigenicity of iPSCs, raising serious clinical precautions. When we are aiming to the regeneration potency of PSC and produce an implant that will rebuild the damaged tissue, iPSCs' immune evasion is intended. Although, when this "invisibility" is shed upon a cancer stem cell we unfortunately can induce an un-controlled tumor growth and development. Normal cells that start to aberrantly proliferate towards tumorigenesis are developing mechanisms to become invisible for the immune system "through three stages of immunoediting: elimination, equilibrium, and escape" [81, 82].

It is very interesting how tumour cells are endowed with complex strategies to evade the immune system action, strategies that resemble in some ways the ones developed by normal stem cells: MHC class I down-regulation and/or alterations [83, 84], FasL over-expression for hindering apoptosis induced-mechanisms [85], TCR inhibition [86] and activation of immune-suppressive population of Tregs [87]. iPSCs induce the same types of processes, but after an eventual transplantation, a reduced immunogenicity of the cells progenitors should be maintained in order to keep the regeneration process without any activation of the immune response.

The iPSC's immunogenicity in clinical advanced cell therapy is one of the main concerns, while tumorigenicity risk associated with therapeutically applied iPSCs is far from being negligible. Tumorigenicity of iPSCs goes hand in hand with pluripotency induction genes that were found in various types of cancer [88].

The line that borders regeneration and tumorigenesis processes is very fine, and several pluripotency-promoting factors were also demonstrated being involved in tumorigenesis. Lately, several studies claimed approaches that limit the tumorigenic potential of iPSCs. A study published in 2014, describes the steps taken to prevail over iPSCs tumorigenic properties. It is interesting that, immune-related products such as the reactive oxygen species (ROS) are critically involved in the endogenous re-programming genes activation (Fig. **2**). Therefore two stages are described, the first one is sustained by exogenous pluripotent factors that induce DNA damage and metabolic stresses activating p53, increased tumor suppressor genes expression (p21CIP1 and p16INK4a), mechanism that induces pre-iPSCs generation. The second step aims to surmount the induced senescence or cell-cycle arrest by shutting off tumor suppressor genes, and induction of endogenous stemness genes, the result being a fully functional and probable less tumorigenic committed iPSCs. Thus, ROS can be one of the inductors of reprogramming genes *via* epigenetic changes [83, 89].

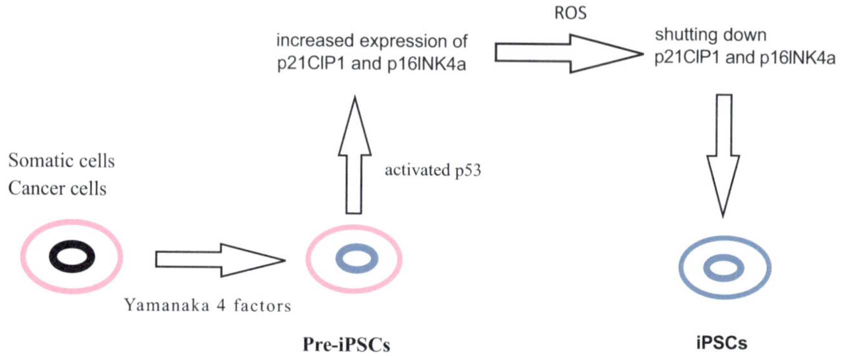

Fig. (2). Mechanism proposed for tumorigenicity risk decrease in iPSC generation. Somatic cells, cancer cells can follow an initial stage of reprogramming by exogenous Yamanaka 4 factors *via* reprogramming-induced senescence, thus obtaining Pre-iPSCs. These pre-iPSCS have p53 activation that leads to increased expression of tumor suppressor genes (p21CIP1 and p16INK4a). Reactive oxygen species (ROS) produced during oxidative stress shuts off the function of p16INK4a, p21CIP1 and p53, process that commits cells toward iPSCs.

In 2013, gene expression microarray reports have shown that pluripotency genes and oncogenic transformation genes are processes that overlap [90]. iPSCs and cancer cells repress a tumor suppressor protein (Lefty) while ESCs have normal levels [91]. From this point of view, one can introduce in iPSC-based therapies also tumorigenic clones, and/or clones that during *in situ* differentiation can reactivate un-intended oncogene [92].

To overcome these important issues several reports have shown the possibilities to decrease tumorigenesis risk. Removing oncogene c-Myc [93], replacing viral-dependent methods that can induce mutagenesis and tumorigenesis [94], using specific protein inhibitors [95], microRNAs [96] and so on, just to enumerate some reported options.

CANCER STEM CELLS – AN IMMUNOLOGICAL DISTINCT POPULATION OF STEM CELLS?

Intense research and debates have as subject, the mere existence of, the so called, cancer stem cells (CSC). However, there is a distinct subpopulation in the tumour mass that has high levels of anti-apoptotic proteins (bcl-2, bcl-xL or survivin) [97], activated PI3K/Akt pathway [98], that sustain a resistance of this subpopulation to chemotherapy and lately proved, also against immune-therapy mediated by the action of T cytotoxic lymphocytes and NK cells [99].

HER2/neu, oncogenic growth factor receptor activates PI3K/Akt signalling [100] and hinders also antigen processing and presentation [101, 102]. The expression of this oncogenic growth factor receptor is involved in keeping up CSC in luminal breast cancer [103]. Several years ago it was reported that CSCs with the phenotype $CD44^{high}/CD24^{low}$ can selectively escape from the cytotoxicity inflicted by NK cells and by ADCC-dependent immune action [104].

Tumor-associated antigen MART-1 (T lymphocyte recognized antigen) is expressed both in cutaneous melanoma A375 cell line and in differentiated melanoma cells. This antigen is not expressed on cancer stem cells (MMICs) and thus cannot be eliminated by MART-1-specific T lymphocytes [105].

In other type of solid cancers, CSCs were demonstrated to secrete immune-related molecules. Breast CSCs [106] and glioblastoma stem cells [107] synthesize increased levels of TGF-beta when compared to tumor cells. Interleukin 4 was shown to be secreted by colon cancer CSCs [108], this cytokine inducing the hindering of specific anti-tumor immune-reaction [109]. Moreover, CSCs express CD200 [110], a membrane protein functioning in immune evasion, probably by inhibiting effector immune cells and helping the tumour escape from the immune system's attack [111, 112].

There are significant immunological features of CSC that are involved in anti-tumoral therapy resistance [113]. Tumour initiating cells, with characteristics that involve mainly self-renewal and "immune invisibility" are maintained through tumour microenvironment molecules and mechanisms that can promote metastasis. When developing new anti-tumoral therapies depicting antigens that induce anti-tumoral immune response is of outmost importance as suppressive factors can be new therapy targets and background for new translational immune-therapies.

Immune suppression can be achieved also by the deficiency of some factors from the cellular milieu [114]. The classical example is tryptophan withdrawal. IDO catabolises L-tryptophan, a physiological process in mammalian foetal allograft. By these mechanisms, the foetus avoids maternal rejection suppressing hence a possible T –dependent reaction [115 - 118]. IDO is not expressed constitutively by MSC but its expression can be induced by IFN-γ [119], leading to T lymphocyte inhibition [119, 120]. This suppression mechanism is not the only one, because it is known that supplementing tryptophan and/or directly inhibiting IDO did not completely abrogate immune-suppression [121, 122]. Other factors, such as hepatocyte growth factor [123], secreted by MSCs can induce immune populations to be tolerogenic (mainly DCs and Tregs) [124], or even flare down autoimmune reactions [125].

CLINICAL TRIALS- HOW FAR FROM THE BEDSIDE APPLICATION OF STEM CELLS?

It is not surprising that the first clinical trial using iPSCs on humans is performed on *immune privileged sites* like the eye. The first trial, launched in Japan, started at the beginning of 2013 for macular degeneration using iPSC harvested from adults. Currently available drugs do not repair the actual damage, thus this type of intervention was long-awaited for. Dr. Takahashi's pilot study aims to obtain autologous iPSCs that will be differentiated into retinal pigment epithelium (RPE), and transplant them in monolayer cell sheets on the damaged site, after excision of the damaged RPE. Post-transplantation, functional integration and adverse reactions will be monitored for one year and then followed-up for three more years [126].

Another scientific breakthrough in this domain was shown in Diamond-Blackfan anaemia (DBA). In this case there is a possibility of developing iPSC cells that are reprogrammed from the harvested skin fibroblasts [127]. DBA is a disorder where bone marrow is dysfunctional and fails to produce sufficient red blood cells, hence the diagnosed anaemia. Moreover, patients' diagnosed with DBA have an increased risk of developing myelodysplastic syndrome (MDS), cancers from the acute myeloid leukaemia family (AML) and osteosarcoma [128]. The disease is actually characterized by mutations encoding ribosomal proteins that append to small (RPS) or large (RPL) subunit [129].The following therapeutical scenario for

DBA was published: harvesting cells from patient's skin, inducing the cells to the pluripotency state, repairing the panel of mutations, and afterwards regenerating the malfunctioning tissue. This possible therapeutical endeavour lacks any immunological deleterious effects. Although the scenario seems applicable, the actual clones that can be selected from these patients should be the cells with no oncogenic potential, although the clones endowed with oncogenic characteristics grow faster and un-controlled [130]. On the contrary, non-nucleated cells, red cells [131] or platelets [132], when producing differentiated iPSCs, tumorigenesis is decreased [133].

There are overall several issues related to the immediate application of iPSC in therapy: methodology of acquiring terminal/functional cells, and iPSC derived from haematological diseases can suffer foetal haematopoiesis, while seldom adult haematopoiesis. These limitations are still intense research subjects [134].

There is a clear lack of robust immunological data from clinical trials using stem cells, although there are numerous reports on *in vitro* studies and some *in vivo* studies in murine models. In terms of clinical trials, implementation of standardized immune monitoring in clinical trials using any type of stem cells would be of outmost importance for the future of this therapeutical approach [135].

CONCLUSIVE REMARKS

The scientists agree upon the fact that stem cells have a reduced immunogenicity. The fact that MHC class I and II reduced as expression in hESCs and iPSCs is beneficial in terms of immune recognition and an eventual damaging immune response. These cells are endowed with cellular mechanisms involved in the immune response modulation, protecting them against immune-mediated destruction. Experimental results have shown that mouse ESCs and human MSCs can represent well tolerated cell grafts in immuno-competent rats and sheep [136]. There are already clinical studies that take advantage of these cells using their potential to reduce graft-*versus*-host disease and sustain the allograft survival [129, 137].

When a transplanted stem cell starts to differentiate *in vivo* it enhances its immunogenicity as well, and, if this modification can be circumvented *in vitro*, probably *in vivo* immuno-suppressive classical drugs could help graft acceptance [4].

Immunologically wise, allogeneic or xenogenic stem cells are rather "invisible". These cells harbour potentially oncogenic mutations and are not likely to generate an immune-response. Thus, stem cells, or even worse cancer stem cell, can build up a tumoral tissue instead of a normal regenerated one [133, 138]. The issues that are still under investigation focus on the clones that retain the best genetic/epigenetic characteristics in order to generate the intended type of tissue [56].

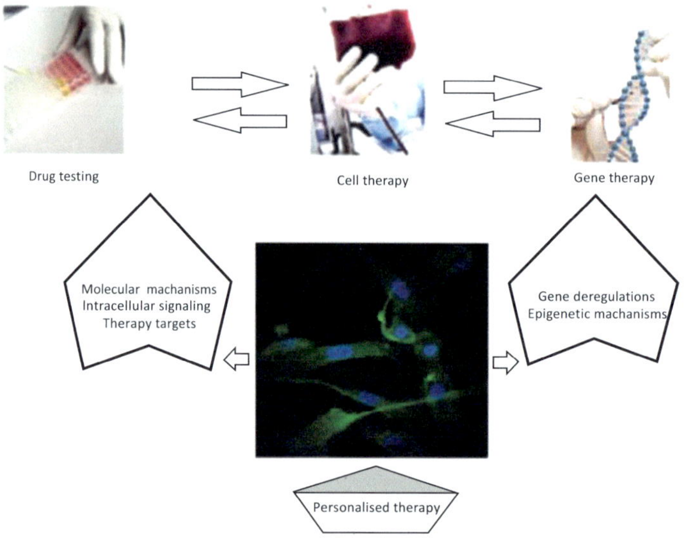

Fig. (3). Clinical pathways for stem cell applications. Stem cell cultures can be the starting point for developing clinical applications. Deciphering molecular mechanisms can depict new therapy targets used further for drug development and testing. Gene deregulations and epigenetic mechanisms can represent the base for developing gene therapy. Advanced therapies using stem cells can dwell on gene therapy results and/or on drug discoveries. Moreover results obtained during cell therapy can furnish valuable data for both new drug therapy development and new insights in cell's genome. (Central image represents cell cultures derived from glioblastoma specimens [139] OncoTargets and Therapy by Dove Medical Press. Reproduced with permission of Dove Medical Press in the format Republish in a book *via* Copyright Clearance Centre. Confirmation Number: 11272901 Order Detail ID: 65885573).

There are still controversies regarding the equivalency of iPSCs and ESCs in terms of immunogenicity, therefore, complex immunome description will have to unveil intimate differences between these cells and their progenitors. Different reprogramming methods, whether *in vitro* or *in vivo* should be scrutinized regarding the stem cells potency to generate an immune response in view of future therapy tools. Clinical pathways for stem cell applications are depicted in Fig. (**3**).

The immunological consequences of stem cell therapies have overridden the knowledge of stem cell biology. Immune system in the context of stem cells that are able to regenerate the planned tissue or build up a tumor, has its seminal involvement in maintaining the balance toward the intended beneficial clinical outcome.

CONFLICT OF INTEREST

The authors confirm that they have no conflict of interest to declare for this publication.

ACKNOWLEDGEMENTS

This work was partially supported by PN-II-PT-PCCA-2013-4-1407 (Project nr. 190), PN-II-PT-PCCA-2013-4-1386 (Project nr. 185) and PN-II-ID-PCE-2-11-3-0918.

REFERENCES

[1] Haworth R, Sharpe M. Editorial Special Focus Issue: Immunological challenges & opportunities. Regen Med 2015; 10(3): 231-4.
[http://dx.doi.org/10.2217/rme.14.50] [PMID: 25933230]

[2] Shlomchik WD. Graft-versus-host disease. Nat Rev Immunol 2007; 7(5): 340-52. [Review].
[http://dx.doi.org/10.1038/nri2000] [PMID: 17438575]

[3] Takahashi K, Yamanaka S. Induction of pluripotent stem cells from mouse embryonic and adult fibroblast cultures by defined factors. Cell 2006; 126(4): 663-76.
[http://dx.doi.org/10.1016/j.cell.2006.07.024] [PMID: 16904174]

[4] Tan Y, Ooi S, Wang L. Immunogenicity and tumorigenicity of pluripotent stem cells and their derivatives: genetic and epigenetic perspectives. Curr Stem Cell Res Ther 2014; 9(1): 63-72.
[http://dx.doi.org/10.2174/1574888X113086660068] [PMID: 24160683]

[5] Stem cells: What are they? Where do they come from? Why are they here? When do they go wrong? Where are they going? In: Sell S, Ed. Stem cells handbook. Humana Press Inc 2004; pp. 1-18.

[6] Bifari F, Pacelli L, Krampera M. Immunological properties of embryonic and adult stem cells. World J Stem Cells 2010; 2(3): 50-60.

[http://dx.doi.org/10.4252/wjsc.v2.i3.50] [PMID: 21607122]

[7] Koch CA, Geraldes P, Platt JL. Immunosuppression by embryonic stem cells. Stem Cells 2008; 26(1): 89-98.
[http://dx.doi.org/10.1634/stemcells.2007-0151] [PMID: 17962705]

[8] de Almeida PE, Ransohoff JD, Nahid A, Wu JC. Immunogenicity of pluripotent stem cells and their derivatives. Circ Res 2013; 112(3): 549-61.
[http://dx.doi.org/10.1161/CIRCRESAHA.111.249243] [PMID: 23371903]

[9] Israel MA, Yuan SH, Bardy C, *et al.* Probing sporadic and familial Alzheimer's disease using induced pluripotent stem cells. Nature 2012; 482(7384): 216-20.
[PMID: 22278060]

[10] Tian C, Liu Q, Ma K, *et al.* Characterization of induced neural progenitors from skin fibroblasts by a novel combination of defined factors. Sci Rep 2013; 3: 1345.
[http://dx.doi.org/10.1038/srep01345] [PMID: 23439431]

[11] Mayshar Y, Ben-David U, Lavon N, *et al.* Identification and classification of chromosomal aberrations in human induced pluripotent stem cells. Cell Stem Cell 2010; 7(4): 521-31.
[http://dx.doi.org/10.1016/j.stem.2010.07.017] [PMID: 20887957]

[12] Mullally A, Ritz J. Beyond HLA: the significance of genomic variation for allogeneic hematopoietic stem cell transplantation. Blood 2007; 109(4): 1355-62.
[http://dx.doi.org/10.1182/blood-2006-06-030858] [PMID: 17008540]

[13] Raveh-Amit H, Berzsenyi S, Vas V, Ye D, Dinnyes A. Tissue resident stem cells: till death do us part. Biogerontology 2013; 14(6): 573-90.
[http://dx.doi.org/10.1007/s10522-013-9469-9] [PMID: 24085521]

[14] Brack AS, Rando TA. Tissue-specific stem cells: lessons from the skeletal muscle satellite cell. Cell Stem Cell 2012; 10(5): 504-14.
[http://dx.doi.org/10.1016/j.stem.2012.04.001] [PMID: 22560074]

[15] Zheng J, Song C, Zhang CC. A new chapter: hematopoietic stem cells are direct players in immunity. Cell Biosci 2011; 1: 33.
[http://dx.doi.org/10.1186/2045-3701-1-33] [PMID: 21978817]

[16] Jaiswal S, Chao MP, Majeti R, Weissman IL. Macrophages as mediators of tumor immunosurveillance. Trends Immunol 2010; 31(6): 212-9.
[http://dx.doi.org/10.1016/j.it.2010.04.001] [PMID: 20452821]

[17] Zou W, Chen L. Inhibitory B7-family molecules in the tumour microenvironment. Nat Rev Immunol 2008; 8(6): 467-77.
[http://dx.doi.org/10.1038/nri2326] [PMID: 18500231]

[18] Gschweng E, De Oliveira S, Kohn DB. Hematopoietic stem cells for cancer immunotherapy. Immunol Rev 2014; 257(1): 237-49.
[http://dx.doi.org/10.1111/imr.12128] [PMID: 24329801]

[19] Sutton Morgan T , Bonfield. Tracey L . Stem Cells: Innovations in clinical applications stem cells international 2014; 2014 Article ID 516278, 9 pages.

[20] Gebler A, Zabel O, Seliger B. The immunomodulatory capacity of mesenchymal stem cells. Trends Mol Med 2012; 18(2): 128-34.
[http://dx.doi.org/10.1016/j.molmed.2011.10.004] [PMID: 22118960]

[21] Rasmusson I. Immune modulation by mesenchymal stem cells. Exp Cell Res 2006; 312(12): 2169-79.
[http://dx.doi.org/10.1016/j.yexcr.2006.03.019] [PMID: 16631737]

[22] Carrion F, Nova E, Ruiz C, *et al.* Autologous mesenchymal stem cell treatment increased T regulatory cells with no effect on disease activity in two systemic lupus erythematosus patients. Lupus 2010; 19(3): 317-22.
[http://dx.doi.org/10.1177/0961203309348983] [PMID: 19919974]

[23] Bassi EJ, Aita CA, Câmara NO. Immune regulatory properties of multipotent mesenchymal stromal cells: Where do we stand? World J Stem Cells 2011; 3(1): 1-8.
[http://dx.doi.org/10.4252/wjsc.v3.i1.1] [PMID: 21607131]

[24] Dominici M, Le Blanc K, Mueller I, *et al.* Minimal criteria for defining multipotent mesenchymal stromal cells. The International Society for Cellular Therapy position statement. Cytotherapy 2006; 8(4): 315-7.
[http://dx.doi.org/10.1080/14653240600855905] [PMID: 16923606]

[25] Singer NG, Caplan AI. Mesenchymal stem cells: mechanisms of inflammation. Annu Rev Pathol 2011; 6: 457-78.
[http://dx.doi.org/10.1146/annurev-pathol-011110-130230] [PMID: 21073342]

[26] Wang H, Wu M, Liu Y. Are mesenchymal stem cells major sources of safe signals in immune system? Cell Immunol 2012; 272(2): 112-6.
[http://dx.doi.org/10.1016/j.cellimm.2011.10.010] [PMID: 22138499]

[27] Nasef A, Mathieu N, Chapel A, *et al.* Immunosuppressive effects of mesenchymal stem cells: involvement of HLA-G. Transplantation 2007; 84(2): 231-7.
[http://dx.doi.org/10.1097/01.tp.0000267918.07906.08] [PMID: 17667815]

[28] Fitzsimons CP, van Bodegraven E, Schouten M, *et al.* Epigenetic regulation of adult neural stem cells: implications for Alzheimer's disease. Mol Neurodegener 2014; 9: 25.
[http://dx.doi.org/10.1186/1750-1326-9-25] [PMID: 24964731]

[29] Van den Hove DL, Kompotis K, Lardenoije R, *et al.* Epigenetically regulated microRNAs in Alzheimer's disease. Neurobiol Aging 2014; 35(4): 731-45.
[http://dx.doi.org/10.1016/j.neurobiolaging.2013.10.082] [PMID: 24238656]

[30] Kalladka D, Muir KW. Brain repair: cell therapy in stroke. Stem Cells Cloning 2014; 7: 31-44.
[PMID: 24627643]

[31] Game DS, Lechler RI. Pathways of allorecognition: implications for transplantation tolerance. Transpl Immunol 2002; 10(2-3): 101-8.
[http://dx.doi.org/10.1016/S0966-3274(02)00055-2] [PMID: 12216939]

[32] Janeway CA Jr, Travers P, Walport M, *et al.* The major histocompatibility complex and its functions in Immunobiology: The Immune System in Health and Disease. 5th ed., New York: Garland Science 2001.

[33] Boyd AS, Rodrigues NP, Lui KO, Fu X, Xu Y. Concise review: Immune recognition of induced pluripotent stem cells. Stem Cells 2012; 30(5): 797-803.
[http://dx.doi.org/10.1002/stem.1066] [PMID: 22419544]

[34] Ayala García MA, González Yebra B, López Flores AL, Guaní Guerra E. The major histocompatibility complex in transplantation. J Transplant 2012; 2012: 842141.
[http://dx.doi.org/10.1155/2012/842141] [PMID: 22778908]

[35] Streilein JW. Regional immunity and ocular immune privilege In: Immune responses and the eye, Chemical Immunology 73. Karger AG, Basel 1999.
[http://dx.doi.org/10.1159/000058741]

[36] Wu DC, Boyd AS, Wood KJ. Embryonic stem cells and their differentiated derivatives have a fragile immune privilege but still represent novel targets of immune attack. Stem Cells 2008; 26(8): 1939-50.
[http://dx.doi.org/10.1634/stemcells.2008-0078] [PMID: 18535153]

[37] Drukker M, Katz G, Urbach A, *et al.* Characterization of the expression of MHC proteins in human embryonic stem cells. Proc Natl Acad Sci USA 2002; 99(15): 9864-9.
[http://dx.doi.org/10.1073/pnas.142298299] [PMID: 12114532]

[38] Boyd AS, Wood KJ. Variation in MHC expression between undifferentiated mouse ES cells and ES cell-derived insulin-producing cell clusters. Transplantation 2009; 87(9): 1300-4.
[http://dx.doi.org/10.1097/TP.0b013e3181a19421] [PMID: 19424028]

[39] Okamura RM, Lebkowski J, Au M, Priest CA, Denham J, Majumdar AS. Immunological properties of human embryonic stem cell-derived oligodendrocyte progenitor cells. J Neuroimmunol 2007; 192(1-2): 134-44.
[http://dx.doi.org/10.1016/j.jneuroim.2007.09.030] [PMID: 17996308]

[40] Suárez-Alvarez B, Rodriguez RM, Calvanese V, *et al.* Epigenetic mechanisms regulate MHC and antigen processing molecules in human embryonic and induced pluripotent stem cells. PLoS One 2010; 5(4): e10192.
[http://dx.doi.org/10.1371/journal.pone.0010192] [PMID: 20419139]

[41] Kim J, Kim H. Recruitment and biological consequences of histone modification of H3K27me3 and H3K9me3. ILAR J 2012; 53(3-4): 232-9.
[http://dx.doi.org/10.1093/ilar.53.3-4.232] [PMID: 23744963]

[42] Favier B, LeMaoult J, Carosella ED. Functions of HLA-G in the immune system. Tissue Antigens 2007; 69(1) (Suppl. 1): 150-2.
[http://dx.doi.org/10.1111/j.1399-0039.2006.763_6.x] [PMID: 17445191]

[43] Verloes A, Van de Velde H, LeMaoult J, *et al.* HLA-G expression in human embryonic stem cells and preimplantation embryos. J Immunol 2011; 186(4): 2663-71.
[http://dx.doi.org/10.4049/jimmunol.1001081] [PMID: 21248264]

[44] Zhao T, Zhang Z-N, Rong Z, Xu Y. Immunogenicity of induced pluripotent stem cells. Nature 2011; 474(7350): 212-5.
[http://dx.doi.org/10.1038/nature10135] [PMID: 21572395]

[45] Swijnenburg R-J, Tanaka M, Vogel H, *et al.* Embryonic stem cell immunogenicity increases upon differentiation after transplantation into ischemic myocardium. Circulation 2005; 112(9) (Suppl.):

1166-72.
[PMID: 16159810]

[46] Takahashi K, Tanabe K, Ohnuki M, *et al.* Induction of pluripotent stem cells from adult human fibroblasts by defined factors. Cell 2007; 131(5): 861-72.
[http://dx.doi.org/10.1016/j.cell.2007.11.019] [PMID: 18035408]

[47] Dhodapkar KM, Feldman D, Matthews P, *et al.* Natural immunity to pluripotency antigen OCT4 in humans. Proc Natl Acad Sci USA 2010; 107(19): 8718-23.
[http://dx.doi.org/10.1073/pnas.0915086107] [PMID: 20404147]

[48] Kroczek RA, Mages HW, Hutloff A. Emerging paradigms of T-cell co-stimulation. Curr Opin Immunol 2004; 16(3): 321-7.
[http://dx.doi.org/10.1016/j.coi.2004.03.002] [PMID: 15134781]

[49] Krampera M, Cosmi L, Angeli R, *et al.* Role for interferon-gamma in the immunomodulatory activity of human bone marrow mesenchymal stem cells. Stem Cells 2006; 24(2): 386-98.
[http://dx.doi.org/10.1634/stemcells.2005-0008] [PMID: 16123384]

[50] Aggarwal S, Pittenger MF. Human mesenchymal stem cells modulate allogeneic immune cell responses. Blood 2005; 105(4): 1815-22.
[http://dx.doi.org/10.1182/blood-2004-04-1559] [PMID: 15494428]

[51] Mills JA, Wang K, Paluru P, *et al.* Clonal genetic and hematopoietic heterogeneity among human-induced pluripotent stem cell lines. Blood 2013; 122(12): 2047-51.
[http://dx.doi.org/10.1182/blood-2013-02-484444] [PMID: 23940280]

[52] Araki R, Uda M, Hoki Y, *et al.* Negligible immunogenicity of terminally differentiated cells derived from induced pluripotent or embryonic stem cells. Nature 2013; 494(7435): 100-4.
[http://dx.doi.org/10.1038/nature11807] [PMID: 23302801]

[53] Jacobs SA, Roobrouck VD, Verfaillie CM, Van Gool SW. Immunological characteristics of human mesenchymal stem cells and multipotent adult progenitor cells. Immunol Cell Biol 2013; 91(1): 32-9.
[http://dx.doi.org/10.1038/icb.2012.64] [PMID: 23295415]

[54] Pearl JI, Wu JC. The immunogenicity of ES cells and their differentiated progeny. In: Fairchild PJ, Ed. The Immunological Barriers to Regenerative Medicine. New York, USA: Springer 2013; pp. 37-48.
[http://dx.doi.org/10.1007/978-1-4614-5480-9_2]

[55] Bar-Nur O, Russ HA, Efrat S, Benvenisty N. Epigenetic memory and preferential lineage-specific differentiation in induced pluripotent stem cells derived from human pancreatic islet beta cells. Cell Stem Cell 2011; 9(1): 17-23.
[http://dx.doi.org/10.1016/j.stem.2011.06.007] [PMID: 21726830]

[56] Sanchez-Freire V, Lee AS, Hu S, *et al.* Effect of human donor cell source on differentiation and function of cardiac induced pluripotent stem cells. J Am Coll Cardiol 2014; 64(5): 436-48.
[http://dx.doi.org/10.1016/j.jacc.2014.04.056] [PMID: 25082575]

[57] Vaskova EA, Stekleneva AE, Medvedev SP, Zakian SM. "Epigenetic memory" phenomenon in induced pluripotent stem cells. Acta Naturae 2013; 5(4): 15-21.
[PMID: 24455179]

[58] Kim K, Doi A, Wen B, *et al.* Epigenetic memory in induced pluripotent stem cells. Nature 2010;

467(7313): 285-90.
[http://dx.doi.org/10.1038/nature09342] [PMID: 20644535]

[59] Kim K, Zhao R, Doi A, *et al.* Donor cell type can influence the epigenome and differentiation potential of human induced pluripotent stem cells. Nat Biotechnol 2011; 29(12): 1117-9.
[http://dx.doi.org/10.1038/nbt.2052] [PMID: 22119740]

[60] Liu P, Chen S, Li X, *et al.* Low immunogenicity of neural progenitor cells differentiated from induced pluripotent stem cells derived from less immunogenic somatic cells. PLoS One 2013; 8(7): e69617.
[http://dx.doi.org/10.1371/journal.pone.0069617] [PMID: 23922758]

[61] Dressel R, Nolte J, Elsner L, *et al.* Pluripotent stem cells are highly susceptible targets for syngeneic, allogeneic, and xenogeneic natural killer cells. FASEB J 2010; 24(7): 2164-77.
[http://dx.doi.org/10.1096/fj.09-134957] [PMID: 20145206]

[62] Boyd AS, Wood KJ. Characteristics of the early immune response following transplantation of mouse ES cell derived insulin-producing cell clusters. PLoS One 2010; 5(6): e10965.
[http://dx.doi.org/10.1371/journal.pone.0010965] [PMID: 20532031]

[63] Hanna J, Wernig M, Markoulaki S, *et al.* Treatment of sickle cell anemia mouse model with iPS cells generated from autologous skin. Science 2007; 318(5858): 1920-3.
[http://dx.doi.org/10.1126/science.1152092] [PMID: 18063756]

[64] Stephens HA. MICA and MICB genes: can the enigma of their polymorphism be resolved? Trends Immunol 2001; 22(7): 378-85.
[http://dx.doi.org/10.1016/S1471-4906(01)01960-3] [PMID: 11429322]

[65] Malarkannan S, Shih PP, Eden PA, *et al.* The molecular and functional characterization of a dominant minor H antigen, H60. J Immunol 1998; 161(7): 3501-9.
[PMID: 9759870]

[66] Tabayoyong WB, Salas JG, Bonde S, Zavazava N. HOXB4-transduced embryonic stem cell-derived Lin-c-kit+ and Lin-Sca-1+ hematopoietic progenitors express H60 and are targeted by NK cells. J Immunol 2009; 183(9): 5449-57.
[http://dx.doi.org/10.4049/jimmunol.0901807] [PMID: 19828634]

[67] Yen M-L, Hou C-H, Peng K-Y, *et al.* Efficient derivation and concise gene expression profiling of human embryonic stem cell-derived mesenchymal progenitors (EMPs). Cell Transplant 2011; 20(10): 1529-45.
[http://dx.doi.org/10.3727/096368910X564067] [PMID: 21396155]

[68] Frenzel LP, Abdullah Z, Kriegeskorte AK, *et al.* Role of natural-killer group 2 member D ligands and intercellular adhesion molecule 1 in natural killer cell-mediated lysis of murine embryonic stem cells and embryonic stem cell-derived cardiomyocytes. Stem Cells 2009; 27(2): 307-16.
[http://dx.doi.org/10.1634/stemcells.2008-0528] [PMID: 18988711]

[69] Swijnenburg R-J, Schrepfer S, Govaert JA, *et al.* Immunosuppressive therapy mitigates immunological rejection of human embryonic stem cell xenografts. Proc Natl Acad Sci USA 2008; 105(35): 12991-6.
[http://dx.doi.org/10.1073/pnas.0805802105] [PMID: 18728188]

[70] Kreisel D, Krupnick AS, Gelman AE, *et al.* Non-hematopoietic allograft cells directly activate CD8+ T cells and trigger acute rejection: an alternative mechanism of allorecognition. Nat Med 2002; 8(3):

233-9.
[http://dx.doi.org/10.1038/nm0302-233] [PMID: 11875493]

[71] Kim E-M, Manzar G, Zavazava N. Human iPS cell-derived hematopoietic progenitor cells induce T-cell anergy in *in vitro*-generated alloreactive CD8(+) T cells. Blood 2013; 121(26): 5167-75. [http://dx.doi.org/10.1182/blood-2012-11-467753] [PMID: 23687092]

[72] Trivedi HL, Vanikar AV, Modi PR, *et al.* Allogeneic hematopoietic stem-cell transplantation, mixed chimerism, and tolerance in living related donor renal allograft recipients. Transplant Proc 2005; 37(2): 737-42. [http://dx.doi.org/10.1016/j.transproceed.2005.01.028] [PMID: 15848518]

[73] Yachimovich-Cohen N, Even-Ram S, Shufaro Y, Rachmilewitz J, Reubinoff B. Human embryonic stem cells suppress T cell responses *via* arginase I-dependent mechanism. J Immunol 2010; 184(3): 1300-8. [http://dx.doi.org/10.4049/jimmunol.0804261] [PMID: 20032293]

[74] Mohib K, AlKhamees B, Zein HS, Allan D, Wang L. Embryonic stem cell-derived factors inhibit T effector activation and induce T regulatory cells by suppressing PKC-θ activation. PLoS One 2012; 7(3): e32420. [http://dx.doi.org/10.1371/journal.pone.0032420] [PMID: 22412871]

[75] Mohib K, Allan D, Wang L. Human embryonic stem cell-extracts inhibit the differentiation and function of monocyte-derived dendritic cells. Stem Cell Rev 2010; 6(4): 611-21. [http://dx.doi.org/10.1007/s12015-010-9185-7] [PMID: 20711689]

[76] Tang Q, Bluestone JA. The Foxp3+ regulatory T cell: a jack of all trades, master of regulation. Nat Immunol 2008; 9(3): 239-44. [http://dx.doi.org/10.1038/ni1572] [PMID: 18285775]

[77] Robertson NJ, Brook FA, Gardner RL, Cobbold SP, Waldmann H, Fairchild PJ. Embryonic stem cell-derived tissues are immunogenic but their inherent immune privilege promotes the induction of tolerance. Proc Natl Acad Sci USA 2007; 104(52): 20920-5. [http://dx.doi.org/10.1073/pnas.0710265105] [PMID: 18093946]

[78] Lui KO, Boyd AS, Cobbold SP, Waldmann H, Fairchild PJ. A role for regulatory T cells in acceptance of ESC-derived tissues transplanted across an major histocompatibility complex barrier. Stem Cells 2010; 28(10): 1905-14. [http://dx.doi.org/10.1002/stem.506] [PMID: 20737577]

[79] Honey K, Cobbold SP, Waldmann H. CD40 ligand blockade induces CD4+ T cell tolerance and linked suppression. J Immunol 1999; 163(9): 4805-10. [PMID: 10528180]

[80] Zhou Y, Singh AK, Hoyt RF Jr, *et al.* Regulatory T cells enhance mesenchymal stem cell survival and proliferation following autologous cotransplantation in ischemic myocardium. J Thorac Cardiovasc Surg 2014; 148(3): 1131-1137, 1117. [http://dx.doi.org/10.1016/j.jtcvs.2014.06.029] [PMID: 25052825]

[81] Vesely MD, Schreiber RD. Cancer immunoediting: antigens, mechanisms, and implications to cancer immunotherapy. Ann N Y Acad Sci 2013; 1284: 1-5. [http://dx.doi.org/10.1111/nyas.12105] [PMID: 23651186]

[82] Lakshmi Narendra B, Eshvendar Reddy K, Shantikumar S, Ramakrishna S. Immune system: a double-edged sword in cancer. Inflamm Res 2013; 62(9): 823-34.
[http://dx.doi.org/10.1007/s00011-013-0645-9] [PMID: 23868500]

[83] Paulson KG, Tegeder A, Willmes C, *et al.* Downregulation of MHC-I expression is prevalent but reversible in Merkel cell carcinoma. Cancer Immunol Res 2014 Aug.; 12 pii: canimm.0005.2014
[http://dx.doi.org/10.1158/2326-6066.CIR-14-0005]

[84] Yang FQ, Liu M, Yang FP, *et al.* Matrix metallopeptidase 2 (MMP2) mediates MHC class I polypeptide-related sequence A (MICA) shedding in renal cell carcinoma. Actas Urol Esp 2014; 38(3): 172-8.
[http://dx.doi.org/10.1016/j.acuro.2013.09.015] [PMID: 24461475]

[85] Houston AM, Michael-Robinson JM, Walsh MD, *et al.* The "Fas counterattack" is not an active mode of tumor immune evasion in colorectal cancer with high-level microsatellite instability. Hum Pathol 2008; 39(2): 243-50.
[http://dx.doi.org/10.1016/j.humpath.2007.06.010] [PMID: 17961631]

[86] Simpson-Abelson MR, Loyall JL, Lehman HK, *et al.* Human ovarian tumor ascites fluids rapidly and reversibly inhibit T cell receptor-induced NF-κB and NFAT signaling in tumor-associated T cells. Cancer Immun 2013; 13: 14-23.
[PMID: 23882159]

[87] Yin Y, Cai X, Chen X, *et al.* Tumor-secreted miR-214 induces regulatory T cells: a major link between immune evasion and tumor growth. Cell Res 2014; 24(10): 1164-80.
[http://dx.doi.org/10.1038/cr.2014.121] [PMID: 25223704]

[88] Lin YC, Murayama Y, Hashimoto K, *et al.* Role of tumor suppressor genes in the cancer-associated reprogramming of human induced pluripotent stem cells. Stem Cell Res Ther 2014; 5(2): 58-68.
[http://dx.doi.org/10.1186/scrt447] [PMID: 25157408]

[89] Banito A, Gil J. Induced pluripotent stem cells and senescence: learning the biology to improve the technology. EMBO Rep 2010; 11(5): 353-9.
[http://dx.doi.org/10.1038/embor.2010.47] [PMID: 20379220]

[90] Riggs JW, Barrilleaux BL, Varlakhanova N, Bush KM, Chan V, Knoepfler PS. Induced pluripotency and oncogenic transformation are related processes. Stem Cells Dev 2013; 22(1): 37-50.
[http://dx.doi.org/10.1089/scd.2012.0375] [PMID: 22998387]

[91] Saito A, Ochiai H, Okada S, Miyata N, Azuma T. Suppression of Lefty expression in induced pluripotent cancer cells. FASEB J 2013; 27(6): 2165-74.
[http://dx.doi.org/10.1096/fj.12-221432] [PMID: 23407711]

[92] Lee AS, Tang C, Rao MS, Weissman IL, Wu JC. Tumorigenicity as a clinical hurdle for pluripotent stem cell therapies. Nat Med 2013; 19(8): 998-1004.
[http://dx.doi.org/10.1038/nm.3267] [PMID: 23921754]

[93] Fang I-M, Yang C-M, Yang C-H, Chiou S-H, Chen M-S. Transplantation of induced pluripotent stem cells without C-Myc attenuates retinal ischemia and reperfusion injury in rats. Exp Eye Res 2013; 113: 49-59.
[http://dx.doi.org/10.1016/j.exer.2013.05.007] [PMID: 23726881]

[94] Stadtfeld M, Nagaya M, Utikal J, Weir G, Hochedlinger K. Induced pluripotent stem cells generated without viral integration. Science 2008; 322(5903): 945-9. [http://dx.doi.org/10.1126/science.1162494] [PMID: 18818365]

[95] Silva J, Barrandon O, Nichols J, Kawaguchi J, Theunissen TW, Smith A. Promotion of reprogramming to ground state pluripotency by signal inhibition. PLoS Biol 2008; 6(10): e253. [http://dx.doi.org/10.1371/journal.pbio.0060253] [PMID: 18942890]

[96] Lin S-L, Ying S-Y. Mechanism and method for generating tumor-free iPS cells using intronic microRNA miR-302 induction. Methods Mol Biol 2013; 936: 295-312. [http://dx.doi.org/10.1007/978-1-62703-083-0_23] [PMID: 23007517]

[97] Abdullah LN, Chow EK. Mechanisms of chemoresistance in cancer stem cells. Clin Transl Med 2013; 2(1): 3-27. [http://dx.doi.org/10.1186/2001-1326-2-3] [PMID: 23369605]

[98] Ni J, Cozzi P, Hao J, *et al.* Cancer stem cells in prostate cancer chemoresistance. Curr Cancer Drug Targets 2014; 14(3): 225-40. [http://dx.doi.org/10.2174/1568009614666140328152459] [PMID: 24720286]

[99] Hamaï A, Meslin F, Benlalam H, *et al.* ICAM-1 has a critical role in the regulation of metastatic melanoma tumor susceptibility to CTL lysis by interfering with PI3K/AKT pathway. Cancer Res 2008; 68(23): 9854-64. [http://dx.doi.org/10.1158/0008-5472.CAN-08-0719] [PMID: 19047166]

[100] Knuefermann C, Lu Y, Liu B, *et al.* HER2/PI-3K/Akt activation leads to a multidrug resistance in human breast adenocarcinoma cells. Oncogene 2003; 22(21): 3205-12. [http://dx.doi.org/10.1038/sj.onc.1206394] [PMID: 12761490]

[101] Seliger B, Kiessling R. The two sides of HER2/neu: immune escape versus surveillance. Trends Mol Med 2013; 19(11): 677-84. [http://dx.doi.org/10.1016/j.molmed.2013.08.003] [PMID: 24035606]

[102] Mimura K, Shiraishi K, Mueller A, *et al.* The MAPK pathway is a predominant regulator of HLA-A expression in esophageal and gastric cancer. J Immunol 2013; 191(12): 6261-72. [http://dx.doi.org/10.4049/jimmunol.1301597] [PMID: 24244023]

[103] Ithimakin S, Day KC, Malik F, *et al.* HER2 drives luminal breast cancer stem cells in the absence of HER2 amplification: implications for efficacy of adjuvant trastuzumab. Cancer Res 2013; 73(5): 1635-46. [http://dx.doi.org/10.1158/0008-5472.CAN-12-3349] [PMID: 23442322]

[104] Reim F, Dombrowski Y, Ritter C, *et al.* Immunoselection of breast and ovarian cancer cells with trastuzumab and natural killer cells: selective escape of CD44high/CD24low/HER2low breast cancer stem cells. Cancer Res 2009; 69(20): 8058-66. [http://dx.doi.org/10.1158/0008-5472.CAN-09-0834] [PMID: 19826050]

[105] Schatton T, Frank MH. Antitumor immunity and cancer stem cells. Ann N Y Acad Sci 2009; 1176: 154-69. [http://dx.doi.org/10.1111/j.1749-6632.2009.04568.x] [PMID: 19796244]

[106] Shipitsin M, Campbell LL, Argani P, *et al.* Molecular definition of breast tumor heterogeneity. Cancer

Cell 2007; 11(3): 259-73.
[http://dx.doi.org/10.1016/j.ccr.2007.01.013] [PMID: 17349583]

[107] Lottaz C, Beier D, Meyer K, *et al.* Transcriptional profiles of CD133+ and CD133- glioblastoma-derived cancer stem cell lines suggest different cells of origin. Cancer Res 2010; 70(5): 2030-40. [http://dx.doi.org/10.1158/0008-5472.CAN-09-1707] [PMID: 20145155]

[108] Todaro M, Alea MP, Di Stefano AB, *et al.* Colon cancer stem cells dictate tumor growth and resist cell death by production of interleukin-4. Cell Stem Cell 2007; 1(4): 389-402. [http://dx.doi.org/10.1016/j.stem.2007.08.001] [PMID: 18371377]

[109] Olver S, Groves P, Buttigieg K, *et al.* Tumor-derived interleukin-4 reduces tumor clearance and deviates the cytokine and granzyme profile of tumor-induced CD8+ T cells. Cancer Res 2006; 66(1): 571-80. [http://dx.doi.org/10.1158/0008-5472.CAN-05-1362] [PMID: 16397274]

[110] Kawasaki BT, Mistree T, Hurt EM, Kalathur M, Farrar WL. Co-expression of the tolerogenic glycoprotein, CD200, with markers for cancer stem cells. Biochem Biophys Res Commun 2007; 364(4): 778-82. [http://dx.doi.org/10.1016/j.bbrc.2007.10.067] [PMID: 17964286]

[111] Kawasaki BT, Farrar WL. Cancer stem cells, CD200 and immunoevasion. Trends Immunol 2008; 29(10): 464-8. [http://dx.doi.org/10.1016/j.it.2008.07.005] [PMID: 18775673]

[112] Kretz-Rommel A, Qin F, Dakappagari N, *et al.* CD200 expression on tumor cells suppresses antitumor immunity: new approaches to cancer immunotherapy. J Immunol 2007; 178(9): 5595-605. [http://dx.doi.org/10.4049/jimmunol.178.9.5595] [PMID: 17442942]

[113] Maccalli C, Volontè A, Cimminiello C, Parmiani G. Immunology of cancer stem cells in solid tumours. A review. Eur J Cancer 2014; 50(3): 649-55. [http://dx.doi.org/10.1016/j.ejca.2013.11.014] [PMID: 24333096]

[114] Ryan JM, Barry FP, Murphy JM, Mahon BP. Mesenchymal stem cells avoid allogeneic rejection. J Inflamm (Lond) 2005; 2: 8-32. [http://dx.doi.org/10.1186/1476-9255-2-8] [PMID: 16045800]

[115] Munn DH, Mellor AL, Rossi M, Young JW. Dendritic cells have the option to express IDO-mediated suppression or not. Blood 2005; 105(6): 2618-20. [http://dx.doi.org/10.1182/blood-2005-01-0122] [PMID: 15746089]

[116] Munn DH, Mellor AL. IDO and tolerance to tumors. Trends Mol Med 2004; 10(1): 15-8. [http://dx.doi.org/10.1016/j.molmed.2003.11.003] [PMID: 14720581]

[117] Mellor AL, Munn DH. Tryptophan catabolism prevents maternal T cells from activating lethal anti-fetal immune responses. J Reprod Immunol 2001; 52(1-2): 5-13. [http://dx.doi.org/10.1016/S0165-0378(01)00118-8] [PMID: 11600174]

[118] Mellor AL, Munn DH. IDO expression by dendritic cells: tolerance and tryptophan catabolism. Nat Rev Immunol 2004; 4(10): 762-74. [http://dx.doi.org/10.1038/nri1457] [PMID: 15459668]

[119] Meisel R, Zibert A, Laryea M, Göbel U, Däubener W, Dilloo D. Human bone marrow stromal cells

inhibit allogeneic T-cell responses by indoleamine 2,3-dioxygenase-mediated tryptophan degradation. Blood 2004; 103(12): 4619-21.
[http://dx.doi.org/10.1182/blood-2003-11-3909] [PMID: 15001472]

[120] Frumento G, Rotondo R, Tonetti M, Damonte G, Benatti U, Ferrara GB. Tryptophan-derived catabolites are responsible for inhibition of T and natural killer cell proliferation induced by indoleamine 2,3-dioxygenase. J Exp Med 2002; 196(4): 459-68.
[http://dx.doi.org/10.1084/jem.20020121] [PMID: 12186838]

[121] Barry FP, Murphy JM, English K, Mahon BP. Immunogenicity of adult mesenchymal stem cells: lessons from the fetal allograft. Stem Cells Dev 2005; 14(3): 252-65.
[http://dx.doi.org/10.1089/scd.2005.14.252] [PMID: 15969620]

[122] Tse WT, Pendleton JD, Beyer WM, Egalka MC, Guinan EC. Suppression of allogeneic T-cell proliferation by human marrow stromal cells: implications in transplantation. Transplantation 2003; 75(3): 389-97.
[http://dx.doi.org/10.1097/01.TP.0000045055.63901.A9] [PMID: 12589164]

[123] Bai L, Lennon DP, Caplan AI, *et al.* Hepatocyte growth factor mediates mesenchymal stem cell–induced recovery in multiple sclerosis models. Nat Neurosci 2012; 15(6): 862-70.
[http://dx.doi.org/10.1038/nn.3109] [PMID: 22610068]

[124] Benkhoucha M, Santiago-Raber ML, Schneiter G, *et al.* Hepatocyte growth factor inhibits CNS autoimmunity by inducing tolerogenic dendritic cells and CD25+Foxp3+ regulatory T cells. Proc Natl Acad Sci USA 2010; 107(14): 6424-9.
[http://dx.doi.org/10.1073/pnas.0912437107] [PMID: 20332205]

[125] Maitra B, Szekely E, Gjini K, *et al.* Human mesenchymal stem cells support unrelated donor hematopoietic stem cells and suppress T-cell activation. Bone Marrow Transplant 2004; 33(6): 597-604.
[http://dx.doi.org/10.1038/sj.bmt.1704400] [PMID: 14716336]

[126] Assawachananont J, Mandai M, Okamoto S, *et al.* Transplantation of embryonic and induced pluripotent stem cell-derived 3D retinal sheets into retinal degenerative mice. Stem Cell Rep 2014; 2(5): 662-74.
[http://dx.doi.org/10.1016/j.stemcr.2014.03.011] [PMID: 24936453]

[127] Garçon L, Ge J, Manjunath SH, *et al.* Ribosomal and hematopoietic defects in induced pluripotent stem cells derived from Diamond Blackfan anemia patients. Blood 2013; 122(6): 912-21.
[http://dx.doi.org/10.1182/blood-2013-01-478321] [PMID: 23744582]

[128] Lipton JM, Ellis SR. Diamond-Blackfan anemia: diagnosis, treatment, and molecular pathogenesis. Hematol Oncol Clin North Am 2009; 23(2): 261-82.
[http://dx.doi.org/10.1016/j.hoc.2009.01.004] [PMID: 19327583]

[129] Le Blanc K, Frassoni F, Ball L, *et al.* Mesenchymal stem cells for treatment of steroid-resistant, severe, acute graft-versus-host disease: a phase II study. Lancet 2008; 371(9624): 1579-86.
[http://dx.doi.org/10.1016/S0140-6736(08)60690-X] [PMID: 18468541]

[130] Mason PJ, Perdigones N, Bessler M. Using iPSCs to study Diamond Blackfan anemia: an outlook on the clinical possibilities. Expert Rev Hematol 2013; 6(6): 627-9.
[http://dx.doi.org/10.1586/17474086.2013.859521] [PMID: 24219546]

[131] Ebihara Y, Ma F, Tsuji K. Generation of red blood cells from human embryonic/induced pluripotent stem cells for blood transfusion. Int J Hematol 2012; 95(6): 610-6.
[http://dx.doi.org/10.1007/s12185-012-1107-9] [PMID: 22648827]

[132] Gekas C, Graf T. Induced pluripotent stem cell-derived human platelets: one step closer to the clinic. J Exp Med 2010; 207(13): 2781-4.
[http://dx.doi.org/10.1084/jem.20102428] [PMID: 21173109]

[133] Di Tomaso T, Mazzoleni S, Wang E, *et al.* Immunobiological characterization of cancer stem cells isolated from glioblastoma patients. Clin Cancer Res 2010; 16(3): 800-13.
[http://dx.doi.org/10.1158/1078-0432.CCR-09-2730] [PMID: 20103663]

[134] Kennedy M, Awong G, Sturgeon CM, *et al.* T lymphocyte potential marks the emergence of definitive hematopoietic progenitors in human pluripotent stem cell differentiation cultures. Cell Reports 2012; 2(6): 1722-35.
[http://dx.doi.org/10.1016/j.celrep.2012.11.003] [PMID: 23219550]

[135] Consentius C, Reinke P, Volk H-D. Immunogenicity of allogeneic mesenchymal stromal cells: what has been seen *in vitro* and *in vivo*? Regen Med 2015; 10(3): 305-15.
[http://dx.doi.org/10.2217/rme.15.14] [PMID: 25933239]

[136] Niemeyer P, Schönberger TS, Hahn J, *et al.* Xenogenic transplantation of human mesenchymal stem cells in a critical size defect of the sheep tibia for bone regeneration. Tissue Eng Part A 2010; 16(1): 33-43.
[http://dx.doi.org/10.1089/ten.tea.2009.0190] [PMID: 19604039]

[137] Roemeling-van Rhijn M, Weimar W, Hoogduijn MJ. Mesenchymal stem cells: application for solid-organ transplantation. Curr Opin Organ Transplant 2012; 17(1): 55-62.
[http://dx.doi.org/10.1097/MOT.0b013e32834ee676] [PMID: 22186092]

[138] Schatton T, Frank MH. Cancer stem cells and human malignant melanoma. Pigment Cell Melanoma Res 2008; 21(1): 39-55.
[http://dx.doi.org/10.1111/j.1755-148X.2007.00427.x] [PMID: 18353142]

[139] Cruceru ML, Enciu AM, Popa AC, *et al.* Signal transduction molecule patterns indicating potential glioblastoma therapy approaches. Onco Targets Ther 2013; 6: 1737-49.
[PMID: 24348050]

CHAPTER 9

Current Proteomic Studies for New Concept in Stem Cell Biology

Cristiana Pistol Tanase[1,2,*], **Elena Codrici**[1], **Ionela Daniela Popescu**[1], **Simona Mihai**[1], **Laura Necula**[1,3] and **Radu Albulescu**[1,2,4]

[1] *"Victor Babes" National Institute of Pathology, Biochemistry-Proteomics Department, Bucharest, Romania*

[2] *"Titu Maiorescu" University, Faculty of Medicine, Bucharest, Romania*

[3] *"Stefan S. Nicolau" Institute of Virology, Cellular and Molecular Pathology, Bucharest, Romania*

[4] *National Institute for Chemical Pharmaceutical R&D, Bucharest, Romania*

Abstract: Recent studies on stem cells and protein interactions using proteomics approaches have yielded novel perception on processes regulating development and stem cell biology. The development of stem cell approaches has evolved in the post-genomic era and the implementation of proteomic applications represents a great challenge.

Understanding the mechanism that controls stem cell pluripotency, self-renewal and differentiation will also improve the ability to develop stem cell-based therapies.

Current proteomics studies of stem cell signaling pathway can lead to the discovery of molecular mechanisms that govern cell-cell interaction and/or with stem cell niche.

Stem cells represent an important potential therapeutical approach, both in regenerative medicine and tumor pathology, allowing the understanding of the mechanisms that underlie the biology of these cells.

Keywords: Cancer stem cells, Interactomics, Proteomics, Regenerative medicine, Secretome, Signaling pathways, Stem cell biology, Stem cells therapy.

* **Corresponding author Cristiana Pistol Tanase:** "Victor Babes" National Institute of Pathology, no 99-101 Splaiul Independentei, 050096, Sector 5, Bucharest, Romania; Tel/Fax: 0040213194528; E-mail: bioch@vbabes.ro.

Cristiana Tanase & Monica Neagu (Eds.)

INTRODUCTION

The domain of stem cells is of outstanding importance in life sciences; a great amount of resources is invested in the development of stem cell based technologies for regenerative medicine. However, there is a "dark" counter-part, represented by cancer stem cells, which are increasingly demonstrated in tumorigenesis and tumor progression.

Regenerative medicine holds the promise of engineering damaged tissues and organs by means of stem cells of both embryonic and adult origins, due to their unique properties of unlimited self-renewal and differentiation once they receive appropriate signals. Proteomics rely on state-of-the-art technology platforms interacting with advancements in mass spectrometry and bioinformatics to develop a large-scale comprehensive study of a specific proteome, including information on protein abundances, their variations and modifications and studies of post-translational modifications and protein-protein interactions. Development of stem cell approaches has evolved in the post-genomic era and the implementation of proteomic applications represent the great challenge. Recent studies on stem cells and protein interactions using proteomics approaches have yielded novel perception on processes regulating development and stem cell pluripotency [1].

Transcriptome alone provides an incomplete and unfairly interpretation of stem cell biology [2, 3]; however, transcriptome mapping using DNA microarray technology is commonplace in the current post-genomic era [4, 5]. Transcriptional profiling approach has some limitations regarding the yet unidentified "stemness" genes that are not present on the microarray and the mRNA level that may not be properly correlated with changes in protein expression [6]. Furthermore, protein complex formation and degradation, post-translational modificationsα (PTMs) highly influence protein-protein and protein-DNA interactions; therefore, it is rather difficult to predict the functional outcome of these systems based only on gene expression and/or genomic data.

The term "proteome" was used for the first time by Wilkins *et al.* [7] and refers to the entire complement of proteins expressed in a given population. Proteomics encompasses a wide variety of advanced techniques capable to provide invaluable insights into the proteome at a genome wide level: from yeast two-hybrid screens for identifying protein-protein interactions [8], antibody-based protein chips for identifying proteins [9] and high throughput crystallography screens [10]; the most widely utilized group of techniques center around mass spectrometry (MS).

Both embryonic stem cells (ESCs) and adult stem cells, the two main groups of stem cells, have been intensively and widely explored to understand the molecular mechanisms such as self-renewal, proliferation, differentiation and the promising results were of great use in many clinical applications [11]. The overall understanding of biological features of stem cells is the great goal of the so-called "omics" technologies, which comprise an increasingly wide range of biology fields to denote studies undertaken on a large or genome-wide scale. The "omics" range from genomics, transcriptomics, proteomics to epigenomics and are based on advanced technologies responsible for the great opportunities and advances in the postgenomic biology and medicine [12, 13], bringing new insights in the field of stem cell biology. The investigation of stem cells using high-throughput screening techniques provided large amounts of information facilitating systemic understanding of relationship between molecular components [14].

It is assumed that the cancer stem cells (CSCs) are the underlying cells which can cause the heterogeneous lineage of cancer cells, also responsible for tumor progression: metastasis, recurrence and chemoresistance [15 - 18]. CSCs were first discovered in acute myeloid leukemia [19], but there are studies that report their existence in a wide variety of cancers, including gastric cancer [20]. The discovery of CSCs markers, which selectively target this specific small population of cells, may arise novel therapeutic opportunities [21, 22]. Some of these CSCs markers: aldehyde dehydrogenase 1 (ALDH1) [23], CD44 [24, 25] and CD133 [26] have been recently reported, even if they are not unique to CSCs. Quantitative protein expression profiling currently gives the most accurate and reproducible differential expression values for proteins [27]. A powerful proteomic instrument useful for tumor biomarkers is iTRAQ (isobaric tags for relative and absolute quantitation) combined with multidimensional liquid chromatography (LC) and tandem mass spectrometry (LC-MS/MS) analysis [28].

Using proteomics technology by comparing the proteomes among parent cells and stem cell-like SP cells containing a rich CSC population, Tamami Morisaki *et al.* identified new gastric CSCs biomarkers useful in clinical diagnosis [29].

Stem cells can be classified into three major categories: embryonic, fetal and adult, according to the occurrence of the ontogenesis stage. During embryogenesis, embryonic stem cells generate the entire body of the organism, while adult stem cells are involved in the regeneration of damaged cells or tissue [30 - 32]. Their unique features qualify stem cells as promising tools for clinical applications and regenerative medicine [31].

PROTEOMICS *VERSUS* TRANSCRIPTOMICS IN STEM CELLS

In contrast with the genome, which is characterized by its stability, the transcriptome actively changes and is a crucial step in discovering the functional components of the genome, revealing developmental processes and mechanisms underlying diseases [33]. High levels of transcriptional activity is representative to pluripotent stem cells, due to their plasticity, while loss of pluripotent nature leads to considerable decrease in transcription of few small parts of the genome [34]. Several studies identified and verified a number of genes of great importance for maintaining the pluripotency, expressed exclusively or predominantly in ESCs such as Oct4, Sox2, Nanog, Rex-1, UTF1, Tert, ABCG2, NODAL, TDGF1, LEFTB, BEX1 and some molecules involved in signaling pathways such as FGF, WNT, and BMP [35 - 37].

MicroRNAs and LncRNAs

The regulatory effects of microRNAs (miRNAs) on transcription are still an issue in debate, although there are many studies, which extensively explore the functions of many protein-coding genes. miRNA can bind to the 3′nontranslational region (3′UTR) of target mRNAs to regulate their expression, giving novel insights into the internal mechanisms of stem cells [38]. There are several studies regarding various miRNAs expressed specifically in ESCs to maintain stem cell self-renewal, such as miR-368, miR-200c, miR-154, miR-371, miR-372, miR-373, and miR-373. Crucial roles in stem cell differentiation are played by miR-301, miR-374, miR-21, miR-29, and miR-29b, whose expression increases after the induction of differentiation [39, 40]. Another remarkable class of potential regulatory RNAs in transcriptome, with important roles in stem cell pluripotency and differentiation, refers to long noncoding RNAs (lncRNAs), which are non-protein coding transcripts longer than 200 nucleotides [41]. Cellular regulation, development and disease are processes in which non-coding RNAs play important roles [42].

The Characteristics of Stem Cell Proteomics

After genomics and transcriptomics, proteomic analysis comes to improve the relative quantitation of proteins and peptides, identification of proteins, their subcellular localization, and identifies protein-protein interactions and posttranslational modifications (PTMs). Proteomics provide valuable new insight into the molecular processes underlying stem cell self-renewal, proliferation and differentiation to specific lineages [41, 43].

The discovery of novel key proteins during different stages of stem cell development to provide a protein signature is the main goal of proteomics [44]. A panel of transcription factors such as Oct4 and UTF-1, alkaline phosphatase, and other molecules including nidogen 2, hepatoma derived growth factor (HDGF), cadherin 1, catenin α1, transgelin, and disabled homolog 2 were identified as markers of ESC [45]. Changes in expression of novel key proteins involved in transcription regulation, cell proliferation (CDV3, RCN1, PCNP and homolog), Ras signaling (G3BP and TTC1), and chromatin remodeling (RUVBL1 and HDGF) were observed by comparing monkey ESCs during proliferation and at different stages of spontaneous differentiation [44]. Furthermore, the key proteins and biomarkers of cancer stem cells were also intensively studied; recent studies detected DAC2 and CTNNB1 as prognostic markers in the malignant transformation of hESCs [46]. Using proteomics strategy, an important role in cancer development by regulating key steps of glycolysis in colon cancer stem cells is assigned to molecule p63 [47]. Assuming that a number of approximate 3400 genes are involved in encoding single-pass transmembrane or secreted proteins in mammalian cells [48], it is of great importance to study the physiological activities of the extracellular proteome during stem cell self-renewal, proliferation and differentiation. It was suggested that activation of pigment epithelium derived factor (PEDF) receptor-Erk1/2 signaling pathway by PEDF is sufficient to maintain the self-renewal of undifferentiated hESCs, in a complete study of hESCs extracellular proteome [49]. Another potential pathway involved in tumorigenesis of cancer stem cells in osteosarcoma is represented by ERK1/2, showing a surface protein signature distinct from differentiated cells [41, 50].

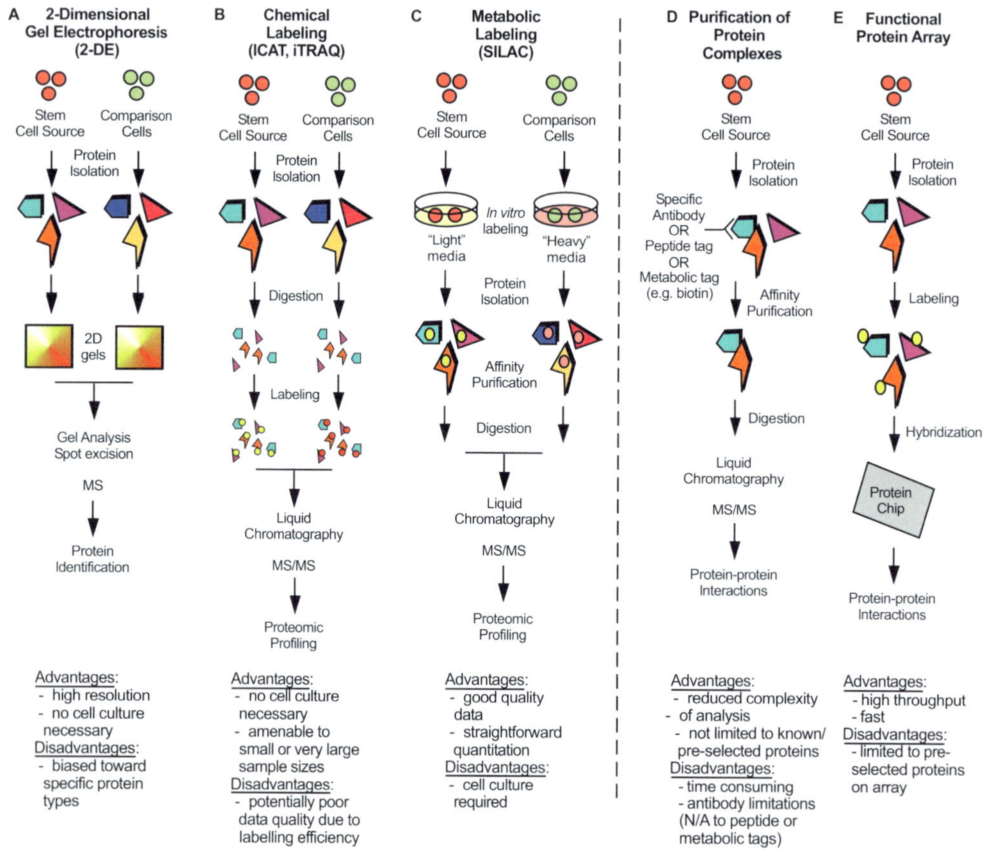

Fig. (1). Comparison between different proteomic approaches in the study of stem cells. Reproduced with permission from [1]. **A:** Bidimensional gel electrophoresis (2-DE) coupled with mass spectrometry (MS); **B:** Chemical labelling (ITAC – isotopic coded affinity labeling) or ITRAQ + Liquid chromatography and MS/MS; **C:** Metabolic L labeling (SILAC) + LC-MS/MS; **D:** Purification of Protein Complexes (Epitopes, Peptide or Metabolic Tags) + LC-MS/MS; **E:** Functional Protein Arrays.

Different proteomics technologies, such as two-dimensional polyacrylamide gel electrophoresis (2D-PAGE), mass spectrometry (MS), and liquid chromatography (LC)-MS/MS, are used to create the proteome signature, based on the abundance of proteins and peptides, stability, subcellular localization, PTMs, and their interactions [43]. New advanced proteomic methods, such as isotope coded affinity tag (ICAT), stable isotope labeling with amino acids in cell culture (SILAC), and iTRAQ are further preferred to enhance sensitivity [14]. A global view of stem cell proteome dynamics can be also generated by protein microarrays, a high-throughput method used to track the interactions and activities of proteins [14, 51] which can provide important signs in stem cell progression and differentiation processes [41, 52] (Fig. **1**).

EMBRYONIC STEM CELLS PROTEOMICS

The establishment of human embryonic stem cells (hESCs) [53] bring new insights into regenerative medicine field, where tissue repair or replacement are currently explored. In addition to transcriptomic analyses of ESCs that establish a genome wide RNA expression signature of stemness [4, 5], the protein pluripotency signature are provided by stem cell proteomics, an outstanding instrument to characterize ESCs at protein level. The proteomic analysis supply both undifferentiated and differentiated ESCs using MS-based protein profiling. Recent studies identified a number of 191 proteins, many of which uncharacterized, exclusively identified in both human and mouse ESCs and which represent potential novel ESC-specific markers or functional proteins [54], all assembled in a catalogue, containing various components derived from ES-specific and stemness genes defined by the transcriptome analysis [5] and components expressed specifically in ESCs, such as Oct4 and UTF1. In a recent study, Graumann *et al.* identified and quantified more than 5000 distinct proteins using LTQ-Orbitrap (linear ion trap-orbitrap) instrument [55], one of the largest reported quantified proteome, which was compared with a recently published map of chromatin states of promoters in ESCs [56]; a great correlation was established between protein expression and the presence of active and repressive chromatin marks.

In order to better understand the underlying stem cell biology, large-scale transcriptomic and proteomic analyses of ESCs are complementary to each other. Nevertheless, there are missing links, such as gene transcription, that may not be directly indicative of, or proportional to protein translational readout (expression); consequently, protein expression and multiprotein complexes do not specify target gene regulation of the protein(s) themselves. A thorough understanding of the establishment of the pluripotent state in ESCs requires the construction of an expanded transcriptional regulatory network in which many key transcription factors besides Nanog, Oct4 and Sox2 and their interaction partners [57] bind directly to their target genes. Some studies started to elucidate the transcriptional networks around the three core ESC transcriptional factors Nanog, Oct4, and Sox2 that operate to control ESC biology. Utilizing ChIP-chip analysis (chromatin immunoprecipitation followed by microarray hybridization to identify binding sites on a genome wide scale), it was shown that Oct4, Sox2 and Nanog co-work to regulate hESC pluripotency and self-renewal through autoregulatory and feedforward loops [58]. There are several other core pluripotency factors within this network, alongside Nanog, Sox2 and Oct4. For example, Klf4 represents an upstream regulator of larger feedforward loops containing Nanog, Sox2 and Oct4 as well as c-Myc. Also, c-Myc, Nanog, Oct4 and Sox2 play different parts in ESCs, according to the regulatory network; c-Myc appears to be involved in the stimulation of cell proliferation and regulation of chromosomal accessibility, while the other three positively regulate ESC factors and negatively regulate differentiation [59].

A potential mechanism that might account for the differential regulation of transcription factor targets in ESCs and offer new insights into the 4-factor (Oct4, Klf4, Sox2, and c-Myc) mediated somatic cell reprogramming [60], paving the way for highly efficient, large-scale proteomic studies in ESCs.

ADULT STEM CELLS PROTEOMICS

Adult somatic stem cells are limited in their capacity to generate cell types within a defined lineage, unlike ESCs. In the last two decades, many studies have been focused on defining this type of cells, developing rigorous isolation strategies, characterizing the *in vitro* and *in vivo* functions, as well as establishing transcriptional profiles. The above-mentioned studies have provided a better understanding of the mechanisms that regulate self-renewal and potency within adult stem cells, requiring integration of multiple high-throughput platforms that assess transcriptomes, proteomes and protein interactomes. There are fewer studies regarding proteomic profiling and protein interaction mapping of adult stem cells compared to ESCs. Most of them focus on hematopoietic stem cells (HSCs), neural stem cells (NSCs) and mesenchymal stem cells (MSCs). The first two will be discussed in different chapters of the present book. Proteomic studies on adult stem cells are difficult and challenging to conduct, since most adult stem cell types are virtually impossible to maintain or expand in culture, without modifying their potency. However, NSCs may be expanded *in vitro* without loss of stem cell properties. Therefore, the available input cells are limited; in addition, currently there is no effective protein amplification method, unlike the development of global nucleic acid amplification for transcriptional profiling. 2-dimensional gel electrophoresis (2-DE) as a front-end fractionation method prior to mass spectrometry (MS) analysis was used as an initial proteomic analysis of *in vitro* expanded adult stem cells, showing that the greatest conserved group of proteins in adult stem cells plays its part in energy metabolism [43]. Nevertheless, these data sets are heavily influenced by the methodology used; subsequently, they might only represent the most abundant proteins broadly expressed among these cell types. Consequently, however, new and more sophisticated proteomic techniques, presenting the advantage of being unbiased, have been used to gain new insights into adult stem cell biology. Sensitive iTRAQ methodology combined with MS analysis represents one of the techniques that allow the comparison of purified populations of hematopoietic stem and progenitor cells.

In contrast to their more differentiated progenitor counterparts, HSCs seem to be adapted for anaerobic environments [61]. However, these differences were not observed at transcriptome level of these populations [61], suggesting that transcriptional profiling by itself would not have led to the deduction of this novel aspect of HSC biology. In addition, iTRAQ has effectively characterized a poorly described population of hematopoietic progenitor cells (Lineage− c-Kit+ Sca-1−) as mainly erythroid in nature [62]. 2-dimensional liquid chromatography (2D LC) or LC/LC fractionation followed by tandem MS/MS has been used in the MSC field, in order to prove that osteogenic differentiation comes from the focusing of gene expression in functional clusters rather than simply from the induced expression of new genes [63].

PTMs have been proven to significantly influence adult stem cell fate decisions. A quantitative phosphoproteomics approach, facilitated by SILAC technology, has been utilized to study the influence of growth factor signaling on MSC differentiation. Namely, the mechanism by which two related growth factors (EGF and PDGF) differentially influenced MSC differentiation was reported to be mediated by tyrosine phosphorylation [64]. Proteomics approach to identify critical protein-protein interactions regulating self-renewal and differentiation has been utilized in order to develop significant new concepts in processes regulating development and stem cell pluripotency [57, 59]. An essential change in subunit composition of a SWI/SNF-like chromatin remodeling complex during differentiation of NSCs to post-mitotic neurons has been characterized, using antibody-based purification of protein complexes [65]. Subunit proteins BAF45a and BAF53a are expressed by neural stem and progenitor cells as part of the SWI/SNF chromatin remodeling complex, being replaced by BAF45b, BAF45c and BAF53b while progenitors exit the cell cycle and are functionally validated. The state-of-the-art pair constituted by proteomic profiling and protein interactome studies of adult stem cells emphasizes the fact that these techniques can provide new perspectives into the underlying cell biology that different methodologies would otherwise be unable to provide.

A great challenge concerning adult stem cell proteomics, such as cell heterogeneity can be paradoxically solved by proteomic work itself.

For instance, in HSCs, the identification of SLAM family of cell surface markers was enabled by transcriptome profiling [66], improving the means of HSC isolation. Given the ability of transcriptome data to achieve moderate success to this end, there is great potential to identify new biomarkers through membrane proteomics.

Furthermore, the isolation of more homogeneous cell populations will be achieved by the use of lineage-specific fluorescent reporters. This strategy has been previously used in proteomic studies, with a high success rate, in order to examine the differentiation of mESCs to mesodermal/hemangioblast lineages with subsequent profiling using iTRAQ [1, 67].

Proteomics of Mesenchymal Stem Cells

Proteomic characterization of different stem cell processes are some of the most promising approaches for understanding the molecular mechanisms involved in stem cell self-renewal, proliferation, and differentiation. An approach in most of the stem cell proteomics studies is the characterization of cytoplasmic protein content, in order to identify novel key proteins and protein interaction maps during different stages of stem cell development. Proteomic analyses include the relative quantitation of proteins and peptides, identification of proteins and their subcellular localization, analysis of protein-protein interactions and posttranslational modifications [41, 44]. The main techniques applied in this field are 2D-PAGE, MS, and (LC)-MS/MS. Proteome analyses are complicated especially because the total number of human protein products has been estimated to be close to 1 million while the human genome harbors roughly 30,000 protein-encoding genes. Moreover, studies that have analyzed early differentiation phases showed that about 50% of the changes in protein expression do not have a corresponding change in mRNA expression [43, 68].

Mass spectrometry-based techniques are frequently used to analyze human stem cells. For example, a comparison of published data sets showed that 32 intracellular proteins and 16 plasma membrane proteins are present in hESCs lines but not in differentiated cells [69].

Using two-dimensional LC and MS/MS analyses, Chaerkady *et al.* have compared pluripotent hESCs and human embryonic carcinoma cells and the results showed that approximately 200 proteins present more than two fold differences in abundance. Among them could be found proteins expressed in early development or factors associated with pluripotency [70].

Another example of proteomic techniques is iTRAQ labeling combined with two dimensional capillary chromatography, coupled with tandem mass spectrometry (μLC-MS/MS); Novak *et al.* used these techniques to achieve comparative proteomics of human embryonic stem cells, starting from undifferentiated stage to

3 days after initiation of differentiation. The results revealed that many processes like chromatin remodeling, and heterochromatin spreading are initiated and later reversed; a decrease in transcription and translation, a decrease in glycolytic proteins and cytoskeleton remodeling, and a decrease in focal and cell adhesion have also been observed [52].

Proteome profiling can be used to define markers for cell differentiation; Osman A *et al.* have compared embryonic stem cells (ESC) with differentiating ESC and the results showed that 57 proteins were downregulated and 42 proteins were upregulated in ESC. Downregulated proteins are linked to cardiomyocyte functionality while pluripotency markers, chaperones and ribosomal proteins have increased expression in ESC. Moreover, the study has identified an increased expression of proteins/enzymes involved in DNA mismatch repair and in methylation reactions [71].

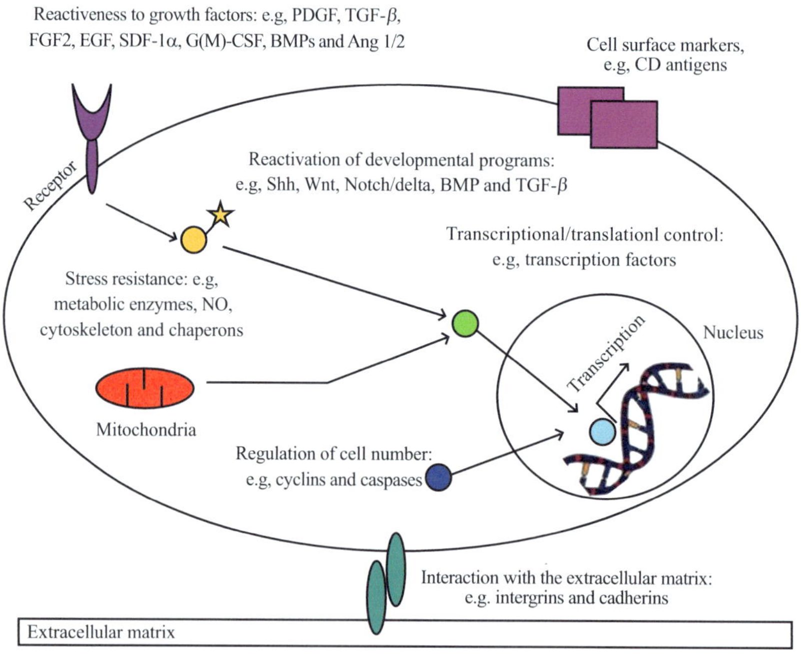

Fig. (2). Protein profiles involved in MSCs mechanisms: specific cell surface markers, growth factors pathway, developmental program pathways, interaction with the extracellular matrix, transcriptional/translational control, and resistance to cellular stress (reproduced with permission from [72]).

Mesenchymal stem cells (MSC), representing pluripotent cells isolated from the bone marrow and various other organs, are characterized by a distinct set of cell surface markers CD13, CD29, CD44, CD73, CD90, CD105, and CD166. The group of proteins and their expression pattern is specific for MSC and includes proteins involved in many processes like growth, stress resistance, regulation of cell number, interaction with extracellular matrix (see Fig. **2**) [72].

SIGNALING PATHWAYS IN STEM CELLS

Proteomic studies of intercellular signaling are being intensely carried out. In order to design physiologically relevant proteomics experiments, the role of MS-based proteomics needs to be significantly promoted, as to better understand these accurately controlled signaling events, especially in their specific micro-environment.

A systems-level approach has been delivered by MS-based proteomics, in order to analyze these signaling networks on posttranslational level, by characterizing dynamic PTMs and protein complexes. A great deal of different PTMs, such as phosphorylation, ubiquitination, and methylation, may be identified using MS instrumentation, specific enriching antibodies, as well as advanced bioinformatics analysis. Different proteomic technologies, such as SILAC and MRM, facilitate the characterization of intercellular signaling dynamics.

MS proteomics has been successfully applied for studying cell–cell communication.

The characterization of stem cells interaction with various niche components, *i.e.* supporting cells, growth factors, and matrix proteins is essential. MS proteomics can be applied for characterization of stem cell proteome and phosphoproteome as response to different extracellular stimuli [73 - 76]. Protein interaction network of key transcription factors that control stem cell pluripotency were also highlighted by IP-MS approach [77, 78].

Plasma membrane proteome profiling and secreted proteins in stem cell niche are crucial in terms of stem cell fate regulation by autocrine or paracrine modes [79 - 81] .

Cancer stem cells present similar properties to normal stem cells and seem to use many of the same signaling pathways that are found in normal stem cells, such as Wnt, Notch, and Hedgehog. JAK/STAT, MAP-Kinase/ERK, PI3K/AKT, NFkB, and TGFβ pathways are also linked to embryonic development and cancer [82]. More studies suggest that developmental signaling pathways play an important role in normal tissue homeostasis through the regulation of stem cell fate but also are involved in maintaining cancer stem cells and thereby the tumor itself. The involvement of these pathways in cancers seems to be directly linked to cancer stem cells. The main problem in the development of agents that target critical molecules of Wnt, Notch, and Hedgehog pathways is signaling cross-talk [82 - 84] (Fig. **3**).

Embryonic development is controlled by extracellular key signal transduction processes, involving Wnt, Hedgehog, Notch, fibroblast growth factors (FGF), and transforming growth factor (TGF)-β proteins. Following these signals, cellular phenotypic changes in self-renewal, differentiation, proliferation or migration occur at stem cells or differentiated cells levels [85].

Aberrant Wnt signaling was observed in many types of cancer, including colorectal cancer [81], and seems to play an important role in brain and bone metastasis during lung adenocarcinoma progression [86]. This association between Wnt signaling and metastasis is explained by the involvement of Wnt in mesenchymal development and the epithelial-mesenchymal transition during embryonic development [87]. Taken together, these studies suggest that Wnt signaling could be considered a candidate therapeutic target to prevent the distal spread of cancer [88].

Another signaling pathway active in stem cells and cancer stem cells is Hedgehog signaling which is required for the maintenance of hematopoietic cancer stem cells, for example [89]. Thereby, several antagonists of this pathway are currently being tested in clinical trials as potential treatments for medulloblastoma, pancreatic cancer, ovarian cancer and hematopoietic cancers [90].

Oncogenic potential of Notch signaling was reported in breast epithelial tumors [81], or T-cell acute leukemia [91] but some carcinomas from skin, head/neck, and lung present loss-of-function mutations in Notch, suggesting that the involvement of Notch in carcinogenesis is cell and tissue context dependent [92, 93].

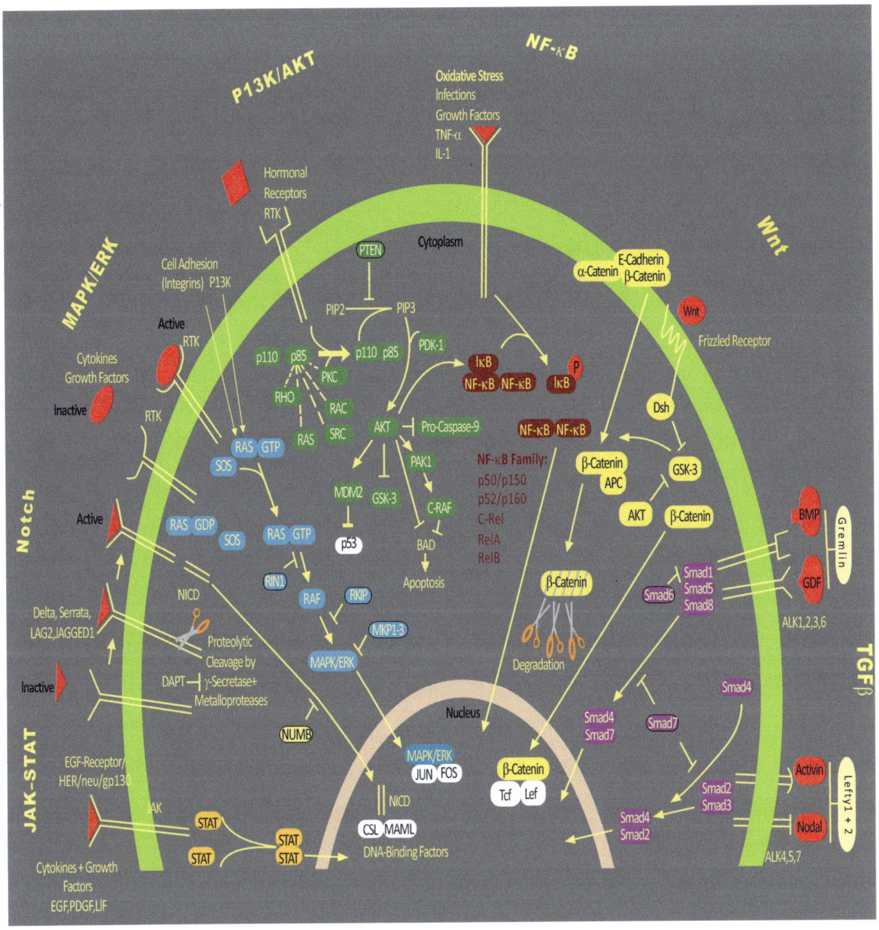

Fig. (3). Representation of seven signaling pathways associated with carcinogenesis and stem cells maintenance. JAK/STAT (orange), Notch, MAPK/ERK (blue), PI3K/AKT (green), NFκB (brown), Wnt (yellow), TGF-β (purple), ligands (red), transcription factors (white), signaling pathway inhibitory factors (black line), cellular membrane (green), nuclear membranes (grey). Modified protein expressions result in perturbation of signaling pathways that control embryonic development, stem cell maintenance and normal tissue homeostasis; aberrant cell signaling affect important cellular processes such as apoptosis, proliferation, angiogenesis and invasion (reproduced with permission from [82]).

Cancer stem cells have been associated with drug resistance and relapse in most cancer types emphasizing the importance of discovery of new targets and drugs for these signaling pathways involved in stem cells and cancer stem cells maintenance [94].

To identify biologically relevant proteins important for stem cell self renewal and pluripotency, the extensive catalogue and benchmark of protein databases are not sufficient. Several biochemical processes are influences in PTMs such as phosphorylation rather than by changes in abundance of proteins themselves. The pluripotent nature of ESCs and differentiation are strongly regulated by epigenetic mechanisms, such as covalent modifications of histones and DNA methylation [81]. Similarly to epigenome, the term "epiproteome" has emerged and it refers to a protein landscape of PTMs and histone variants [1, 95].

Cell-fate determination is also regulated by protein phosphorylation, a critical determinant of cell signaling [73, 75]. Phosphorylation status exhibits dynamic changes during the differentiation period. Four recent phosphoproteomic analyses of hESCs, using different cell culture conditions and different technologies, have identified 3067, 2546, 11 995, and 23 522 protein phosphorylation sites [73, 96, 97].

Approximately 50% of these sites presented dynamic changes in the phosphorylation status during 24 h of differentiation [73, 97]. Among the dynamically phosphorylated proteins, CDK1/2 was identified as a central factor in controlling stem cell self-renewal and lineage specification [73]. Brill *et al.* also discovered that 389 proteins contained more phosphorylation site identifications in undifferentiated hESCs, whereas 540 proteins contained more such identifications in differentiated derivatives [75]. Moreover, numerous phosphoproteins in receptor tyrosine kinase (RTK) signaling pathways were present in hESCs [75].

Phosphorylation is a PTM with an essential role in modulating protein function. In order to provide a better understanding of intracellular signals governing ESC self-renewal and differentiation, a multivariate systems analysis of proteomic data obtained from combinatorial stimulation of mESCs (line CCE) by fibronectin, laminin, LIF and fgf4 was performed [98]. Western blot has been used to achieve the phosphorylation states of 31 intracellular signaling network components across 16 different stimulus conditions at three time points. Likewise, computer modeling determined which components were most strongly correlated with cell proliferation and differentiation rate constants. The study revealed a group of signaling network components highly associated with differentiation and proliferation of undifferentiated as well as differentiated cells [1].

Understanding the phosphorylation landscape that controls stem cell pluripotency, self-renewal and differentiation will also improve our ability to develop stem cell-based therapies [41].

INTERACTOMICS IN STEM CELLS

In spite of advances reached in stem cell biology, great efforts are made for the exact characterization of the mechanisms by which pluripotency and self-renewal are established and maintained. A better understanding of these processes may lead to discovery of more reliable applications of stem cells in therapy. Proteome analysis of ESCs is an important approach to elucidate the molecular mechanisms that regulate the behavior of these cells [43, 99]. The expression-based analyses of ESC proteome and epiproteome offer a valuable insight for proteins and their PTMs; some may be used as ESC markers. The identification of individual polypeptides that form a proteome is not sufficient enough to completely understand the biological processes. Vital cellular functions involve the synchronized action of a great number of proteins that are gathered into an array of complexes of different composition and structure. The investigation of protein complexes and intricate protein-protein interaction (PPI) networks is crucial for stem cells protein function and regulation understanding [57, 100, 101].

Identifying several network proteins, which are not specific to ESCs and cannot be identified by transcriptional profiling, emphasizes the significance and benefit of proteomic studies in ESCs.

One of the major goals of functional proteomics in stem cells is to decrypt the molecular function by characterizing all molecular mechanisms, their roles in preserve stem cell features, their reactions to external stimuli during differentiation, and their interconnectivities.

Moreover, it offers a context for investigating the combinations of factors that may allow ideal reprogramming of differentiated cells to an ES cell state [1, 102, 103].

The investigation of protein-protein interaction (PPI) networks is a new area with which to explore complicated biological systems. Applying proteomics approaches to investigate interactomes that control self-renewal, differentiation and plasticity will offer an important perspective on the manner in which these interactions are involved in inducing differentiation of ESCs to specific lineages.

These PPI networks of ESCs can lead to a mechanistic comprehension of self-renewal and pluripotency. Yousefi *et al.* assess various proteomics methodologies that have been used to study PPI networks. According to the study, the biological findings and technical challenges found with interactome studies of pluripotent stem cells; foreseeing a potential way of development of interactomics.

As the behavior of proteins in cells may differ from what is measured *in vitro*, introduction of methods that enable visualization of in-cell PPI is necessary.

Some techniques enable screening of a large number of proteins in a cell, such as yeast two hybrid (Y2H). The currently available methods such as Förster/fluorescence resonance energy transfer (FRET), bimolecular fluorescence complementation (BiFC) and proximity ligation assay (PLA) are based on proximity of interacting partners [104]. Recent studies have shown the role of protein interaction networks of several major transcription factors such as Nanog, Oct4 and Sox2 [57, 77, 78].

Novel evidence for Oct4 interaction with UBC9, an E2 conjugation enzyme for sumoylation has been found, which suggested an important role of post-translational modification in modulating Oct4 function in ESCs [105]. Dejosez *et al.* applied a similar approach to identify the functional significance of interaction between Ronin and HCF-1 [106].

Introduction of tandem affinity purification (TAP) which uses two steps for purification is a breakthrough in the identification of interactions [107].

A number of previously identified factors (*e.g.*, Oct4, ESRRB, RIF1 and SALL4) and a number of novel interacting partners (*e.g.*, SMARCA2, SMARCA4 and BAF180) which interacted with Nanog were identified. Introduction of epitope tagging circumvents the need for a distinctive antibody for each bait protein and an individual protocol for the washing steps [107].

Among the studies on protein interaction networks in ESCs that used epitope tagging, Van den Berg *et al.* used flag-tagged Oct4 as an entry point to the protein interaction network for pluripotency and the acquired network was extended by tagged Oct4 partners SALL4, TCFCP2l1, DAX1 and ESRRB. A total of 166 proteins were identified as interacting partners of these 5 factors [78]. One- and two-step purifications were undertaken on total cell extract and analyzed by nanoliquid chromatography/tandem mass spectrometry (LC-MS/MS). As a result, 92 proteins were determined to be Oct4 interacting partners [77].

FRET is used to detect the interaction between two candidate proteins, a donor and acceptor and BiFC is the most popular technique for detecting interactions based on fragments of one molecule for fragments of autofluorescent proteins [108].

Several PPI such as Myc-Max, HER2-EGFR and SMAD3-SMAD4 have been successfully detected by using proximity ligation assay (PLA) [109]. The interactions during differentiation were investigated. Their results have shown that interactions of Oct4 and Nanog with Npm1 diminish during differentiation because Nanog and Oct4 were mostly present in stem cells, whereas the interaction between Sox2 and Npm1 was present during differentiation using retinoic acid [110].

Protein microarrays have been developed for determination of binary interactions (protein-protein, protein–lipid, protein–ligand and enzyme–substrate), drug development and clinical diagnostics [111].

Many new interactions for EGFR were identified, using protein microarrays, with the majority of Src homology 2 (SH2) and phosphotyrosine binding (PTB) domains encoded in the human genome [112].

In addition to the experimental detection of interactions, a wide range of computational methods exist that have become more powerful in the identification of potential interacting proteins and protein domains. These methods are also useful to infer functional associations between proteins. The recent developments in methods for extraction of protein interactions are one of the most exciting developments in molecular biology. Major computational approaches that study proteome interactions are discussed [113].

A powerful tool for studying diseases and designing drug targets has been developed by gaining a full understanding of the details of protein interactions in a proteome interaction network such as domain interaction analysis [114].

However, many domain interactions, especially those that participate in cell regulatory systems, are highly specific. It is possible that two proteins that contain potentially interacting domains do not interact with each other, since they may be located at different cell compartments, or may be expressed at different times during the cell cycle. Both structures and environmental conditions influence PPIs. For instance, two proteins with the same domain structure may interact differently with other proteins. In accordance with recent developments in high-throughput technologies, protein functions are predicted by utilizing interaction data. PPI networks show the relation between proteome and cell phenotype in different situations. They are highly dynamic and change with varying physiological conditions. Large-scale direct physical interactomes have been determined for several model organisms and experimental methods to identify interacting proteins are sometimes dependent on or complement each other [115]. Analyzing the changes that take place during ESCs differentiation can provide highly useful data to better understand the process, enabling more reliable use of these cells in therapy [116]. Before pluripotent stem cells can be effectively used in regenerative medicine, a thorough comprehension of the molecular mechanisms involved in maintaining their undifferentiated and pluripotent nature is crucial. Nevertheless, in order to fully comprehend the processes that take place within cells, it is not enough to merely identify the individual polypeptides that form the proteome. PPI identification and characterization are important steps towards a complex insight from an integrative point of view regarding the networks of interacting proteins that regulate hESC behavior [103].

CANCER STEM CELLS AND PROTEOMICS

Cancer Stem Cells Involvement in Tumorigenesis

CSCs, identified in a large number of human malignancies, represent 0.2%-0.8% of cancer cells and are considered responsible for tumor growth, invasion, metastasis and recurrence [117]. Currently, there are two models that explain tumor development [118 - 120]. The "stochastic model" states that every cancer cell has the ability to initiate and promote tumoral growth. The "cancer stem cell hypothesis", proposes that tumor evolution is based on stem cells with a "deregulated" self-renewal pathway. A recent and rapidly growing body of research shows solid evidence sustaining cancer stem cell model [121, 122]. The American Association for Cancer Research defines CSCs as cells within a tumor able to generate the heterogeneous cancer cell lineages found in the tumor and that possesses the capacity to self-renew. CSCs also share several other important attributes: active telomerase expression, drug resistance to harming agents, the activation of anti-apoptotic pathways, the ability to migrate and to metastasize and increased membrane transporter activity. To date, CSCs have been isolated and characterized only from a relatively small number of tumor types: breast, brain, pancreas, colon, blood and head and neck [123, 124].

Several studies argue that cancer stem cells cannot be eradicated by current therapy and thus are responsible for tumor relapse and metastasis [125]. Many studies have demonstrated that multiple critical genes, including K-ras, p53 and p16, and key signaling kinases, such as PI-3K, mTOR, NF-κB, epidermal growth factor receptor (EGFR) and SHH, play important roles in tumorigenesis (Fig. **4**) [125, 126].

Recently, a progressive alteration of CSCs concept has emerged, describing them as cells more or less differently characterized by a peculiar genotypic and phenotypic plasticity. It is the selective pressure of a specific microenvironment that influences these cells to assume some stem cell-like features and/or attain the capacity to go through an epithelial–mesenchymal transition (EMT), consequently colonizing distant organs and/or tissues [127].

The acquisition of knowledge on these intriguing cancer cell subpopulations depends strictly on the particular methodological approach adopted [128]. A proteomic approach must be the answer because proteomic analyses could significantly contribute to solving the predominant problem related to these particular cancer cells, *i.e.*, their exact definition and, thus, their identification [127]. Proteome-based methods could enhance knowledge about deregulation of pathways as a result of altered expression and activities of proteins in CSCs [129].

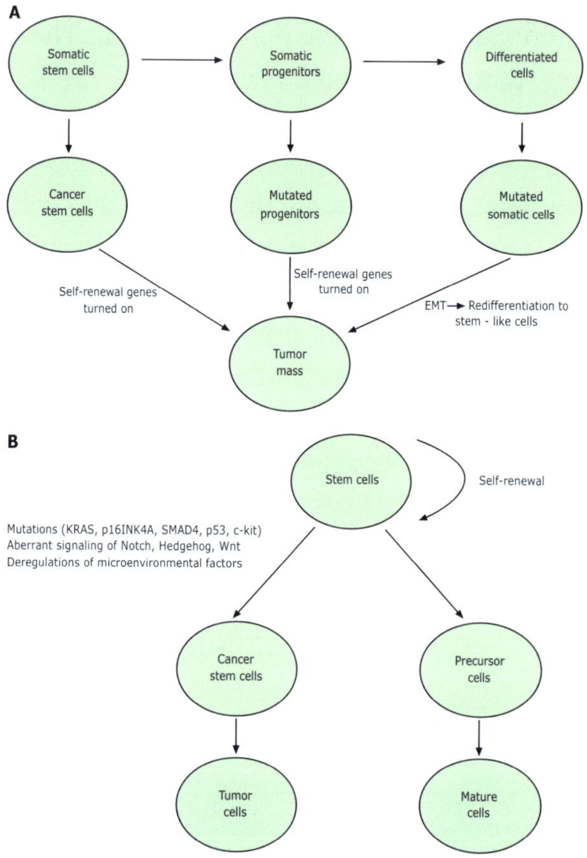

Fig. (4). Models that explain tumor development. Aberrant protein expression, mainly as result of gene mutations (*e.g.* KRAS, p16INK4A, SMAD4, p53, c-kit), and deregulation of signaling pathways (*e.g.* Notch, Hedgehog, Wnt) can contribute to malignant transformation of stem cells (reproduced with permission of [126]).

The capacity to properly identify CSCs has been inaccurate. According to the original remarks on leukemic stem cells, several markers were proposed for various CSCs, but no marker has established to bear an important specificity for

CSCs. Some markers that were proposed for various CSCs: CD133, CD44, CD24, and CD138; ABCG2, ABCB5, epithelial cell adhesion molecule (EpCAM), N-Cadherin, vimentin, integrin αVβ6, GSK-3β, β-Catenin, Sox10, Snail family, twist-related protein 1 (Twist1) and aldehyde-dehydrogenase-1 (ALDH1); CXCR4; several signalling pathways – Hedgehog, Notch, Wnt/ β-catenin, BM1, and PTEN; and different microRNAs [127].

In spite of the up to date technology, a major challenge for proteomics research is represented by large amounts of starting materials required. A sufficient number of CSCs for proteomic analyses can be obtained in a variety of ways: fluorescence (FACS) and magnetic (MACS) activated cell sorting, laser cell capture microdissection, and three-dimensional spheroid/organoid cell culture. These methods to enrich and isolate CSCs can be performed either with or without staining using antibodies against currently known CSC-specific cell surface molecules, such as CD44, CD24, CD133, EpCAM, ALDH1, *etc.* The most important limitation of using antibody-based staining of CSC is that we still do not possess definitive CSCs surface markers. Proteomic methods could be used to study protein profiling of CSCs and to identify novel CSC-specific biomarkers and therapeutic targets [129].

Proteomic methods can be used effectively to clarify which pathways are involved in regulating various intratumoral processes, including activation of CSCs. Based on this point of view, searching and identification of single molecule as biomarkers or therapeutic targets could become possible when CSC-associated pathways are well described and clearly understood due to detailed investigation of the protein patterns in pre-clinical models and clinical samples [129].

There are hundreds of reports regarding the proteomic discoveries in CSCs, but most of the recent reports focus on breast CSCs, digestive cancers and intracranial tumors. CSCs came into the spot light as major players for tumor recurrence, metastasis and resistance to therapy.

Breast Cancer

As it is the leading death cause for women worldwide, CSCs in this type of cancer were subject to intense research. Comparative proteomics performed recently on breast cancer cell line BrCA-MZ-01 has shown that there are differences between cells with ALDH+ phenotype and ALDH-hence mature cancer cell populations. Using 2-Dimensional Difference In-Gel Electrophoresis (2D-DIGE) differentially-expressed spots were obtained and subjected to identification using MALDI-ToF MS. 2D-DIGE has shown that poly(ADP-ribose) polymerase 1 (PARP1) was found overexpressed in ALDH+ cells, CSC type. Moreover, ALDH+ cells with the highest level of PARP1 overexpression were endowed with resistance to olaparib, a specific PARP1 inhibitor. This recent report highlights that upregulated PARP1 in CSCs breast cancer can contribute to clinical resistance to PARP inhibitors [130].

For intimate molecular quantification, a recent report published a nanochannel-based sensor for specific proteomic biomarkers associated with breast cancer CSCs and their potency to metastasize. Four breast cancer cell lines (SUM159 invasive, SUM159 noninvasive, HC1143 and DCIS cell lines) were tested and PDGFR, ALDH1A1 and ALDH1A3 were quantified. The sensor could detect as low as 0.1ng/mL PDGFR, correlating to the detection of unit stem cell count [131].

Using isotope reductive dimethylation and large-scale MS-based quantitative phosphoproteome analysis, it was recently reported the pattern of protein phosphorylation induced by SDF-1/CXCR4 signaling in breast CSCs. Using this proteomic method, over 11000 phosphorylation sites were identified in 2500 phosphoproteins. While over 85% were unchanged as response to SDF-1/CXCR4 stimulation, there were also phosphosites that were found to statistically increase, while other was found to decrease. SDF-1/CXCR4 was reported as increasing phosphorylation in proteins related to migration and invasion. System-based analyses that has examined SDF-1/CXCR4-mediated phosphoproteome, has identified a new pathway - SDF-1/CXCR4-PKA-MAP2K2-ERK. This new pathway has a feedback regulation on MEK, ERK1/2, δ-catenin, and PPP1Cα in SDF-1/CXCR4 signaling in breast CSCs. This study brings besides a new molecular proteomic insight in CSCs, possible new therapy targets [132].

The early detection of breast cancer is a prerequisite for positive clinical outcome, thus recently extracellular vesicles studies have shown the importance of this inter-communicating particles for early diagnostic. Using proteomic approach for primary breast enriched for stem/progenitor cells cultured as mammospheres several proteins contained in these vesicles were identified, their expression was affected by hormonal changes. Secreted extracellular vesicles and their protein content can be potential diagnostic and/or prognostic markers [133].

Another study on extracellular vesicles on cancer breast cell lines (MCF-7 and MDA-MB 231) has identified proteomic profiling of vesicles using liquid chromatography-mass spectrometry (LC-MS/MS) has shown different protein profiles of the vesicles in these cell lines. 27 proteins were commonly found in the vesicles from both cell lines, but vesicles from MDA contain higher amount of matrix-metalloproteinases, in relation to the higher metastatic potential of this cell line. Microarray analysis of extracellular vesicles has shown several oncogenic miRNA in these vesicles. Extracellular vesicles are possible pathways through which signals from malignant cells reach to non-malignant cells and thus could promote tumor progression and metastasis [134].

It has recently been demonstrated that breast CSCs are regulated by complex interactions with components of the tumour microenvironment, including mesenchymal stem cells (MSCs), adipocytes, tumour-associated fibroblasts, endothelial cells, and immune cells, through networks of cytokines and growth factors. Specifically, tumour-associated fibroblasts, macrophages and MSCs secrete IL-6, IL-8 and CXCL7 to activate STAT3/NF-κB signaling, which in turn induces the self-renewal of breast CSCs [127, 135].

Digestive Cancers

The second cancers, in frequency of published studies regarding proteomics of CSCs, are the digestive cancers. Using proteomics biomarkers for gastric CSCs were recently published. CSC-like cell lines (OCUM-12/SP cells, OCUM-2MD3/SP cells, and their parent OCUM-12 cells and OCUM-2MD3) were lysed and analyzed with QSTAR Elite Liquid Chromatography - Tandem Mass Spectrometry, coupled with isobaric tags. After obtaining the candidate proteins, they were validated by immunohistochemical analysis on actual tumors from 300 gastric cancers. Eight proteins were highlighted (RBBP6, GLG1, VPS13A,

DCTPP1, HSPA9, HSPA4, ALDOA, and KRT18) as up-regulated in OCUM-12/SP cells and OCUM-2MD3/SP cells in comparison to parent cells. All these proteins were associated with invasion, lymph node metastasis, distant metastasis, or clinical stage. RBBP6, DCTPP1, HSPA4, and ALDOA were associated in clinical samples with a poor prognosis, while RBBP6 was pinpointed as an independent prognostic factor. These 8 proteins could be new markers for CSC in gastric cancer, as RBBP6 is a promising prognostic biomarker [29].

In colorectal cancer, comparative expression was performed on CD133+ and CD133- cell populations from primary and metastatic tumors. In CD133+ cells, the study reports the up-regulation of 17 genes in comparison to CD133-cells (RNASE2, PRB2, IL-4, MGC27382, CLEC4C, SALL3, GIMAP1, ISG15, LOC728875, ZIK1, ICAM2, CCDC7, CDYL2, LRRC2, ZEB1, OSTF1 and CCDC144B). Interestingly IL-4, as survivin inducer was found up-regulated in colon cancer CSCs. After evaluating survivin in colon cancer samples, over 40% showed an increase associated with primary tumor sites, lymph node metastasis, advanced stages and reduced disease-free survival. CSCs in colon cancer with CD133+ phenotype have a distinct expression pattern where survivin, up-regulated by IL-4, can be a candidate biomarker for recurrence prediction and survival [136].

For hepatocellular carcinoma cells - two clones of Hep3B cell lines were functionally characterized as control or CSC-like cells, based on properties including spheroid formation, drug resistance, and tumor initiation. The protein expression profiles were investigated by iTRAQ, and from a total of 1,127 proteins identified and quantified, only 50 proteins were analyzed by GeneGo being associated with liver neoplasms, hepatocellular carcinoma (HCC), and liver diseases. Finally, the expressions of S100P, S100A14, and vimentin were validated in several HCC cell lines, and their expressions showing a correlation with tumorigenicity in HCC cell lines [137].

Signaling Pathways Involved in Pancreatic Cancer Stem Cells

Several signaling pathways are modified in CSCs and EMT like cells in pancreatic cancer: Hedgehog, Notch, Wnt, AKT and NF-κB. Hedgehog, Notch and Wnt have been shown to be of particular importance in pancreatic cancer stem cells, due to their function in pancreatic embryonic development and differentiation [138]. These signaling pathways play important roles in the self-renewal of CSCs, tumor growth, invasion, metastasis and resistance to therapy [139]. miRNAs were lately considered to have an important role in the regulation of CSCs [117, 126].

Intracranial Tumours

Glioblastoma CSCs are lately accounted for the tumour resistance to radiation and chemotherapy and probably are the main cells that induce rapid tumor recurrence. In glioblastoma studies, neurospheres enriched in CSCs are used for the discovery of new drugs and new therapy targets. Recently a proteomic analysis of primary glioblastoma cell lines as adherent cells or organized as neurospheres was performed. Nano-LC/Q-TOF MS technology showed that over 340 proteins were identified only in neurosphere, proteins involved in metabolic and cellular processes, while over 100 proteins were found only in adherent variants and mainly linked to cell adhesion. In adherent cells a 10 proteins signature was reported; proteins mostly involved in cell adhesion. The proteomic signature of neurospheres was linked to cell cycle and protein metabolism. The identified proteins can have prognostic relevance for glioblastoma and investigating these types of cells, new complementary tools for drug screening are highlighted [140].

Membrane proteomics in CSC of glioblastoma origin was recently implied for identifying new surface antigens in order to recognize, isolate, and target CSCs [141].

In another study, using iTRAQ - tagging combined with two dimensional liquid chromatography tandem MS analysis has observed differentially expressed proteins between glioma stem cells and differentiated tumor cells isolated from the human glioma cell line, U251. S100A9 levels have been found to increase in glioma stem cells than differentiated cells [27].

Glioblastoma stem like cells (GSCs) have some distinctive characteristics such as self-renewal, cellular hierarchy, resistance to current therapies and present similar molecular markers with NSCs which make them different from the other cells within the tumor. The proteome profile of GSCs was evaluated by 2D-DIGE coupled to tandem MS [142]. HDGF was found to be overexpressed in GSCs compared to the whole tissue cells and NSCs, confirmed by blotting analysis. Additionally, *in vivo* experiments showed that GSCs secretome induced neoangiogenesis and this effect was blocked by HDGF targeting siRNAs [142]. This data suggested a significant role for the secreted HDGF in GSCs in tumor formation and progression [143].

Other Cancers

In oral cancer, CSCs were studied using mass spectrometry-based quantitative proteomics. Oral cancer cell lines with high invasive potential (UM1) and reduced invasive (UM2) were subjected to sphere-formation assays. Quantitative proteomic analysis was performed using tandem mass tagging and two-dimensional liquid chromatography with Orbitrap mass spectrometry. Stem cell markers (Sox2, Oct4, Sox9 and CD44) were found up-regulated, as HIF-1 alpha and PGK-1 down-regulated in UM1cells *versus* non-invasive cells. The two cell variants had different expression in many proteins involved in cell cycle, metabolism, G protein signal transduction, translational elongation, development, and RNA splicing pathways. In particular, authors mention CREB-1-binding protein (CBP) and phosphorylated CREB-1 as over-expressed in CSC-like UM1 cells [144].

CSCs targeting agents could become a new approach for the treatment of pancreatic cancer with better outcome. To sum up, specific targeting of pancreatic CSCs and EMT-type cells would definitely open advanced perspectives towards better treatment of this pathology, for which there is no curative therapy [117].

STEM CELLS SECRETOME

Novel methodologies, proteome and secretome approaches can be essential in comprehending the molecular processes regarding the function of stem cells [145].

Secretome represents a complex panel of molecules secreted from cells or shed from the cell surface, being encoded by almost 10% of the human genome [146,

147]. Secretome constitutes a challenging area, especially because of the difficulties in the collection and preparation. Nevertheless, the number of studies increased over the last years, aiming to describe its features; this has been allowed by technological progress in proteomics [148]. There is an increasing effort to analyze stem cell secretome using proteomic technologies, targeting the characterization of various molecules secreted by stem cells and therefore getting insight into the processes of immunomodulation, angiogenesis, cell survival, differentiation and recruitment [149, 150] and generally cellular state at different stages and conditions [151].

We resume the progresses made on secretome analyses of human embryonic stem cells, adult stem cells and cancer stem cells. A comprehensive characterization of the hESC secretome was obtained by specific enhanced MS approach in combination with different fractionation methods resulting in the identification of over 10-12 fold increased extracellular proteins compared to previously published studies [152]. A comparison of the secretome from undifferentiated human Bone Marrow-derived Mesenchymal Stem Cells (hBMSCs) and differentiated osteoblasts, by LC-MS/MS resulted in the identification of 315 proteins [153]. Osteoblasts secretome showed 177 proteins with increased and 88 proteins with decreased level compared to the hBMSCs secretome, results were consequently validated by quantitative RT-PCR and western blot analysis.

Another approach of stem cell secretome analyzed by two-dimensional gel electrophoresis and tandem mass spectrometry, in human subcutaneous adipose-derived stem cells (hASC), which revealed a panel of adipokines such as adiponectin and plasminogen activator inhibitor 1 (PAI-1) [154]. The detection of multiple serine protease inhibitors (serpins) results in a biomarkers panel: PAI-1, pigment epithelium-derived factor, placental thrombin inhibitor, pregnancy zone protein, and protease C1 inhibitor [155]. The analysis of hASC secretome, by LC-MS/MS in relation to TNF-α induction, identified 187 proteins, such as IL-6, IL-8, CXCL6 and MCP-1. Analysis of human multipotent adipose-derived stem cells (hMADS) secretome at an initial stage of commitment to adipocytes and osteoblasts led to the identification of 73 proteins [156], an 8 proteins set being validated by western blot analysis.

The CSCs inside the tumor are able to self-renew and exhibit the capacity to generate heterogeneous populations of cancer cells [143]. Glioblastoma stem like cells (GSCs) present unique features such as self-renewal, cellular hierarchy, resistance to current therapies, thus leading to a difference from the other cells within the tumor [142, 157, 158]. Common molecular markers are similar with NSCs. A comparative proteome profile of GSCs from tumor tissues and hNSCs was analyzed by 2D-DIGE coupled to tandem MS [142], HDGF being upregulated in GSCs compared to the tumor tissue cells and NSCs. HDGF was found exclusively in the secretome of GSCs, validated by western blot. *In vivo* experiments further showed that GSCs secretome induced neoangiogenesis, this effect being blocked by HDGF targeting siRNAs [142]; therefore, an important role for the secreted HDGF in GSCs involvement in tumor formation and progression has been established [143].

Stem cell secretome including cytokines, chemokines and growth factors, became increasingly popular over the last years due to its various implications in regenerative medicine [159, 160]. Stem cells and particularly MSCs regulate immune response and have a contribution to regenerative processes in several pathological conditions; their major mechanism in tissue repair is related to their paracrine activity. The recent studies point that the regenerative process is mediated by secretome in combination with secreted vesicles [159]. Progress in proteomics technology has improved secretome analysis [148], by identifying secreted factors particularly at nanoscale levels [152], as well as adoption of an alternative methodology targeting the organelles of the secretory pathway. It holds promise for overcoming subsequent technical difficulties in secretome analyses (for instance, the need for serum-free culture conditions) [161]. The existing evidence shows great promise and it contributes to the area of regenerative medicine [162].

BIOINFORMATICS – OMICS DATA ANALYSIS

Databases for stem cell omics data encourage researchers all over the world to present their experimental stem cell data given the progress of high-throughput approaches, such as RNA sequencing and protein studies, resulting in increasing developments in bioinformatics for further analysis. Computational methods and commercial software packages are continuously developed. As a consequence, an increasingly comprehensive view of the molecular processes underlying self-renewal, proliferation, and differentiation of stem cells and other kinds of cell lines at different regulatory levels is obtained based on comparisons of omics data [41].

The future of genomic and proteomic technologies holds tremendous expectations. Yet, computational integration is needed so that transcriptomic and proteomic data achieve their optimal potential. Computational algorithms and software packages including multiple-dimension experimental data sets for stem cells to construct their models and regulation networks have been recently developed. Integrated mRNA and microRNA expression profiles represent the general strategy for correlation analysis. There are various versions of software that perform statistical tests to predict mRNA targets for each differentially expressed miRNA and to identify whether the number of predicted target mRNAs that are differentially expressed is higher than that expected by chance ($P<0.01/0.05$). More, bioinformatics databases and software are used in gene ontology (GO) analysis and network analysis [163].

Comprehensive quality data analysis of proteome and transcriptome is used increasingly more, the two data sets being complementary rather than comparable [164, 165]. Mapping for short reads of transcriptome raw data and amino acid sequences from proteome data to the reference genome is still needed. Presence and qualities of transcripts and their corresponding protein products depend on a series of post-transcriptional regulation and metabolic processes [164]. So the overlaps of the two data sets are not expected too much. Where the differences are present, we may find the post-transcriptional regulation and metabolic processes occur and the actions between transcription and translation will be figured out in the near future. Moreover, Unwin *et al.* found the proteome and transcriptome tend to shift in same direction by a comparison of information on great numbers of mRNA transcripts and the levels of expression of their corresponding proteins in dynamic systems of primary hematopoietic stem cells [166]. Several laboratories have used other omics data sets to perform a comprehensive analysis of stem cell data. To analyze transcriptome and epigenomic data altogether, Xu *et al.* developed a classifier to predict self-renewal and pluripotency of mESCs stemness membership genes, using support vector machines [167]. The Stem Cell Discovery Engine (SCDE), a new platform for analysis of multiomics data, has allowed the users to consistently describe, share, and compare multiomics information at the gene and pathway level [41, 168].

CONCLUSION

Omics research of stem cells has substantially progressed by the development and improvement of experimental techniques and computational methods, leading to major discoveries in stem cell biology and emphasizing the potential of stem cells for biomedicine, as well as the consciousness that proteomics can accelerate these advances and might open unexplored areas of research.

The proteomics analyses of embryonic, adult and/or cancer stem cells will improve the description of these cellular phenotypes.

The integration of proteomic studies into the entire spectrum of biological and biomedical research of these cells in this field represents a true challenge.

Depicting the proteome and interactome of stem cells by proteome profile identification, protein quantitation and protein interaction networks will provide a valuable insight in the involvement of stem cell both in regeneration and in tumorigenesis.

Combined advanced technologies in stem cell biology: proteomics, signalomics, secretome analysis, interactomics, bio-informatics will provide a perspective view for multilayered omics data in understanding the processes regulating stem cell self-renewal and potency.

In conclusion, we anticipate that all of these efforts will lead to the comprehension of the underlying biology of stem cells for full clinical therapeutic advantages of these unique cells, either in regeneration or in tumor pathology.

CONFLICT OF INTEREST

The authors confirm that they have no conflict of interest to declare for this publication.

ACKNOWLEDGEMENTS

This work was partially supported by PNII 93/2012, PNII 192/2014, 4SEE/2014, PNII 265/2014, PN 16.22.04.01, PN-II-RU-TE-2014-4-1534, SMIS P_40_197.

REFERENCES

[1] Wang J, Trowbridge JJ, Rao S, *et al.* Proteomic studies of stem cells. Cambridge, MA: StemBook 2008.

[2] Evsikov AV, Solter D. Comment on " 'Stemness': transcriptional profiling of embryonic and adult stem cells" and "a stem cell molecular signature". Science 2003; 302(5644): 393. [http://dx.doi.org/10.1126/science.1082380] [PMID: 14563991]

[3] Fortunel NO, Otu HH, Ng HH, *et al.* Comment on " 'Stemness': transcriptional profiling of embryonic and adult stem cells" and "a stem cell molecular signature". Science 2003; 302(5644): 393. [http://dx.doi.org/10.1126/science.1086384] [PMID: 14563990]

[4] Ivanova NB, Dimos JT, Schaniel C, Hackney JA, Moore KA, Lemischka IR. A stem cell molecular signature. Science 2002; 298(5593): 601-4. [http://dx.doi.org/10.1126/science.1073823] [PMID: 12228721]

[5] Ramalho-Santos M, Yoon S, Matsuzaki Y, Mulligan RC, Melton DA. "Stemness": transcriptional profiling of embryonic and adult stem cells. Science 2002; 298(5593): 597-600. [http://dx.doi.org/10.1126/science.1072530] [PMID: 12228720]

[6] Gygi SP, Rist B, Gerber SA, Turecek F, Gelb MH, Aebersold R. Quantitative analysis of complex protein mixtures using isotope-coded affinity tags. Nat Biotechnol 1999; 17(10): 994-9. [http://dx.doi.org/10.1038/13690] [PMID: 10504701]

[7] Wilkins MR, Sanchez JC, Gooley AA, *et al.* Progress with proteome projects: why all proteins expressed by a genome should be identified and how to do it. Biotechnol Genet Eng Rev 1996; 13: 19-50. [http://dx.doi.org/10.1080/02648725.1996.10647923] [PMID: 8948108]

[8] Rual JF, Venkatesan K, Hao T, *et al.* Towards a proteome-scale map of the human protein-protein interaction network. Nature 2005; 437(7062): 1173-8. [http://dx.doi.org/10.1038/nature04209] [PMID: 16189514]

[9] MacBeath G. Protein microarrays and proteomics. Nat Genet 2002; 32 (Suppl.): 526-32. [http://dx.doi.org/10.1038/ng1037] [PMID: 12454649]

[10] Stevens RC, Yokoyama S, Wilson IA. Global efforts in structural genomics. Science 2001; 294(5540): 89-92. [http://dx.doi.org/10.1126/science.1066011] [PMID: 11588249]

[11] Ahn SM, Simpson R, Lee B. Genomics and proteomics in stem cell research: the road ahead. Anat Cell Biol 2010; 43(1): 1-14. [http://dx.doi.org/10.5115/acb.2010.43.1.1] [PMID: 21190000]

[12] Callinan PA, Feinberg AP. The emerging science of epigenomics. Hum Mol Genet 2006; 15(Spec No

1): R95-R101.
[http://dx.doi.org/10.1093/hmg/ddl095] [PMID: 16651376]

[13] Schneider MV, Orchard S. Omics technologies, data and bioinformatics principles. Methods Mol Biol 2011; 719: 3-30.
[http://dx.doi.org/10.1007/978-1-61779-027-0_1] [PMID: 21370077]

[14] Stanton LW, Bakre MM. Genomic and proteomic characterization of embryonic stem cells. Curr Opin Chem Biol 2007; 11(4): 399-404.
[http://dx.doi.org/10.1016/j.cbpa.2007.05.029] [PMID: 17646122]

[15] Yashiro M, Nishii T, Hasegawa T, *et al.* A c-Met inhibitor increases the chemosensitivity of cancer stem cells to the irinotecan in gastric carcinoma. Br J Cancer 2013; 109(10): 2619-28.
[http://dx.doi.org/10.1038/bjc.2013.638] [PMID: 24129235]

[16] Nishii T, Yashiro M, Shinto O, Sawada T, Ohira M, Hirakawa K. Cancer stem cell-like SP cells have a high adhesion ability to the peritoneum in gastric carcinoma. Cancer Sci 2009; 100(8): 1397-402.
[http://dx.doi.org/10.1111/j.1349-7006.2009.01211.x] [PMID: 19493275]

[17] Glinsky GV, Berezovska O, Glinskii AB. Microarray analysis identifies a death-from-cancer signature predicting therapy failure in patients with multiple types of cancer. J Clin Invest 2005; 115(6): 1503-21.
[http://dx.doi.org/10.1172/JCI23412] [PMID: 15931389]

[18] Clarke MF, Dick JE, Dirks PB, *et al.* Cancer stem cells--perspectives on current status and future directions: AACR Workshop on cancer stem cells. Cancer Res 2006; 66(19): 9339-44.
[http://dx.doi.org/10.1158/0008-5472.CAN-06-3126] [PMID: 16990346]

[19] Bonnet D, Dick JE. Human acute myeloid leukemia is organized as a hierarchy that originates from a primitive hematopoietic cell. Nat Med 1997; 3(7): 730-7.
[http://dx.doi.org/10.1038/nm0797-730] [PMID: 9212098]

[20] Takaishi S, Okumura T, Wang TC. Gastric cancer stem cells. J Clin Oncol 2008; 26(17): 2876-82.
[http://dx.doi.org/10.1200/JCO.2007.15.2603] [PMID: 18539967]

[21] Stojnev S, Krstic M, Ristic-Petrovic A, Stefanovic V, Hattori T. Gastric cancer stem cells: therapeutic targets. Gastric Cancer 2014; 17(1): 13-25.
[http://dx.doi.org/10.1007/s10120-013-0254-x] [PMID: 23563919]

[22] Hartgrink HH, Jansen EP, van Grieken NC, van de Velde CJ. Gastric cancer. Lancet 2009; 374(9688): 477-90.
[http://dx.doi.org/10.1016/S0140-6736(09)60617-6] [PMID: 19625077]

[23] Charafe-Jauffret E, Ginestier C, Iovino F, *et al.* Aldehyde dehydrogenase 1-positive cancer stem cells mediate metastasis and poor clinical outcome in inflammatory breast cancer. Clin Cancer Res 2010; 16(1): 45-55.
[http://dx.doi.org/10.1158/1078-0432.CCR-09-1630] [PMID: 20028757]

[24] Al-Hajj M, Wicha MS, Benito-Hernandez A, Morrison SJ, Clarke MF. Prospective identification of tumorigenic breast cancer cells. Proc Natl Acad Sci USA 2003; 100(7): 3983-8.
[http://dx.doi.org/10.1073/pnas.0530291100] [PMID: 12629218]

[25] Li C, Heidt DG, Dalerba P, *et al.* Identification of pancreatic cancer stem cells. Cancer Res 2007;

67(3): 1030-7.
[http://dx.doi.org/10.1158/0008-5472.CAN-06-2030] [PMID: 17283135]

[26] Ricci-Vitiani L, Lombardi DG, Pilozzi E, *et al.* Identification and expansion of human colon-cance--initiating cells. Nature 2007; 445(7123): 111-5.
[http://dx.doi.org/10.1038/nature05384] [PMID: 17122771]

[27] Chen S, Zhao H, Deng J, Liao P, Xu Z, Cheng Y. Comparative proteomics of glioma stem cells and differentiated tumor cells identifies S100A9 as a potential therapeutic target. J Cell Biochem 2013; 114(12): 2795-808.
[http://dx.doi.org/10.1002/jcb.24626] [PMID: 23836528]

[28] Kakehashi A, Ishii N, Shibata T, *et al.* Mitochondrial prohibitins and septin 9 are implicated in the onset of rat hepatocarcinogenesis. Toxicol Sci 2011; 119(1): 61-72.
[http://dx.doi.org/10.1093/toxsci/kfq307] [PMID: 20935162]

[29] Morisaki T, Yashiro M, Kakehashi A, *et al.* Comparative proteomics analysis of gastric cancer stem cells. PLoS One 2014; 9(11): e110736.
[http://dx.doi.org/10.1371/journal.pone.0110736] [PMID: 25379943]

[30] Morrison SJ, Kimble J. Asymmetric and symmetric stem-cell divisions in development and cancer. Nature 2006; 441(7097): 1068-74.
[http://dx.doi.org/10.1038/nature04956] [PMID: 16810241]

[31] Klimanskaya I, Rosenthal N, Lanza R. Derive and conquer: sourcing and differentiating stem cells for therapeutic applications. Nat Rev Drug Discov 2008; 7(2): 131-42.
[http://dx.doi.org/10.1038/nrd2403] [PMID: 18079756]

[32] Pappa KI, Anagnou NP. Novel sources of fetal stem cells: where do they fit on the developmental continuum? Regen Med 2009; 4(3): 423-33.
[http://dx.doi.org/10.2217/rme.09.12] [PMID: 19438317]

[33] Wang Z, Gerstein M, Snyder M. RNA-Seq: a revolutionary tool for transcriptomics. Nat Rev Genet 2009; 10(1): 57-63.
[http://dx.doi.org/10.1038/nrg2484] [PMID: 19015660]

[34] Efroni S, Duttagupta R, Cheng J, *et al.* Global transcription in pluripotent embryonic stem cells. Cell Stem Cell 2008; 2(5): 437-47.
[http://dx.doi.org/10.1016/j.stem.2008.03.021] [PMID: 18462694]

[35] Chin MH, Mason MJ, Xie W, *et al.* Induced pluripotent stem cells and embryonic stem cells are distinguished by gene expression signatures. Cell Stem Cell 2009; 5(1): 111-23.
[http://dx.doi.org/10.1016/j.stem.2009.06.008] [PMID: 19570518]

[36] Bhattacharya B, Miura T, Brandenberger R, *et al.* Gene expression in human embryonic stem cell lines: unique molecular signature. Blood 2004; 103(8): 2956-64.
[http://dx.doi.org/10.1182/blood-2003-09-3314] [PMID: 15070671]

[37] Zhan M. Genomic studies to explore self-renewal and differentiation properties of embryonic stem cells. Front Biosci 2008; 13: 276-83.
[http://dx.doi.org/10.2741/2678] [PMID: 17981546]

[38] He L, Hannon GJ. MicroRNAs: small RNAs with a big role in gene regulation. Nat Rev Genet 2004;

5(7): 522-31.
[http://dx.doi.org/10.1038/nrg1379] [PMID: 15211354]

[39] Suh MR, Lee Y, Kim JY, *et al.* Human embryonic stem cells express a unique set of microRNAs. Dev Biol 2004; 270(2): 488-98.
[http://dx.doi.org/10.1016/j.ydbio.2004.02.019] [PMID: 15183728]

[40] Gangaraju VK, Lin H. MicroRNAs: key regulators of stem cells. Nat Rev Mol Cell Biol 2009; 10(2): 116-25.
[http://dx.doi.org/10.1038/nrm2621] [PMID: 19165214]

[41] Wang H, Zhang Q, Fang X. Transcriptomics and proteomics in stem cell research. Front Med 2014; 8(4): 433-44.
[http://dx.doi.org/10.1007/s11684-014-0336-0] [PMID: 24972645]

[42] Mattick JS. A new paradigm for developmental biology. J Exp Biol 2007; 210(Pt 9): 1526-47.
[http://dx.doi.org/10.1242/jeb.005017] [PMID: 17449818]

[43] Baharvand H, Fathi A, van Hoof D, Salekdeh GH. Concise review: trends in stem cell proteomics. Stem Cells 2007; 25(8): 1888-903.
[http://dx.doi.org/10.1634/stemcells.2007-0107] [PMID: 17495109]

[44] Nasrabadi D, Rezaei Larijani M, Pirhaji L, *et al.* Proteomic analysis of monkey embryonic stem cell during differentiation. J Proteome Res 2009; 8(3): 1527-39.
[http://dx.doi.org/10.1021/pr800880v] [PMID: 19226164]

[45] Yan L, Yang M, Guo H, *et al.* Single-cell RNA-Seq profiling of human preimplantation embryos and embryonic stem cells. Nat Struct Mol Biol 2013; 20(9): 1131-9.
[http://dx.doi.org/10.1038/nsmb.2660] [PMID: 23934149]

[46] Sun Y, Yang Y, Zeng S, Tan Y, Lu G, Lin G. Identification of proteins related to epigenetic regulation in the malignant transformation of aberrant karyotypic human embryonic stem cells by quantitative proteomics. PLoS One 2014; 9(1): e85823.
[http://dx.doi.org/10.1371/journal.pone.0085823] [PMID: 24465727]

[47] D'Aguanno S, Barcaroli D, Rossi C, *et al.* p63 isoforms regulate metabolism of cancer stem cells. J Proteome Res 2014; 13(4): 2120-36.
[http://dx.doi.org/10.1021/pr4012574] [PMID: 24597989]

[48] Lin H, Lee E, Hestir K, *et al.* Discovery of a cytokine and its receptor by functional screening of the extracellular proteome. Science 2008; 320(5877): 807-11.
[http://dx.doi.org/10.1126/science.1154370] [PMID: 18467591]

[49] Gonzalez R, Jennings LL, Knuth M, *et al.* Screening the mammalian extracellular proteome for regulators of embryonic human stem cell pluripotency. Proc Natl Acad Sci USA 2010; 107(8): 3552-7.
[http://dx.doi.org/10.1073/pnas.0914019107] [PMID: 20133595]

[50] Gemei M, Corbo C, D'Alessio F, Di Noto R, Vento R, Del Vecchio L. Surface proteomic analysis of differentiated *versus* stem-like osteosarcoma human cells. Proteomics 2013; 13(22): 3293-7.
[http://dx.doi.org/10.1002/pmic.201300170] [PMID: 24106197]

[51] Stoevesandt O, Taussig MJ, He M. Protein microarrays: high-throughput tools for proteomics. Expert Rev Proteomics 2009; 6(2): 145-57.

[http://dx.doi.org/10.1586/epr.09.2] [PMID: 19385942]

[52] Novak A, Amit M, Ziv T, *et al.* Proteomics profiling of human embryonic stem cells in the early differentiation stage. Stem Cell Rev 2012; 8(1): 137-49. [http://dx.doi.org/10.1007/s12015-011-9286-y] [PMID: 21732092]

[53] Prelle K, Zink N, Wolf E. Pluripotent stem cells--model of embryonic development, tool for gene targeting, and basis of cell therapy. Anat Histol Embryol 2002; 31(3): 169-86. [http://dx.doi.org/10.1046/j.1439-0264.2002.00388.x] [PMID: 12479360]

[54] Van Hoof D, Passier R, Ward-Van Oostwaard D, *et al.* A quest for human and mouse embryonic stem cell-specific proteins. Mol Cell Proteomics 2006; 5(7): 1261-73. [http://dx.doi.org/10.1074/mcp.M500405-MCP200] [PMID: 16600995]

[55] Graumann J, Hubner NC, Kim JB, *et al.* Stable isotope labeling by amino acids in cell culture (SILAC) and proteome quantitation of mouse embryonic stem cells to a depth of 5,111 proteins. Mol Cell Proteomics 2008; 7(4): 672-83. [http://dx.doi.org/10.1074/mcp.M700460-MCP200] [PMID: 18045802]

[56] Mikkelsen TS, Ku M, Jaffe DB, *et al.* Genome-wide maps of chromatin state in pluripotent and lineage-committed cells. Nature 2007; 448(7153): 553-60. [http://dx.doi.org/10.1038/nature06008] [PMID: 17603471]

[57] Wang J, Rao S, Chu J, *et al.* A protein interaction network for pluripotency of embryonic stem cells. Nature 2006; 444(7117): 364-8. [http://dx.doi.org/10.1038/nature05284] [PMID: 17093407]

[58] Boyer LA, Plath K, Zeitlinger J, *et al.* Polycomb complexes repress developmental regulators in murine embryonic stem cells. Nature 2006; 441(7091): 349-53. [http://dx.doi.org/10.1038/nature04733] [PMID: 16625203]

[59] Kim J, Chu J, Shen X, Wang J, Orkin SH. An extended transcriptional network for pluripotency of embryonic stem cells. Cell 2008; 132(6): 1049-61. [http://dx.doi.org/10.1016/j.cell.2008.02.039] [PMID: 18358816]

[60] Lewitzky M, Yamanaka S. Reprogramming somatic cells towards pluripotency by defined factors. Curr Opin Biotechnol 2007; 18(5): 467-73. [http://dx.doi.org/10.1016/j.copbio.2007.09.007] [PMID: 18024106]

[61] Unwin RD, Smith DL, Blinco D, *et al.* Quantitative proteomics reveals posttranslational control as a regulatory factor in primary hematopoietic stem cells. Blood 2006; 107(12): 4687-94. [http://dx.doi.org/10.1182/blood-2005-12-4995] [PMID: 16507774]

[62] Spooncer E, Brouard N, Nilsson SK, *et al.* Developmental fate determination and marker discovery in hematopoietic stem cell biology using proteomic fingerprinting. Mol Cell Proteomics 2008; 7(3): 573-81. [http://dx.doi.org/10.1074/mcp.M700292-MCP200] [PMID: 18083999]

[63] Salasznyk RM, Klees RF, Westcott AM, Vandenberg S, Bennett K, Plopper GE. Focusing of gene expression as the basis of stem cell differentiation. Stem Cells Dev 2005; 14(6): 608-20. [http://dx.doi.org/10.1089/scd.2005.14.608] [PMID: 16433616]

[64] Kratchmarova I, Blagoev B, Haack-Sorensen M, Kassem M, Mann M. Mechanism of divergent

growth factor effects in mesenchymal stem cell differentiation. Science 2005; 308(5727): 1472-7. [http://dx.doi.org/10.1126/science.1107627] [PMID: 15933201]

[65] Lessard J, Wu JI, Ranish JA, *et al.* An essential switch in subunit composition of a chromatin remodeling complex during neural development. Neuron 2007; 55(2): 201-15. [http://dx.doi.org/10.1016/j.neuron.2007.06.019] [PMID: 17640523]

[66] Kiel MJ, Yilmaz OH, Iwashita T, Yilmaz OH, Terhorst C, Morrison SJ. SLAM family receptors distinguish hematopoietic stem and progenitor cells and reveal endothelial niches for stem cells. Cell 2005; 121(7): 1109-21. [http://dx.doi.org/10.1016/j.cell.2005.05.026] [PMID: 15989959]

[67] Williamson AJ, Smith DL, Blinco D, *et al.* Quantitative proteomics analysis demonstrates post-transcriptional regulation of embryonic stem cell differentiation to hematopoiesis. Mol Cell Proteomics 2008; 7(3): 459-72. [http://dx.doi.org/10.1074/mcp.M700370-MCP200] [PMID: 18045800]

[68] van Hoof D, Krijgsveld J, Mummery C. Proteomic analysis of stem cell differentiation and early development. Cold Spring Harb Perspect Biol 2012; 4(3): a008177. [http://dx.doi.org/10.1101/cshperspect.a008177] [PMID: 22317846]

[69] Van Hoof D, Heck AJ, Krijgsveld J, Mummery CL. Proteomics and human embryonic stem cells. Stem Cell Res (Amst) 2008; 1(3): 169-82. [http://dx.doi.org/10.1016/j.scr.2008.05.003] [PMID: 19383398]

[70] Chaerkady R, Kerr CL, Kandasamy K, Marimuthu A, Gearhart JD, Pandey A. Comparative proteomics of human embryonic stem cells and embryonal carcinoma cells. Proteomics 2010; 10(7): 1359-73. [http://dx.doi.org/10.1002/pmic.200900483] [PMID: 20104618]

[71] Osman AM, van Dartel DA, Zwart E, Blokland M, Pennings JL, Piersma AH. Proteome profiling of mouse embryonic stem cells to define markers for cell differentiation and embryotoxicity. Reprod Toxicol 2010; 30(2): 322-32. [http://dx.doi.org/10.1016/j.reprotox.2010.05.084] [PMID: 20553848]

[72] Maurer MH. Proteomic definitions of mesenchymal stem cells. Stem Cells Int 2011 2011. 2011:704256. Epub 2011/03/26. [http://dx.doi.org/10.4061/2011/704256]

[73] Van Hoof D, Muñoz J, Braam SR, *et al.* Phosphorylation dynamics during early differentiation of human embryonic stem cells. Cell Stem Cell 2009; 5(2): 214-26. [http://dx.doi.org/10.1016/j.stem.2009.05.021] [PMID: 19664995]

[74] Tian R, Wang S, Elisma F, *et al.* Rare cell proteomic reactor applied to stable isotope labeling by amino acids in cell culture (SILAC)-based quantitative proteomics study of human embryonic stem cell differentiation. Mol Cell Proteomics 2011; 10(2) M110 000679. Epub 2010/06/10

[75] Brill LM, Xiong W, Lee KB, *et al.* Phosphoproteomic analysis of human embryonic stem cells. Cell Stem Cell 2009; 5(2): 204-13. [http://dx.doi.org/10.1016/j.stem.2009.06.002] [PMID: 19664994]

[76] Phanstiel DH, Brumbaugh J, Wenger CD, *et al.* Proteomic and phosphoproteomic comparison of

human ES and iPS cells. Nat Methods 2011; 8(10): 821-7.
[http://dx.doi.org/10.1038/nmeth.1699] [PMID: 21983960]

[77] Pardo M, Lang B, Yu L, *et al.* An expanded Oct4 interaction network: implications for stem cell biology, development, and disease. Cell Stem Cell 2010; 6(4): 382-95.
[http://dx.doi.org/10.1016/j.stem.2010.03.004] [PMID: 20362542]

[78] van den Berg DL, Snoek T, Mullin NP, *et al.* An Oct4-centered protein interaction network in embryonic stem cells. Cell Stem Cell 2010; 6(4): 369-81.
[http://dx.doi.org/10.1016/j.stem.2010.02.014] [PMID: 20362541]

[79] Bonardi F, Fusetti F, Deelen P, van Gosliga D, Vellenga E, Schuringa JJ. A proteomics and transcriptomics approach to identify leukemic stem cell (LSC) markers. Mol Cell Proteomics 2013; 12(3): 626-37.
[http://dx.doi.org/10.1074/mcp.M112.021931] [PMID: 23233446]

[80] Kristensen LP, Chen L, Nielsen MO, *et al.* Temporal profiling and pulsed SILAC labeling identify novel secreted proteins during *ex vivo* osteoblast differentiation of human stromal stem cells. Mol Cell Proteomics 2012; 11(10): 989-1007.
[http://dx.doi.org/10.1074/mcp.M111.012138] [PMID: 22801418]

[81] Tian R. Exploring intercellular signaling by proteomic approaches. Proteomics 2014; 14(4-5): 498-512.
[http://dx.doi.org/10.1002/pmic.201300259] [PMID: 24214021]

[82] Dreesen O, Brivanlou AH. Signaling pathways in cancer and embryonic stem cells. Stem Cell Rev 2007; 3(1): 7-17.
[http://dx.doi.org/10.1007/s12015-007-0004-8] [PMID: 17873377]

[83] Karamboulas C, Ailles L. Developmental signaling pathways in cancer stem cells of solid tumors. Biochim Biophys Acta 2013; 1830(2): 2481-95.
[http://dx.doi.org/10.1016/j.bbagen.2012.11.008] [PMID: 23196196]

[84] Takebe N, Harris PJ, Warren RQ, Ivy SP. Targeting cancer stem cells by inhibiting Wnt, Notch, and Hedgehog pathways. Nat Rev Clin Oncol 2011; 8(2): 97-106.
[http://dx.doi.org/10.1038/nrclinonc.2010.196] [PMID: 21151206]

[85] An SM, Ding QP, Li LS. Stem cell signaling as a target for novel drug discovery: recent progress in the WNT and Hedgehog pathways. Acta Pharmacol Sin 2013; 34(6): 777-83.
[http://dx.doi.org/10.1038/aps.2013.64] [PMID: 23708555]

[86] Nguyen DX, Chiang AC, Zhang XH, *et al.* WNT/TCF signaling through LEF1 and HOXB9 mediates lung adenocarcinoma metastasis. Cell 2009; 138(1): 51-62.
[http://dx.doi.org/10.1016/j.cell.2009.04.030] [PMID: 19576624]

[87] Yang J, Weinberg RA. Epithelial-mesenchymal transition: at the crossroads of development and tumor metastasis. Dev Cell 2008; 14(6): 818-29.
[http://dx.doi.org/10.1016/j.devcel.2008.05.009] [PMID: 18539112]

[88] Yu M, Ting DT, Stott SL, *et al.* RNA sequencing of pancreatic circulating tumour cells implicates WNT signalling in metastasis. Nature 2012; 487(7408): 510-3.
[http://dx.doi.org/10.1038/nature11217] [PMID: 22763454]

[89] Zhao C, Chen A, Jamieson CH, *et al.* Hedgehog signalling is essential for maintenance of cancer stem cells in myeloid leukaemia. Nature 2009; 458(7239): 776-9. [http://dx.doi.org/10.1038/nature07737] [PMID: 19169242]

[90] Schweizer L, Zhang L. Enhancing Cancer Drug Discovery through Novel Cell Signaling Pathway Panel Strategy. Cancer Growth Metastasis 2013; 6: 53-9. [http://dx.doi.org/10.4137/CGM.S11134] [PMID: 24665207]

[91] Weng AP, Ferrando AA, Lee W, *et al.* Activating mutations of NOTCH1 in human T cell acute lymphoblastic leukemia. Science 2004; 306(5694): 269-71. [http://dx.doi.org/10.1126/science.1102160] [PMID: 15472075]

[92] Wang NJ, Sanborn Z, Arnett KL, *et al.* Loss-of-function mutations in Notch receptors in cutaneous and lung squamous cell carcinoma. Proc Natl Acad Sci USA 2011; 108(43): 17761-6. [http://dx.doi.org/10.1073/pnas.1114669108] [PMID: 22006338]

[93] An SM, Ding Q, Zhang J, Xie J, Li L. Targeting stem cell signaling pathways for drug discovery: advances in the Notch and Wnt pathways. Sci China Life Sci 2014; 57(6): 575-80. [http://dx.doi.org/10.1007/s11427-014-4665-7] [PMID: 24829106]

[94] Duhagon MA, Crea F, Danesi R, Farrar WL. Embryonic stem cell pathways and chemotherapy response: an unexplored route. Ann Oncol 2013; 24(2): 558-9. [http://dx.doi.org/10.1093/annonc/mds610] [PMID: 23233652]

[95] Dai B, Rasmussen TP. Global epiproteomic signatures distinguish embryonic stem cells from differentiated cells. Stem Cells 2007; 25(10): 2567-74. [http://dx.doi.org/10.1634/stemcells.2007-0131] [PMID: 17641388]

[96] Swaney DL, Wenger CD, Thomson JA, Coon JJ. Human embryonic stem cell phosphoproteome revealed by electron transfer dissociation tandem mass spectrometry. Proc Natl Acad Sci USA 2009; 106(4): 995-1000. [http://dx.doi.org/10.1073/pnas.0811964106] [PMID: 19144917]

[97] Rigbolt KT, Prokhorova TA, Akimov V, *et al.* System-wide temporal characterization of the proteome and phosphoproteome of human embryonic stem cell differentiation. Sci Signal 2011; 4(164): rs3. [http://dx.doi.org/10.1126/scisignal.2001570] [PMID: 21406692]

[98] Prudhomme W, Daley GQ, Zandstra P, Lauffenburger DA. Multivariate proteomic analysis of murine embryonic stem cell self-renewal *versus* differentiation signaling. Proc Natl Acad Sci USA 2004; 101(9): 2900-5. [http://dx.doi.org/10.1073/pnas.0308768101] [PMID: 14978270]

[99] Reiland S, Salekdeh GH, Krijgsveld J. Defining pluripotent stem cells through quantitative proteomic analysis. Expert Rev Proteomics 2011; 8(1): 29-42. [http://dx.doi.org/10.1586/epr.10.100] [PMID: 21329426]

[100] Levchenko A. Proteomics takes stem cell analyses to another level. Nat Biotechnol 2005; 23(7): 828-30. [http://dx.doi.org/10.1038/nbt0705-828] [PMID: 16003369]

[101] Alberts B. The cell as a collection of protein machines: preparing the next generation of molecular biologists. Cell 1998; 92(3): 291-4.

[http://dx.doi.org/10.1016/S0092-8674(00)80922-8] [PMID: 9476889]

[102] Wang J, Orkin SH. A protein roadmap to pluripotency and faithful reprogramming. Cells Tissues Organs (Print) 2008; 188(1-2): 23-30.
[http://dx.doi.org/10.1159/000113532] [PMID: 18196921]

[103] Yousefi M, Hajihoseini V, Jung W, *et al.* Embryonic stem cell interactomics: the beginning of a long road to biological function. Stem Cell Rev 2012; 8(4): 1138-54.
[http://dx.doi.org/10.1007/s12015-012-9400-9] [PMID: 22847281]

[104] Ciruela F. Fluorescence-based methods in the study of protein-protein interactions in living cells. Curr Opin Biotechnol 2008; 19(4): 338-43.
[http://dx.doi.org/10.1016/j.copbio.2008.06.003] [PMID: 18602005]

[105] Zhang Z, Liao B, Xu M, Jin Y. Post-translational modification of POU domain transcription factor Oct-4 by SUMO-1. FASEB J 2007; 21(12): 3042-51.
[http://dx.doi.org/10.1096/fj.06-6914com] [PMID: 17496161]

[106] Dejosez M, Levine SS, Frampton GM, *et al.* Ronin/Hcf-1 binds to a hyperconserved enhancer element and regulates genes involved in the growth of embryonic stem cells. Genes Dev 2010; 24(14): 1479-84.
[http://dx.doi.org/10.1101/gad.1935210] [PMID: 20581084]

[107] Rigaut G, Shevchenko A, Rutz B, Wilm M, Mann M, Séraphin B. A generic protein purification method for protein complex characterization and proteome exploration. Nat Biotechnol 1999; 17(10): 1030-2.
[http://dx.doi.org/10.1038/13732] [PMID: 10504710]

[108] Kerppola TK. Visualization of molecular interactions by fluorescence complementation. Nat Rev Mol Cell Biol 2006; 7(6): 449-56.
[http://dx.doi.org/10.1038/nrm1929] [PMID: 16625152]

[109] Leuchowius KJ, Weibrecht I, Landegren U, Gedda L, Söderberg O. Flow cytometric *in situ* proximity ligation analyses of protein interactions and post-translational modification of the epidermal growth factor receptor family. Cytometry A 2009; 75(10): 833-9.
[http://dx.doi.org/10.1002/cyto.a.20771] [PMID: 19650109]

[110] Johansson H, Simonsson S. Core transcription factors, Oct4, Sox2 and Nanog, individually form complexes with nucleophosmin (Npm1) to control embryonic stem (ES) cell fate determination. Aging (Albany, NY) 2010; 2(11): 815-22.
[http://dx.doi.org/10.18632/aging.100222] [PMID: 21076177]

[111] Smith MG, Jona G, Ptacek J, *et al.* Global analysis of protein function using protein microarrays. Mech Ageing Dev 2005; 126(1): 171-5.
[http://dx.doi.org/10.1016/j.mad.2004.09.019] [PMID: 15610776]

[112] Jones RB, Gordus A, Krall JA, MacBeath G. A quantitative protein interaction network for the ErbB receptors using protein microarrays. Nature 2006; 439(7073): 168-74.
[http://dx.doi.org/10.1038/nature04177] [PMID: 16273093]

[113] Panchenko A, Przytycka TM. Protein-protein Interactions and Networks - Identification, Computer Analysis, and Prediction. Biology C, editor 2008.

[http://dx.doi.org/10.1007/978-1-84800-125-1]

[114] Singhal M, Resat H. A domain-based approach to predict protein-protein interactions. BMC Bioinformatics 2007; 8: 199.
[http://dx.doi.org/10.1186/1471-2105-8-199] [PMID: 17567909]

[115] Pawson T, Nash P. Assembly of cell regulatory systems through protein interaction domains. Science 2003; 300(5618): 445-52.
[http://dx.doi.org/10.1126/science.1083653] [PMID: 12702867]

[116] Bader S, Kühner S, Gavin AC. Interaction networks for systems biology. FEBS Lett 2008; 582(8): 1220-4.
[http://dx.doi.org/10.1016/j.febslet.2008.02.015] [PMID: 18282471]

[117] Li Y, Kong D, Ahmad A, Bao B, Sarkar FH. Pancreatic cancer stem cells: emerging target for designing novel therapy. Cancer Lett 2013; 338(1): 94-100.
[http://dx.doi.org/10.1016/j.canlet.2012.03.018] [PMID: 22445908]

[118] Clevers H. The cancer stem cell: premises, promises and challenges. Nat Med 2011; 17(3): 313-9.
[http://dx.doi.org/10.1038/nm.2304] [PMID: 21386835]

[119] Rasheed ZA, Kowalski J, Smith BD, Matsui W. Concise review: Emerging concepts in clinical targeting of cancer stem cells. Stem Cells 2011; 29(6): 883-7.
[http://dx.doi.org/10.1002/stem.648] [PMID: 21509907]

[120] La Porta CA. Thoughts about cancer stem cells in solid tumors. World J Stem Cells 2012; 4(3): 17-20.
[http://dx.doi.org/10.4252/wjsc.v4.i3.17] [PMID: 22577494]

[121] Valent P, Bonnet D, De Maria R, *et al.* Cancer stem cell definitions and terminology: the devil is in the details. Nat Rev Cancer 2012; 12(11): 767-75.
[http://dx.doi.org/10.1038/nrc3368] [PMID: 23051844]

[122] Rosen JM, Jordan CT. The increasing complexity of the cancer stem cell paradigm. Science 2009; 324(5935): 1670-3.
[http://dx.doi.org/10.1126/science.1171837] [PMID: 19556499]

[123] Hermann PC, Bhaskar S, Cioffi M, Heeschen C. Cancer stem cells in solid tumors. Semin Cancer Biol 2010; 20(2): 77-84.
[http://dx.doi.org/10.1016/j.semcancer.2010.03.004] [PMID: 20371287]

[124] Nguyen LV, Vanner R, Dirks P, Eaves CJ. Cancer stem cells: an evolving concept. Nat Rev Cancer 2012; 12(2): 133-43.
[PMID: 22237392]

[125] Sureban SM, May R, Qu D, *et al.* DCLK1 regulates pluripotency and angiogenic factors *via* microRNA-dependent mechanisms in pancreatic cancer. PLoS One 2013; 8(9): e73940.
[http://dx.doi.org/10.1371/journal.pone.0073940] [PMID: 24040120]

[126] Tanase CP, Neagu AI, Necula LG, *et al.* Cancer stem cells: involvement in pancreatic cancer pathogenesis and perspectives on cancer therapeutics. World J Gastroenterol 2014; 20(31): 10790-801.
[http://dx.doi.org/10.3748/wjg.v20.i31.10790] [PMID: 25152582]

[127] Scatena R, Bottoni P, Giardina B. Circulating tumour cells and cancer stem cells: a role for proteomics in defining the interrelationships between function, phenotype and differentiation with potential

clinical applications. Biochim Biophys Acta 2013; 1835(2): 129-43. [PMID: 23228700]

[128] Scatena R, Bottoni P, Pontoglio A, Giardina B. The proteomics of cancer stem cells: potential clinical applications for innovative research in oncology. Proteomics Clin Appl 2011; 5(11-12): 590-602. [http://dx.doi.org/10.1002/prca.201000142] [PMID: 21956883]

[129] Skvortsov S, Debbage P, Skvortsova I. Proteomics of cancer stem cells. Int J Radiat Biol 2014; 90(8): 653-8. [http://dx.doi.org/10.3109/09553002.2013.873559] [PMID: 24350919]

[130] Gilabert M, Launay S, Ginestier C, *et al.* Poly(ADP-ribose) polymerase 1 (PARP1) overexpression in human breast cancer stem cells and resistance to olaparib. PLoS One 2014; 9(8): e104302. [http://dx.doi.org/10.1371/journal.pone.0104302] [PMID: 25144364]

[131] Selvam AP, Wong J, Flanagan K, Prasad S. Cellular level classification of breast cancer through proteomic markers using nanochannel array sensors. Nanomedicine (Lond) 2013. [http://dx.doi.org/10.2217/nnm.13.152] [PMID: 24237262]

[132] Yi T, Zhai B, Yu Y, *et al.* Quantitative phosphoproteomic analysis reveals system-wide signaling pathways downstream of SDF-1/CXCR4 in breast cancer stem cells. Proc Natl Acad Sci USA 2014; 111(21): E2182-90. [http://dx.doi.org/10.1073/pnas.1404943111] [PMID: 24782546]

[133] Gonzalez E, Piva M, Rodriguez-Suarez E, *et al.* Human mammospheres secrete hormone-regulated active extracellular vesicles. PLoS One 2014; 9(1): e83955. [http://dx.doi.org/10.1371/journal.pone.0083955] [PMID: 24404144]

[134] Kruger S, Abd Elmageed ZY, Hawke DH, *et al.* Molecular characterization of exosome-like vesicles from breast cancer cells. BMC Cancer 2014; 14: 44. [http://dx.doi.org/10.1186/1471-2407-14-44] [PMID: 24468161]

[135] Korkaya H, Liu S, Wicha MS. Breast cancer stem cells, cytokine networks, and the tumor microenvironment. J Clin Invest 2011; 121(10): 3804-9. [http://dx.doi.org/10.1172/JCI57099] [PMID: 21965337]

[136] Kim ST, Sohn I, Do IG, *et al.* Transcriptome analysis of CD133-positive stem cells and prognostic value of survivin in colorectal cancer. Cancer Genomics Proteomics 2014; 11(5): 259-66. [PMID: 25331798]

[137] Ko CH, Cheng CF, Lai CP, *et al.* Differential proteomic analysis of cancer stem cell properties in hepatocellular carcinomas by isobaric tag labeling and mass spectrometry. J Proteome Res 2013; 12(8): 3573-85. [http://dx.doi.org/10.1021/pr4004294] [PMID: 23782096]

[138] Bailey JM, Alsina J, Rasheed ZA, *et al.* DCLK1 marks a morphologically distinct subpopulation of cells with stem cell properties in preinvasive pancreatic cancer. Gastroenterology 2014; 146(1): 245-56. [http://dx.doi.org/10.1053/j.gastro.2013.09.050] [PMID: 24096005]

[139] Castellanos JA, Merchant NB, Nagathihalli NS. Emerging targets in pancreatic cancer: epithelial-mesenchymal transition and cancer stem cells. Onco Targets Ther 2013; 6: 1261-7.

[PMID: 24049451]

[140] Collet B, Avril T, Aubry M, *et al.* Proteomic analysis underlines the usefulness of both primary adherent and stem-like cell lines for studying proteins involved in human glioblastoma. J Proteomics 2014; 110: 7-19. [http://dx.doi.org/10.1016/j.jprot.2014.07.022] [PMID: 25088051]

[141] Pointer KB, Clark PA, Zorniak M, Alrfaei BM, Kuo JS. Glioblastoma cancer stem cells: Biomarker and therapeutic advances. Neurochem Int 2014; 71: 1-7. [http://dx.doi.org/10.1016/j.neuint.2014.03.005] [PMID: 24657832]

[142] Thirant C, Galan-Moya EM, Dubois LG, *et al.* Differential proteomic analysis of human glioblastoma and neural stem cells reveals HDGF as a novel angiogenic secreted factor. Stem Cells 2012; 30(5): 845-53. [http://dx.doi.org/10.1002/stem.1062] [PMID: 22331796]

[143] Makridakis M, Roubelakis MG, Vlahou A. Stem cells: insights into the secretome. Biochim Biophys Acta 2013; 1834(11): 2380-4. [http://dx.doi.org/10.1016/j.bbapap.2013.01.032] [PMID: 23376432]

[144] Misuno K, Liu X, Feng S, Hu S. Quantitative proteomic analysis of sphere-forming stem-like oral cancer cells. Stem Cell Res Ther 2013; 4(6): 156. [http://dx.doi.org/10.1186/scrt386] [PMID: 24423398]

[145] Roubelakis MG, Trohatou O, Anagnou NP. Amniotic fluid and amniotic membrane stem cells: marker discovery. Stem Cells Int 2012 2012. 107836. Epub 2012/06/16 [http://dx.doi.org/10.1155/2012/107836]

[146] Skalnikova H, Motlik J, Gadher SJ, Kovarova H. Mapping of the secretome of primary isolates of mammalian cells, stem cells and derived cell lines. Proteomics 2011; 11(4): 691-708. [http://dx.doi.org/10.1002/pmic.201000402] [PMID: 21241017]

[147] Pavlou MP, Diamandis EP. The cancer cell secretome: a good source for discovering biomarkers? J Proteomics 2010; 73(10): 1896-906. [http://dx.doi.org/10.1016/j.jprot.2010.04.003] [PMID: 20394844]

[148] Makridakis M, Vlahou A. Secretome proteomics for discovery of cancer biomarkers. J Proteomics 2010; 73(12): 2291-305. [http://dx.doi.org/10.1016/j.jprot.2010.07.001] [PMID: 20637910]

[149] Roche S, D'Ippolito G, Gomez LA, *et al.* Comparative analysis of protein expression of three stem cell populations: models of cytokine delivery system *in vivo*. Int J Pharm 2013; 440(1): 72-82. [http://dx.doi.org/10.1016/j.ijpharm.2011.12.041] [PMID: 22285475]

[150] Farina A, D'Aniello C, Severino V, *et al.* Temporal proteomic profiling of embryonic stem cell secretome during cardiac and neural differentiation. Proteomics 2011; 11(20): 3972-82. [http://dx.doi.org/10.1002/pmic.201100063] [PMID: 21770033]

[151] Stastna M, Van Eyk JE. Secreted proteins as a fundamental source for biomarker discovery. Proteomics 2012; 12(4-5): 722-35. [http://dx.doi.org/10.1002/pmic.201100346] [PMID: 22247067]

[152] Bendall SC, Hughes C, Campbell JL, *et al.* An enhanced mass spectrometry approach reveals human

embryonic stem cell growth factors in culture. Mol Cell Proteomics 2009; 8(3): 421-32. [http://dx.doi.org/10.1074/mcp.M800190-MCP200] [PMID: 18936058]

[153] Kim JM, Kim J, Kim YH, *et al.* Comparative secretome analysis of human bone marrow-derived mesenchymal stem cells during osteogenesis. J Cell Physiol 2013; 228(1): 216-24. [http://dx.doi.org/10.1002/jcp.24123] [PMID: 22674502]

[154] Zvonic S, Lefevre M, Kilroy G, *et al.* Secretome of primary cultures of human adipose-derived stem cells: modulation of serpins by adipogenesis. Mol Cell Proteomics 2007; 6(1): 18-28. [http://dx.doi.org/10.1074/mcp.M600217-MCP200] [PMID: 17018519]

[155] Lee MJ, Kim J, Kim MY, *et al.* Proteomic analysis of tumor necrosis factor-alpha-induced secretome of human adipose tissue-derived mesenchymal stem cells. J Proteome Res 2010; 9(4): 1754-62. [http://dx.doi.org/10.1021/pr900898n] [PMID: 20184379]

[156] Chiellini C, Cochet O, Negroni L, *et al.* Characterization of human mesenchymal stem cell secretome at early steps of adipocyte and osteoblast differentiation. BMC Mol Biol 2008; 9: 26. [http://dx.doi.org/10.1186/1471-2199-9-26] [PMID: 18302751]

[157] Cheng L, Bao S, Rich JN. Potential therapeutic implications of cancer stem cells in glioblastoma. Biochem Pharmacol 2010; 80(5): 654-65. [http://dx.doi.org/10.1016/j.bcp.2010.04.035] [PMID: 20457135]

[158] Cruceru ML, Neagu M, Demoulin JB, Constantinescu SN. Therapy targets in glioblastoma and cancer stem cells: lessons from haematopoietic neoplasms. J Cell Mol Med 2013; 17(10): 1218-35. [http://dx.doi.org/10.1111/jcmm.12122] [PMID: 23998913]

[159] Baglio SR, Pegtel DM, Baldini N. Mesenchymal stem cell secreted vesicles provide novel opportunities in (stem) cell-free therapy. Front Physiol 2012; 3: 359. [http://dx.doi.org/10.3389/fphys.2012.00359] [PMID: 22973239]

[160] Ranganath SH, Levy O, Inamdar MS, Karp JM. Harnessing the mesenchymal stem cell secretome for the treatment of cardiovascular disease. Cell Stem Cell 2012; 10(3): 244-58. [http://dx.doi.org/10.1016/j.stem.2012.02.005] [PMID: 22385653]

[161] Sarkar P, Randall SM, Muddiman DC, Rao BM. Targeted proteomics of the secretory pathway reveals the secretome of mouse embryonic fibroblasts and human embryonic stem cells. Mol Cell Proteomics 2012; 11(12): 1829-39. [http://dx.doi.org/10.1074/mcp.M112.020503] [PMID: 22984290]

[162] Heo SC, Jeon ES, Lee IH, Kim HS, Kim MB, Kim JH. Tumor necrosis factor-α-activated human adipose tissue-derived mesenchymal stem cells accelerate cutaneous wound healing through paracrine mechanisms. J Invest Dermatol 2011; 131(7): 1559-67. [http://dx.doi.org/10.1038/jid.2011.64] [PMID: 21451545]

[163] Nielsen JA, Lau P, Maric D, Barker JL, Hudson LD. Integrating microRNA and mRNA expression profiles of neuronal progenitors to identify regulatory networks underlying the onset of cortical neurogenesis. BMC Neurosci 2009; 10: 98. [http://dx.doi.org/10.1186/1471-2202-10-98] [PMID: 19689821]

[164] Hokke CH, Fitzpatrick JM, Hoffmann KF. Integrating transcriptome, proteome and glycome analyses of Schistosoma biology. Trends Parasitol 2007; 23(4): 165-74.

[http://dx.doi.org/10.1016/j.pt.2007.02.007] [PMID: 17336161]

[165] Tarun AS, Peng X, Dumpit RF, *et al.* A combined transcriptome and proteome survey of malaria parasite liver stages. Proc Natl Acad Sci USA 2008; 105(1): 305-10. [http://dx.doi.org/10.1073/pnas.0710780104] [PMID: 18172196]

[166] Unwin RD, Whetton AD. Systematic proteome and transcriptome analysis of stem cell populations. Cell Cycle 2006; 5(15): 1587-91. [http://dx.doi.org/10.4161/cc.5.15.3101] [PMID: 16861929]

[167] Xu H, Lemischka IR, Ma'ayan A. SVM classifier to predict genes important for self-renewal and pluripotency of mouse embryonic stem cells. BMC Syst Biol 2010; 4: 173. [http://dx.doi.org/10.1186/1752-0509-4-173] [PMID: 21176149]

[168] Ho Sui SJ, Begley K, Reilly D, *et al.* The Stem Cell Discovery Engine: an integrated repository and analysis system for cancer stem cell comparisons. Nucleic Acids Res 2012; 40(Database issue): D984-91. [http://dx.doi.org/10.1093/nar/gkr1051] [PMID: 22121217]

CHAPTER 10

Nano and Microtechnology for Monitoring Stem Cell Differentiation

Larisa-Emilia Cheran[1], **Alin Cheran**[2,*], **Andreea-Roxana Lupu**[3] and **Traian Popescu**[4]

[1] *Department of Chemistry, University of Toronto, Toronto, Ontario, Canada*

[2] *Ross University School of Medicine, New Jersey, USA*

[3] *Immunobiology Laboratory, "Victor Babeş" National Institute of Pathology, Bucharest, Romania*

[4] *National Institute of Materials Physics, Magurele, Romania*

Abstract: This chapter gives an overview of the new micro- and nano-technologies designed to monitor SC differentiation in the context of their potential applications in disease modeling, tissue engineering, regenerative medicine, as well as drug screening and toxicology. Representative examples of such applications are presented and the crucial importance of the differentiation processes for the safety of SC therapies is discussed. The roles of the main factors that influence SC differentiation are briefly summarized and the vital need to control and monitor the differentiation process using non-invasive methods is emphasized. The basic principles of new micro- and nano-technologies for monitoring SC differentiation are presented, with special focus on the use of acoustic vibrational fields to characterize SC. Literature studies on SC differentiation, involving methods based on Impedance Sensing (IS), Surface Enhanced Raman Spectroscopy (SERS), Fourier Transformed Infrared (FTIR) spectroscopy, Microelectrode Array (MEA) sensors, Light-addressable Potentiometric Sensors (LAPS), Field-effect Transistors (FET), Surface Plasmon Resonance (SPR) and Quartz Crystal Microbalance (QCM), are briefly described.

Keywords: Differentiation, Monitoring, Nano and microtechnology, Stem cells.

* **Corresponding author Alin Cheran:** Department of Chemistry, University of Toronto, Toronto, Ontario, Canada; E-mail: alin.cheran@gmail.com.

Cristiana Tanase & Monica Neagu (Eds.)

INTRODUCTION

The potential uses of stem cells (SC) in treating a series of neurological, hepatic, hematopoietic, diabetic and skin diseases [1 - 5] rely on their capacities to self-renew and differentiate into multiple cell lineages.

These characteristics promote SC as basic ingredients in the design of efficient tissue repair and vital organ regeneration therapies [3, 6].

Since SC evolution strongly depends on their microenvironment [6 - 8], the safety of SC therapies can only be ensured by studying SC under relevant conditions, mimicking their *in vivo* environments. However, standard cell culture methods are inappropriate for SC culture. SC need sequential delivery of specific growth factors and other soluble molecules, an extracellular scaffold allowing proper cell-cell interactions, lower oxygen concentration than those provided by the majority of cell culture incubators *etc.* [7, 9, 10].

To avoid such inconveniences, efforts have been made to develop nano and micro-technologies and devices capable to replicate the *in vivo* SC microenvironments, as well as to monitor SC behavior and differentiation *in vitro* or to track their location and distribution following their transplantation to the human body. Such multifunctional tools allow for a better understanding of SC biology and a more meaningful extrapolation of results from *in vitro* studies to *in vivo* situations.

Regarding the monitoring methods, they should ideally be based on physical principles that allow them not to interfere with the cellular processes.

While the advances regarding SC tracking have been presented in several comprehensive review articles and books [1, 11 - 16], the monitoring of SC differentiation by means of advanced nano and micro-technologies has been poorly summarized.

This chapter gives an overview of the present technologies used to monitor SC differentiation, classified with respect to their underlying detection/monitoring principles, with special focus on the use of electro-acoustic vibrational fields in sensing and monitoring the differentiation process.

THE NEED TO MONITOR SC DIFFERENTIATION IN THE CONTEXT OF THEIR RESEARCH AND THERAPEUTIC APPLICATIONS

SC – Research

Due to their remarkable properties, SC represent excellent experimental models for both basic as well as applied research.

Basic Research

SC offer valuable information concerning cell biology and tissue formation in embryogenesis and postembryonic development, allowing in-depth studies regarding normal growth and development and/or identification of major causes of birth defects [17].

Applied Research

a) Disease Modeling

Induced pluripotent SC (iPSC)-based research offers both clinical translation opportunities and new approaches to clarify the molecular basis of human diseases [18, 19]. Depending on the desired application, iPS cells can be used to generate disease specific somatic cells or patient specific cells. By linking disease, patient specificity and drug discovery and design, SC-based research allows patients to benefit from the newly emerging personalized medicine [2, 20, 21].

Due to specific similarities between SC and cancer cells (*e.g.* high telomerase activity and unlimited cell division capacity), SC represent key models for the study of abnormal cell division that leads to cancer [5, 22].

b) Tissue Engineering and Regenerative Medicine

Transplanted SC are expected to differentiate in response to tissue specific stimuli, leading to tissue regeneration and repair [23]. Like any other therapy, SC transplantation needs to provide maximum therapeutic benefits to patient, with minimum side effects [4].

Bone marrow transplantation has been traditionally used for many years as replacement therapy in certain cancers (hematogenic tissue destruction following chemotherapy, compromised immune system) or other diseases including genetic diseases (Fanconi anemia, Shwachman-Diamond Syndrome *etc.*) [5, 24 - 27].

Recent developments suggest that umbilical cord blood (UCB) transplantation represent an alternative better way to treat some blood cancerous and genetic diseases in children. UCB contains hematopoietic SC, which are easier and more safe to collect compared to bone marrow SC, this approach causes less frequent graft *versus* host disease cases due to newborn's immune system characteristics [28 - 30]. However, the efficiency of UCB transplantation is restricted to age and blood-related diseases. In case of adults, it appears that the amount of hematopoietic SC needed for transplantation cannot be isolated from a single umbilical cord. For this reason, UCB transplantation was inefficient in adult patients, even after the *ex vivo* expansion of SC. Regarding the SC content of USB, it is not clear yet whether it contains hematopoietic SC only, or if other types of SC (*e.g.* nervous SC) can also be found in its structure. Due to this lack of knowledge, therapies using UCB-derived SC have not yet been developed [30 - 32].

The transplantation of SC or SC-derived chondrocytes represents a new treatment strategy in arthritic cartilage defects [33].

Ischemic conditions in coronary artery disease and stroke lead to severe damage of cardiac and brain tissues. A bioartificial myocardium was developed using SC and nanobiotechnology templates, aiming to become an efficient treatment of myocardial disease [3, 34].

Due to the limited regeneration capabilities of the central nervous system (CNS) [5], a disturbance in the CNS homeostasis leads to neurodegeneration and subsequently compromised neurological functions.

A central player during differentiation of embryonic stem cells (ES) into neural progenitors and finally mature neurons and glial cells is the Notch signaling pathway. The disruption of Notch causes defects in brain and heart development in mice and humans [35 - 37]. Following its activation in normal tissue, the intracellular domain of Notch receptor (Notch ICD) translocates into the nucleus and interacts with recombination signal-sequence binding protein-J (RBP-J) to regulate target genes [35]. As a result of these signaling events, Notch, *via* RBP-J, regulates the maintenance of stem cells as well as their differentiation into neurons [38, 39]. A number of key molecules are known to directly participate in neurogenesis, in particular Bone Morphogenetic Protein 7 (BMP7) and Retinoic Acid (RA), a derivative of vitamin A [40 - 50].

Despite the extensive present knowledge regarding SC biology, the molecular control of stem cell growth and differentiation is only partially understood and significant issues remain unclear, precluding the widespread use of stem cells for applications in regenerative medicine. Most notably, each source of SC presents a degree of heterogeneity and most of the methods available for examining such cells are invasive and render them unusable for subsequent applications. Under these circumstances, the design of novel methods for probing the behavior of stem cells is of great importance in filling this gap of knowledge and enabling basic and clinical stem cell research.

c) Toxicological and Drug Screening Studies

The ability of pluripotent SC to develop into any cell type is very useful for toxicological and drug screening studies (both targeted and phenotypic approaches). Because animal cells and human cells could manifest different sensitivity to tested stimuli, human- derived pluripotent SC represent a valuable modality to reduce and / or replace *in vivo* (animal models) testing [2]. For this purpose, an increased genetic diversity of SC lines is needed to better represent the human population (responders and non-responders against a certain tested compound) [51].

The poor control of SC differentiation into particular cell types (including alteration of genetic profile), low proliferation rates and high costs (still lower than animal models), represent the main disadvantages of using SC in toxicological and drug screening studies [2, 51].

Importance of Monitoring SC Differentiation

SC fate is decided by intrinsic factors (signaling molecules, growth factors *etc.*) and cell microenvironment, with characteristic physical and biochemical properties [1].

The conventional methods used to induce SC differentiation are based on at least one of the following factors.

a) Soluble Factors

Presence of growth factors and small molecules (*e.g.* paracrine factors released by surrounding cells) is needed to activate specific signaling pathways leading to SC differentiation [51]. The time-dependent composition of soluble factors pattern is crucial for SC fate decision [52]. Complex microfluidic systems, allowing the modulation of spatial and temporal gradients of chemical (including growth factors) and biophysical stimuli, were recently developed [7, 53].

b) Cell Microenvironment/Niches

SC niches represent 3D environments which can influence SC evolution (self-renewal or differentiation) through soluble factors, adhesive context and mechanical signals [54 - 56].

Cell microenvironment can act through tensile forces (extracellular matrix, ECM) or through fluid applied forces (interstitial fluid) [53]. ECM is composed of protein fibers (collagen, elastin *etc.*), specifically organized (fiber's diameter, orientation and spacing) depending on tissue functions [6].

External mechanical stimuli regulate cell function by changing cell shape and subsequent mechanotransduction-related modifications. Cell shape changes decide the lineage commitment of mesenchymal SC [54].

ECM stiffness is an important parameter which has the potential to decide the mesenchymal SC differentiation into neuronal, skeletal muscle or osteogenic markers [54, 57, 58]. Mechanical stimuli - activated Wnt/Frizzled/beta catenin signaling also regulates neuronal and skeletal differentiation of mesenchymal SC

[58 - 61].

Changes in shear flow induced reprogramming of gene expression in endothelial cells and promoted endothelial differentiation of mesenchymal SC [54].

Since, SC biology is not yet fully understood and minor manipulation mistakes can lead to dramatic consequences, the experimentally induced SC differentiation (for research and medical purposes) must be highly monitored and controlled. In the absence of these elements, SC transplanted in a wrong state of differentiation or in a wrong place in the body, can become cancerous or form teratomers or ectopic tissues [5, 22, 62] , leading to severe side-effects of SC-based therapy.

To avoid unwanted responses, SC should be triggered to differentiate only in appropriate biological contexts and should be monitored using extremely sensitive technologies, capable of detecting subtile cellular responses to microenvironment [54].

To correctly differentiate, SC need to pass through all differentiation stages (they need to became progenitor, precursor and terminally differentiated cells), under appropriate conditions [5, 63 - 65]. Although sustained efforts are made to introduce workflow standardization (non-automated or automated), it is not yet possible to fully control the differentiation of SC into particular cell phenotypes.

The successful application of SC in regenerative medicine depends on at least two essential aspects: monitoring and control of SC differentiation and their *in vivo* tracking.

Although tracking methods probe the presence of fully differentiated cell phenotypes, they cannot necessarily be used to also monitor SC differentiation.

For instance, functionalized and/or non-functionalized superparamagnetic iron oxide nanoparticles (SPIONs) show promising potential for SC tracking applications [66]. However, SPIONs are not appropriate for monitoring SC differentiation. Chang *et al.* [67] showed that differentiation of human mesenchymal SC is altered by amine-surface-modified SPIONs [67, 68]. Moreover, the detection of SPIONs-labeled cell function and the potential toxicity of SPIONs, represent yet unsolved problems [69]. Particular types of such iron oxide nanoparticles were reported to induce apoptosis through reactive oxygen species (ROS) (especially hydroxyl radicals) – dependent mechanisms [1, 3, 70]. Nanoparticle-treated embryonic SC showed cell viability decrease and impaired cytoskeletal development without effect on the size of nucleus [4].

For an enhanced control of SC differentiation under laboratory conditions, researchers need to identify additional factors required to optimize SC *in vitro* micro environments and engineer the differentiation process.

The monitoring of SC differentiation requires new methods, capable of detecting subtile cellular responses without being toxic or interfering (directly or indirectly- (*e.g.* through feeder cells activation/inhibition)) with the differentiation process. Also, researchers should have the possibility to choose methods for *in vitro* or *in vivo* applications that either keep the sample suitable or render it impracticable for further experimental use.

The recently developed micro- and nano-technologies outlined below bring significant progress in achieving this goal.

TECHNOLOGIES FOR MONITORING SC DIFFERENTIATION

The successful replacement of damaged tissues by means of stem cell therapy strongly depends on the possibility to evaluate stem cell responses to their microenvironment.

For this purpose, innovative technologies, highly sensitive to subtle cellular phenomena occurring at the micro- or nanometer scale, have been developed. Such micro- and nanotechnologies include biosensors designed to detect changes in cell physiology with a high level of spatial and temporal resolution. The output of these biosensors is transduced to signals that allow the real-time monitoring of cellular processes using basic physical methods. An outline of the most important of these methods, with a focus on the use of acoustic physics for the detection of cell processes, is presented below. Given the large variety of devices corresponding to each of the presented methods, device-specific improvements and drawbacks can only be discussed with respect to particular models.

Monitoring SC Differentiation by Spectroscopic Methods

Spectroscopic methods represent important tools in analyzing any type of organic or inorganic material sample. From a large variety of such methods, Impedance spectroscopy (IS), Surface-enhanced Raman spectroscopy (SERS) and Fourier Transformed Infrared (FTIR) spectroscopy have high potential for use in biological applications, including SC analysis. Some of the general advantages and shortcomings of these methods are summarized in (Table **1**). Details regarding the basic functioning principles of these methods and their use for monitoring SC differentiation are described below.

Table 1. Spectroscopic methods used for monitoring SC differentiation.

Spectroscopic methods			
Method	**Advantages**	**Shortcomings**	**Use in monitoring SC differentiation**
Impedance spectroscopy	-non-invasive - real-time monitoring without the use of markers	- applicability to adherent cells only - difficult distinction between cell death and detachment or proliferation reduction processes -lack of single-cell specificity	- intercellular junction formation, morphological changes, cell-substrate interactions
Surface enhanced Raman spectroscopy	- single cell analysis - marker free - high accuracy	- weak signal - requires the use of nanoparticles as signal enhancers	- nucleic acids and proteins contents and mitochondrial activity
FTIR spectroscopy	- non-invasive - single cell analysis - marker free - facile sample preparation	-water background	- complementary to SERS - variations in the IR bands of proteins, lipids and nucleic acids

a) Impedance Spectroscopy

Impedance spectroscopy (IS) is a non-invasive method that allows the real-time monitoring of cells during growth and differentiation without the use of markers.

The experimentally determined quantity is the complex impedance Z^* of a cultured cell system exposed to a conveniently chosen AC voltage. While the frequency of the applied voltage is varied, the amplitude and phase of the resulting AC currents are measured. The presence of cells modulates the measured parameters by acting as dielectric particles that adhere to the surface of the electrodes.

Together with dedicated microelectrode array (MEA) systems and theoretical models, the measured complex impedance (Z^*) variations can be related to cell dielectric properties, cell electrode coverage, cell-to-cell junctions and cell-substrate interactions. By detecting changes in cell number, their spreading and adhesion as well as morphological changes [71, 72] IS measurements can offer valuable information regarding cell morphology and its electrophysiological properties, cytotoxicity, cell death, growth, migration and differentiation [73 - 78].

Impedance spectroscopy is limited by its applicability to adherent cells only. By measuring average impedances over MEA systems covered by cells, it lacks single-cell specificity as well as the possibility to distinguish cell death from detachment or proliferation reduction processes [79].

Impedance changes throughout SC differentiation have been exploited to monitor the differentiation of adipose-derived SC (ADSCs) toward osteoblasts and adipocytes by Bagnaninchi and Drummond (2011) [74]. During early differentiation times (less than twenty four hours after induction), the impedance differences between the two evolving phenotypes occurred as a consequence of intercellular junction formation [74] (Fig. **1**). These differences were significant even at twelve hours post induction, when no clear morphological differences were found between the induced cell lineages. Important differences were also found in the mean cell capacitances, Cm, determined four days after induction: $Cm = 2.25 \pm 0.27\ \mu F/cm^2$ for adipocyte progenitors and $Cm = 1.72 \pm 0.10\ \mu F/cm^2$ for osteoblast progenitors [74]. The distinct dielectric properties of the induced cell lineages suggest different mechanisms for osteo- and adipo-induction [74].

During late differentiation stages, the changes of complex impedance correlated with the emergence of specialized functions, being found in the 1-8 kHz frequency range.

The described study confirmed the potential of impedance sensing as a monitoring method, capable of revealing the real-time evolution of ADSCs differentiation into adipocytes and osteoblasts.

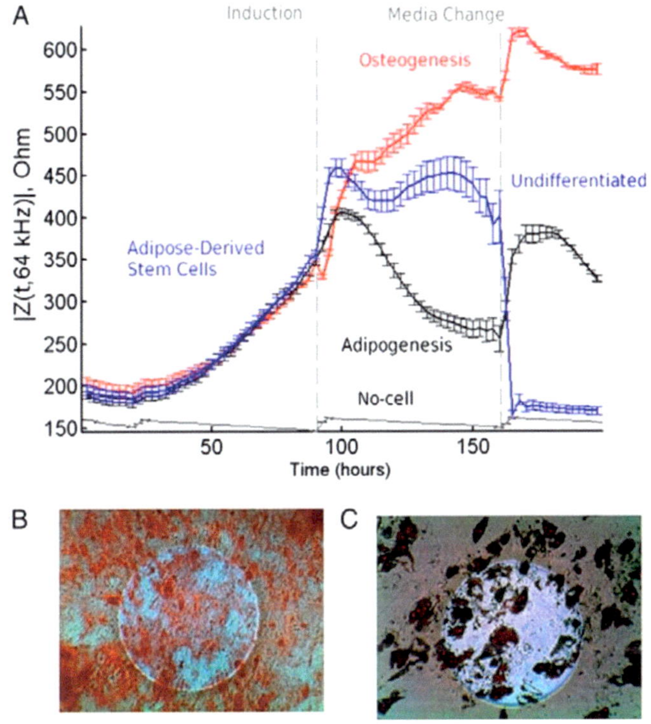

Fig. (1). Real-time and label-free monitoring of ADSCs induced toward osteoblasts and adipocytes. (A) Time-course measurement of mean impedance for different groups throughout early induction; Histochemical end-point staining was performed to assess that cells underwent osteogenesis (B) and adipogenesis (C) Reused with permission from [74].

A promising and relatively new detection and monitoring technology uses impedance measurements to extract cellular information mediated by acoustic vibrational fields.

The use of such electromagnetic and acoustic fields of variable frequencies in cell biology represents the state-of-the art in detection science. Unlike patch-clamp methods which destroy the cell during measurements, methods employing acoustic physics are *non-invasive*, allowing the monitoring of cell activity before and after stimulation with a chemical or pharmaceutical factor, without affecting the integrity of the physical cell [80 - 83]. Optical and electron microscopy methods reveal only physical changes and are plagued by detection limits.

At the center of this new cellular investigation technique is the Thickness Shear Mode device (TSM) [84, 85].

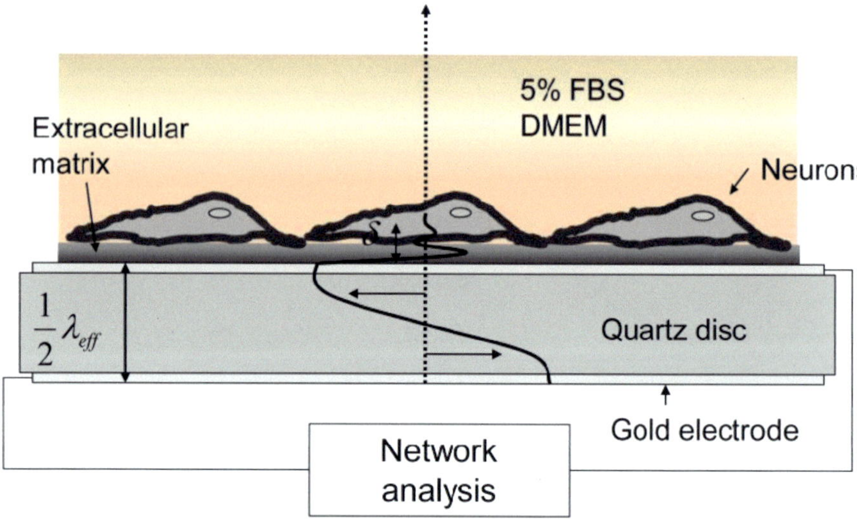

Fig. (2). A schematic of a TSM system for the study of neuron cells. Cells are constantly bathed in a fresh supply of Dublecco's Modified Eagle Medium (DMEM) supplemented with 5% Fetal Bovine Serum (FBS). All test solutions being made up using either 5% FBS DMEM or DMEM as the solvent to prevent adverse effects to cells [80]. *(Reprinted with kind permission of the Royal Society of Chemistry).*

As shown in a typical experiment involving study of neurons in Fig. (**2**) [80], a resonant shear wave is produced in a piezoelectric material (quartz) by gold electrodes placed on each face of the sensor, to which an oscillating electrical potential of 9 MHz is applied. The acoustic wave generated in the device travels through the piezoelectric material with little energy dissipation. The wave is reflected at the interface and a standing wave is maintained in the crystal. When operated in a liquid environment (one face of the device, on which the neurons are grown is exposed to the growth liquid), a portion of the acoustic energy is not reflected at the interface but is dissipated into the liquid. The process is influenced by the liquid density, conductivity, dielectric constant, viscosity, roughness, slip and stress at the liquid-solid interface. Acoustic coupling effects and interfacial effects govern the response of the TSM device, offering a unique mechanism for biomolecular and cellular detection. The propagation of the acoustic shear into the liquid and resultant energy dissipation is detected *via* variations in the series resonant frequency (f_s) and other acoustical parameters such as the motional resistance (R_m) in a flow-injection configuration. These parameters are detected by acoustic network analysis using an impedance analyzer attached to the system which measures the magnitude and the phase of the impedance of the sensor at a set of frequencies under resonant conditions.

The interface between the neurons and the sensor can be modeled by an electrical equivalent circuit containing capacities, inductances and resistances, related to energy accumulation (inertial, elastic storage) and dissipative processes at interface (Fig. **3**).

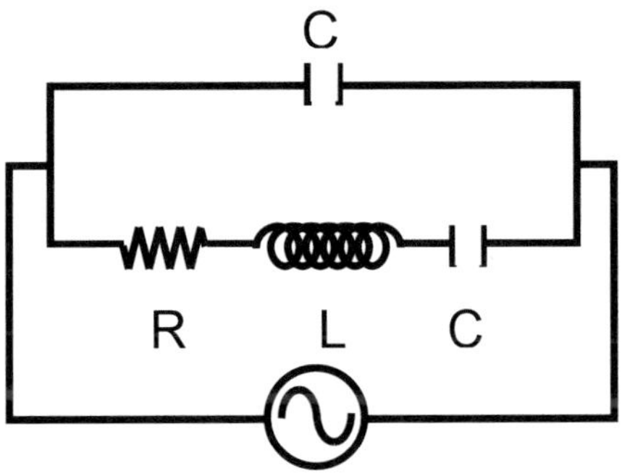

Fig. (3). The electrical model of the liquid-solid interface.

At resonance, the resonant frequency is defined by Fig. (**4**):

The sensor system is incorporated into a temperature-controlled static or flow-though conformation which allows cellular growth and, subsequently, the introduction of various reagents, in order to detect the response of the cells. The latter are kept in a CO_2 environment.

The thickness-shear mode sensor (TSM) with cells on the device electrode surface is placed horizontally in a flow-through system. The TSM device is sandwiched between two Viton o-rings in a cell made of Plexi-glass. The top side of the quartz crystal is exposed to solution while the bottom is flushed with a low flow of nitrogen gas. A network analyzer (HP4195A Network/Spectrum Analyzer,

Hewlett Packard, Colorado Springs, USA) was employed to generate an impedance spectrum of the crystal around its resonant frequency. Characteristic frequencies (f_s) and selected impedance characteristics were recorded and also equivalent circuit parameters were extracted.

$$f_s \approx \frac{1}{2\pi\sqrt{L_m C_m}}$$

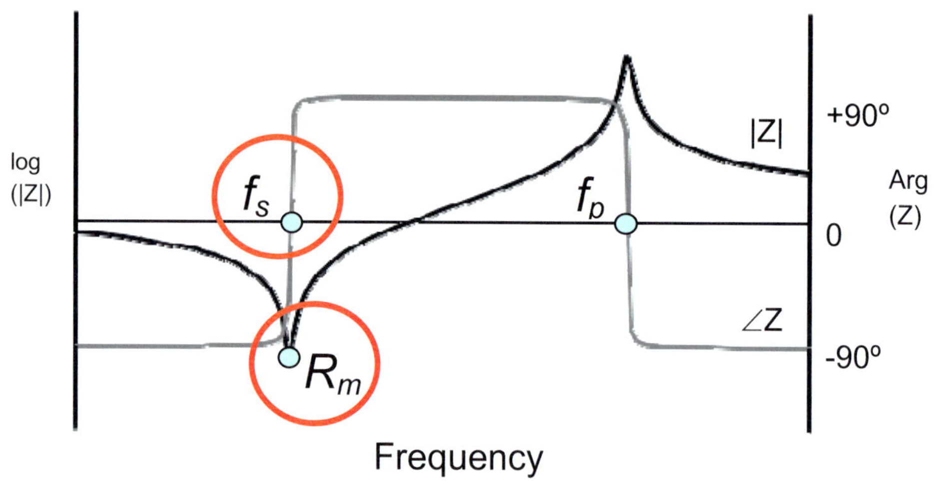

Fig. (4). The impedance spectrum from which the two main characteristics are extracted (motional resistance R_m and resonant frequency f_s.

TSM Device in Exploring Stem Cells Behavior

The TSM method was consequently adapted for the study of SC [98]. In order to be analyzed with the TSM, the cells were processed as follows: Mouse embryonic SC (mES D3 wild type) were purchased from ATCC and the RBPJk knockout and heterozygous mES D3 cells were a gift from Dr Zuniga-Pflucker (University of Toronto). Cells were plated on tissue culture dishes coated with 0.2% gelatin (Sigma Aldrich) in water for 1 hour at room temperature in mES media (DME, (Invitrogen) with 20% knockout serum replacement (Invitrogen) and 0.8mM beta mercaptoethanol (Sigma/Aldrich). To maintain cells in the undifferentiated state leukemia inhibitory factor (LIF, Sigma Aldrich) was added at 20ng/ml. When the cell density reached 80% confluence, cells were trypsinized with 0.25% trypsin/ 0.1% EDTA (Wisent Inc.) suspended gently in media and split into gelatin coated dishes at 1:20 dilution in mES media + LIF.

To demonstrate the competency of ESCs to respond to RA, the cells were trypsinized and plated in mES media (without LIF) at 2 x 10^5 cells/ml in non tissue culture dishes. The media was replaced after 2 days. After an additional 2 days the media was replaced with mES media + 1ug/ml retinoic acid (RA, Sigma/Aldrich) and again 2 days later. Cell spheres were collected, trypsinised with 0.25% trypsin/0.1%EDTA and plated on poly-l lysine/laminin coated dishes containing the sensing TSM crystals at 2 x 10^5 cells/ml. The media was replaced every 2 days. Cells were fixed for immunofluorescence staining. New paragraph mES cells can be differentiated into neurons and glia in response to RA (Fig. **5**), the question to test is: are biophysical cellular responses to such cues detectable in a real-time environment and if these signals can be used to discriminate ESC populations that would be predicted to be deficient in their response on a cellular biology level. To this end wild-type ESC and ESC null were examined for RBP-J, a key component of the Notch signaling pathway and predicted to be necessary for neural differentiation by TSM in the presence of two neural inductive molecules, RA and BMP7.

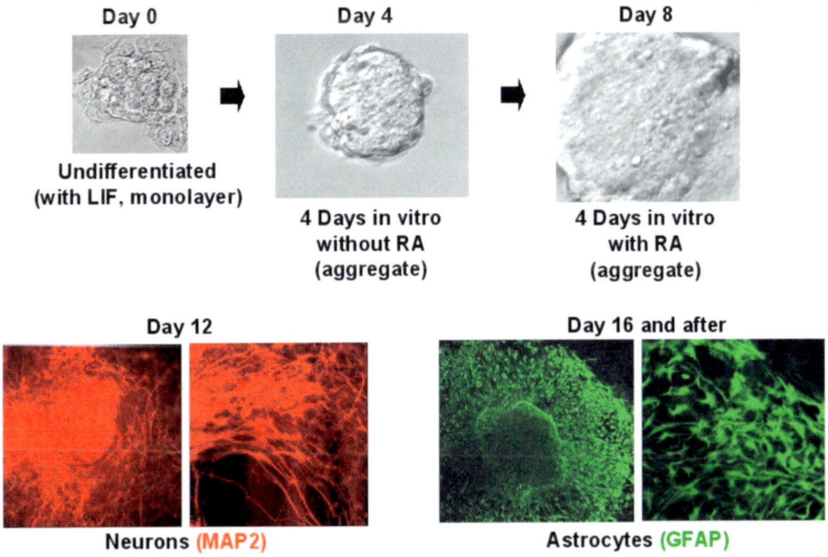

Fig. (5). Neuronal differentiation and glial determination in mouse embryonic stem (ES) cells; neurons (MAP2+) and astrocytes (GFAP+) in WT cultures after 16 days *in vitro*; scale bar: 10 μm [44].

Experiments were conducted to ascertain under what conditions healthy functional cells are maintained. The longevity of signal and amplitude of changes in frequency and motional resistance were monitored. Such details were analyzed to determine the mechanism responsible for altering the acoustic parameters as well as to define the charge or structural perturbation connected to cellular response. Another aspect is the use of the afore-mentioned surfaces with different confluences to establish the level of intercellular communication possible with variable cell packing density.

At the most fundamental level, the energetic and acoustic properties of ES cells can be analyzed and reveal differences between ES cells that are null for Notch signaling (RBP-J-/-) and their wild-type counterparts. In order to test proof this principle undifferentiated WT and KO (-/-) ES cells were subjected to testing by acoustic wave propagation. This was performed in a flow-through configuration at 37°C for introduction of BMP7 and RA with monitoring of the TSM device resonant frequency (Fig. **6A** and **B**), where the vertical dotted line indicates the addition of the factor and the arrow indicates the removal of the factor from the medium).

Clear and discernable responses could be observed between the cell types whereby WT cells (black squares) showed an immediate increase in resonant frequency in response to BMP7 and RA (in the case of BMP this response was significantly sustained) which was attenuated by the removal of each factor. The KO cells (grey diamonds) on the other hand appeared to show a delayed and reduced response to the addition of BMP7 and a significantly reduced response to the withdrawal of RA.

Embryonic stem cells hold significant therapeutic potential, but are also currently subject to investigation by largely invasive methodologies. The experiments presented here were intended to investigate whether at the single cell stage ESCs have biophysical characteristics that may be used to further study their development and differentiation, and possibly be used to discriminate differing stem cell populations.

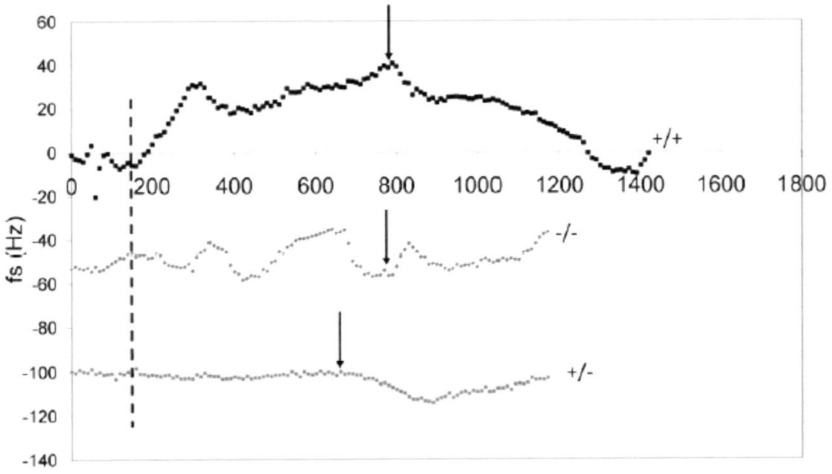

Fig. (6A). Resonant frequency variations at the addition of BMP-7 10 ng/ml to mES wild type (+/+), heterogenous (+/-) and knockout (-/-) cells. Dotted line indicates the time of BMP-7 addition. Arrows indicate the times of wash off.

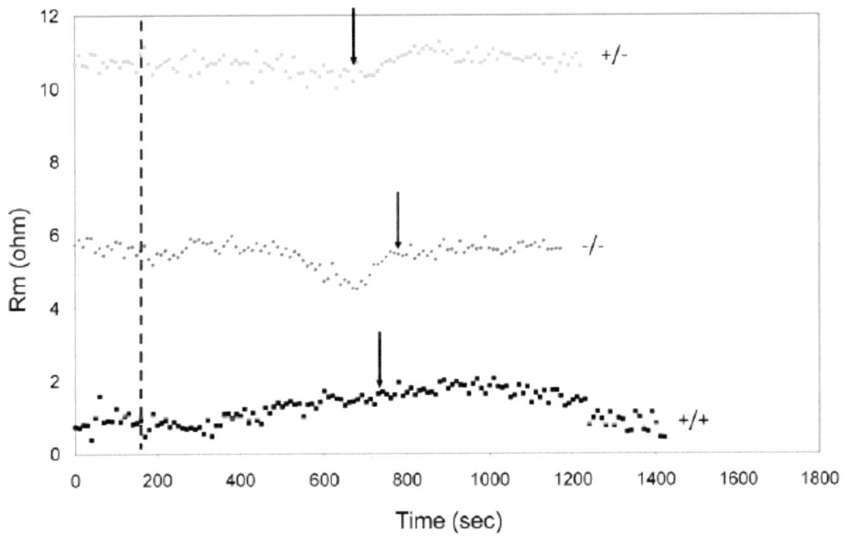

Fig. (6B). Motional resistance variations at the addition of BMP-7 10 ng/ml to mES wild type (+/+), heterogenous (+/-) and knockout (-/-) cells. Dotted line indicates the time of BMP-7 addition. Arrows indicate the times of wash off.

Data indicate a number of significant points: Firstly, ESCs have energetic signatures that can be detected by acoustic wave physics. This was very clearly demonstrated by the fact that a baseline resonant frequency was detected in all cell types across a field of cells. Secondly, the cells are subject to variation in response to external stimuli. Again, this was a very clear and robust finding, whereby all three cell populations (WT, HT, and KO) showed distinct differences in response to both RA and BMP7. Given that at the microscopic level, the cell populations appear phenotypically identical, this would suggest that cells can be discriminated in a label-free manner which otherwise could only be distinguished *via* invasive genotypic analysis, or lengthy cytological analysis.

b) Surface Enhanced Raman Spectroscopy (with Nanoparticles (NPs)

Raman spectroscopy is a non-invasive and label free technique generally used to investigate molecular bonds in a system. The method relies on the detection of characteristic changes in the energy of photons from a laser source, inelastically scattered by molecular vibrations or other excitations in the studied system.

Although Raman microspectroscopy was already used for cell characterization in several studies [86], a major drawback of the method in the study of cells is the weakness of the Raman scattering signal. Surface enhanced Raman scattering (SERS) represents a method by which the scattering signal is greatly enhanced (typically between 10^4-10^6 folds) when the molecules of interest are adsorbed on certain metal surfaces. The enhancement comes from two main sources: electromagnetic effects (formation of surface plasmons, valid for chemisorption and physisorption of molecules) and chemical (only valid for chemisorbed molecules, the process depending on the surface roughness).

Gold nanoparticles (GNP) were used to monitor the differentiation of mouse embryonic SC by SERS, probing different evolution stages: undifferentiated cells, embryoid bodies and fully differentiated cardiomyocytes by [87]. The localization of GNP inside the studied cells was assessed by transmission electron microscopy (TEM). The undifferentiated cells generated SERS peaks mostly associated to nucleic acids, DNA/RNA (O-P-O stretch in DNA: 787 cm^{-1} and O-P-O stretch in RNA: 813 cm^{-1} [87]), as a consequence of their high proliferation rates (Fig. **7** and Table **2**).

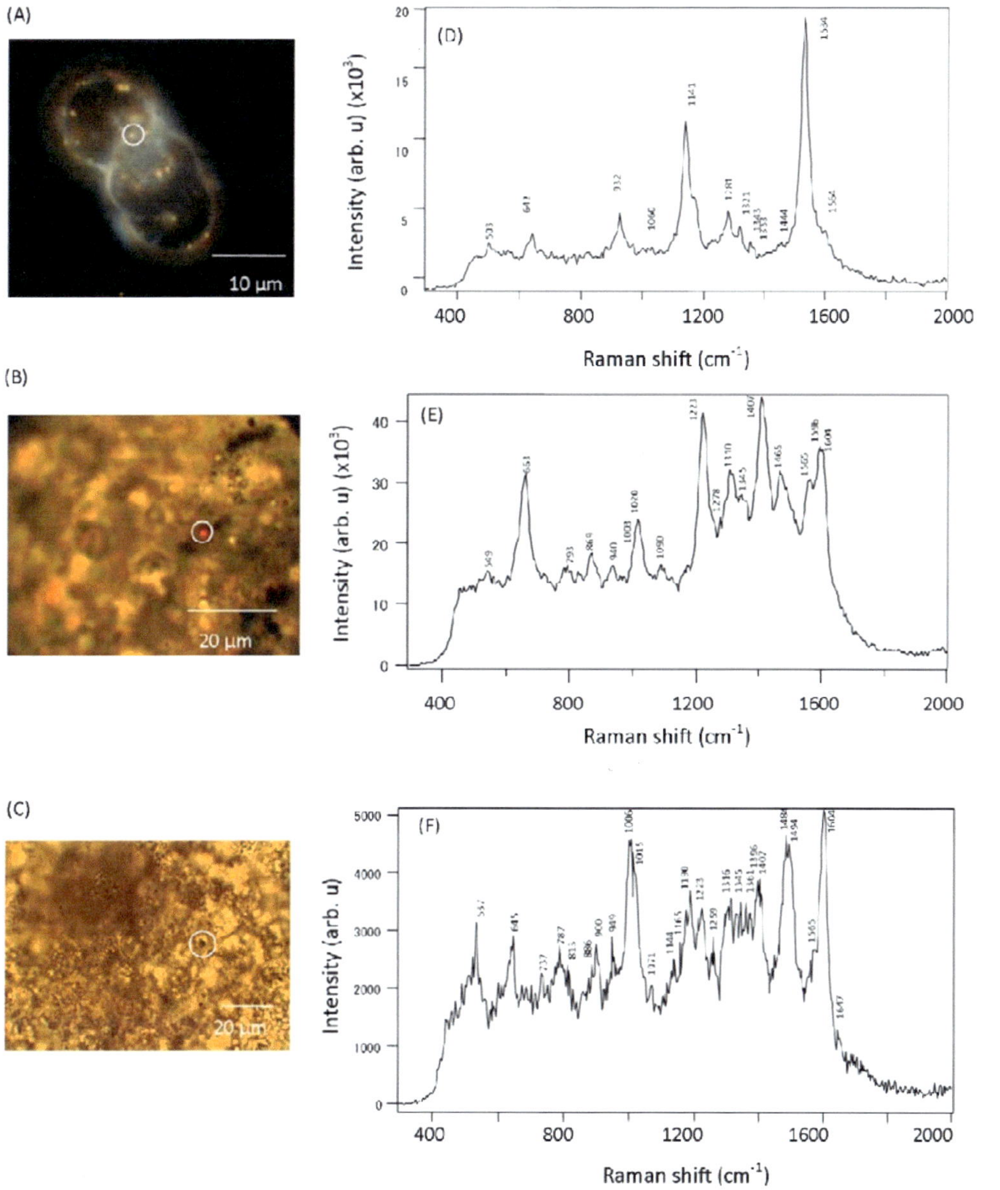

Fig. (7). SERS spectra characteristic of the 3 differentiation stages of mES cells. Reused with permission (under the terms of Creative Commons Attribution License) from [87], (doi:10.1371/journal.pone.0022802.g004).

In the case of embryoid bodies (EB) and mature cardiomyocytes, the obtained SERS peaks were related to protein translation (amide I, II, and III) and mitochondrial metabolic activity (1604 cm^{-1}), illustrating enhanced protein translation and cell signaling processes compared to undifferentiated cells.

SERS was also used by Huefner *et al.* [88], to assess the fingerprints of progenitor and differentiated human neuroblastoma cells.

Table 2. Tentative assignment of SERS peaks derived from undifferentiated ES cells, EBs, and differentiated cardiomyocytes; Reused with permission (under the terms of Creative Commons Attribution License) from [87], (doi:10.1371/journal.pone.0022802.t001).

ES cell (cm^{-1})	Peak assignment	Embryoid body (cm^{-1})	Peak assignment	Cardiomyocyte tissue (cm^{-1})	Peak assignment
642	C-C twist tyrosine O-P-O stretch in DNA	663	T,G (DNA bases) C: ring breathing	645	C-C twist tyrosine C-N stretch in lipid/adenine
787	O-P-O stretch in RNA Histidine	793	C-DNA: phosphodiester stretching ν(C-C), α-helix	737	O-P-O stretch in DNA O-P-O stretch in RNA
813	Proline ring ν(C-C) PO_2^- str (protein)	869	Phenylalanine ring vibration DNA backbone: C-O stretch	787	C-O-C skeletal mode; disaccharide C-C skeletal stretch in protein
918	Ribose phosphate		Histidine phosphate	813	P:C-C skeletal mode (random)
932	P: Amide III (O-sheet) P: Amide III	940 1000–1003	Ribose phosphate P: Amide III(O-sheet)	 886	Phenylalanine PO_2^-
1060	C: base (cytosine) G (guanine), CH def	 1020	P: Amide III C: base (cytosine)		Ribose phosphate C-C stretching in protein
1141–1146	A: base Proteins: CH_3 deformation		A: ring Proteins: γ_T (CH_2-CH_3) A: base	900	Nucleotides: base & Try, Phe P: Amide III (O-sheet)
1220	Alanine, Try, G: base δCH_2	1090–1096	Proteins: CH_3 deformation Proteins: COO- symmetric stretching	 949	P: Amide III C: base (cytosine)
1270–1289	A, G, T P: Amide II C=C (lipid);	1138–1141 1223	A, G, T P: Amide II C=C (lipid);	 1006	Proteins: γ_T (CH_2-CH_3) A: base Proteins: CH_3 deformation
1321	A, C, G	 1275–1283	A, C, G Mitochondria	1071 1144	Alanine, Try, G: base Proteins: COO- symmetric stretching
1335					A, G, T
				1165	P: Amide II
		1310			C=C (lipid);
1353–1367				1190–1196	A, C, G
					Mitochondria
1444		1337			P: Amide I
1484				1223	
1534–1567					
		1394–1415		1256–1275	
				1316–1318	
		1481–1502			
		1520–1565		1329–1340	
		1596–1604		1361	
				1404–1407	
				1484–1502	
				1531–1567	
				1596–1604	
				1647	

Separate SERS spectra were recorded using gold nanoparticles (GNP) from the whole cell and nucleus only. The process of differentiation was mostly visible in the SERS signals from the nuclear region, illustrated by an increase in the DNA/RNA ratio and protein transcription processes (Fig. **8**).

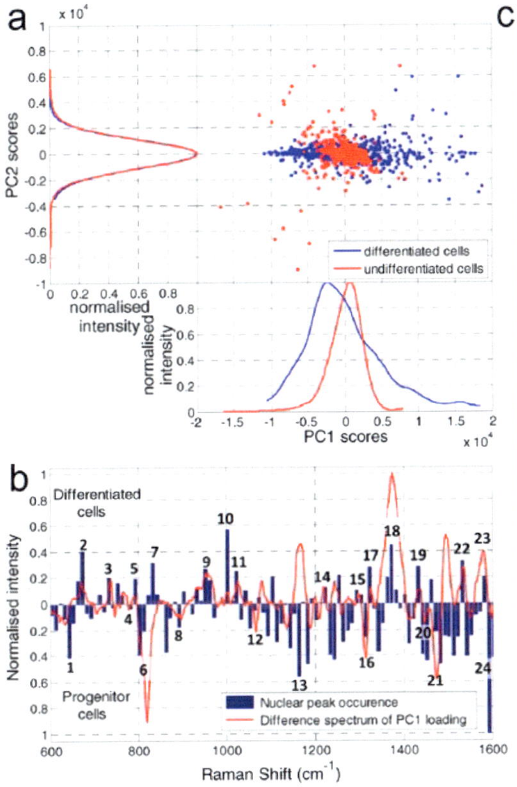

c

No	Peak Position	Assignments [34, 37, 45-51, 53-56]
1	635/642*	Tyr (C-C twist)
2	670	T,G
3	736/732*	Tyr, A
4	780	U,C,T ring br (RNA)
5	792	C,T, DNA bkb(O-P-O str)
6	818/805*	O-P-O RNA
7	834/832*	Tyr, ribose phosphate (DNA)
8	895	ribose phosphate(DNA bkb)
9	952	C-C skeletal
10	1005	Phe
11	1025	C-N str, Phe (C-H in-plane)
12	1060 (UDCs) 1074 (DCs) 1092 (UDCs)	1060-1095 PO_2^- str
13	1188/1185	C,G,A
14	1218/1222	Amide III
15	1295	Amide III
16	1312	CH_2 ring str, CH_3 twist, A
17	1322	G
18	1375	A,T, G
19	1432	A,G
20	1446	CH_2 def
21	1480	A,G
22	1534	A,C,G
23	1580	A,G
24	1595	A

* peak position for PC1 loadings/peak occurrence method; A: adenine, C: cytosine, G: guanine, U: uracil, T:thymine, str: stretch, def: deformation, br: breathing, bkb: backbone, Tyr: tyrosine, Phe: phenylalanine

Fig. (8). Analysis of nuclear SERS spectra; (**a**) Principal component analysis (PCA) - scatter plot and 1D intensity plots of principal components. (**b**) Normalized difference in the spectrum of nuclear peak occurrences (bars) and the PC1 loadings. Bars/peaks pointing to the upper side of the plot correspond to undifferentiated cells (UDCs) while those pointing to the lower side of the plot represent differentiated cells (DCs). (**c**) Assignments of peaks (1-24) marked in (**b**); Reused with permission from [88] (http://pubs.acs.org/doi/pdf/10.1021/nl400448n).

While the undifferentiated cells exhibit distinct SERS peaks for uracil/cytosine at 780 cm^{-1} and symmetric O-P-O stretch in RNA in the region 805-819 cm^{-1}, mature neuroblastoma cells reveal symmetric and asymmetric DNA O-P-O phosphodiester stretch at 793 cm^{-1} and 835 cm^{-1} respectively. The PO_2^- stretch of the DNA backbone is observed in both, undifferentiated and differentiated cells, being associated to the 1060–1092 cm^{-1} bands.

The SERS results presented in [88] reveal an increased protein content (illustrated by the larger number of SERS peaks attributed to proteins) and a tight DNA packaging with histones (most SERS peaks of nucleic acids correspond to DNA and histones) in heterochromatin in differentiated cells, while a looser histone-DNA wrapping, in the form of euchromatin, in undifferentiated cells [88].

The performed studies confirm the potential of SERS to identify Raman spectral characteristics corresponding to distinct stages of SC differentiation and to characterize nuclear modifications occurred during the differentiation process.

Besides signal enhancement, SERS offers in depth molecular information that cannot be recovered from ensemble-averaged conventional Raman spectra [87].

c) Fourier Transformed Infrared Spectroscopy

Fourier transformed infrared spectroscopy (FTIR) functions on similar principles and offers complementary information compared to Raman spectroscopy. It represents a versatile method to obtain information about the molecular fingerprint of a sample which absorbs IR radiation as dictated by its chemical bonds. A comprehensive review regarding the role of FTIR spectroscopy in stem cell research was recently published by Aksoy and Severcan [86].

The value of FTIR determinations consists in their capacity to reveal significant modifications in the IR absorption bands of proteins, lipids and nucleic acids, occurred during the differentiation process.

FTIR spectroscopy was used to determine characteristic markers of murine embryonic stem cell differentiation to neural cell lineages by Tanthanuch *et al.* [89]. Their results revealed increased lipid bands associated to CH_3 and CH_2 stretching vibrations of membrane glycerophospholipids involved in the neural differentiation. Changes were also observed in the protein bands, significant enhancements in α-helix protein secondary structures being detected in differentiated cells. This change was accompanied by a diminished presence of β-sheet secondary structures [89]. These differentiation-induced modifications were attributed to the enhanced expression of the α-helix-rich proteins of the cytoskeleton, a structure with important roles in neural processes [89].

The spontaneous differentiation of murine embryonic SCs was also monitored using FTIR microspectroscopy by. Ami *et al.* [90]. Their results revealed spectral changes in the protein amide I band (1700-1600 cm^{-1}) and nucleic acid absorption region (1050-850 cm^{-1}), indicating the undergoing of mRNA translation and synthesis of new proteins, characteristic to the newly appearing phenotype in the early stages of differentiation (days 4 to 7) [90].

The chemical composition and differentiation into hepatocytes of single hepatic progenitor cells obtained from mouse embryonic stem cells was studied using synchrotron FTIR microspectroscopy by Thumanu K *et al.* [91]. Mature hepatocytes showed an increased intensity of α-helix protein secondary structure, associated to the formation of albumin protein in the differentiation process. The maturation process induced a frequency shift corresponding to the β-sheet secondary structure of the hepatic progenitor, from 1627 cm^{-1} to 1641 cm^{-1}.

At present, the studies concerning the use of FTIR microspectroscopy for SC characterization confirm its potential to offer valuable information and improve our understanding of the differentiation process.

Monitoring SC Differentiation by Amperometric, Voltammetric / Potentiometric and Capacitive Methods

The basic functioning principles of the methods described in this section define classes of devices such as micro-electrode array (MEA) electrochemical biosensors, Light-addressable Potentiometric Sensors (LAPS) and Field-effect Transistor (FET) based sensors. The use of these methods in stem cells differentiation research is outlined below, their main advantages and drawbacks being summarized in Table **3**.

a) Micro-Electrode Array (MEA)

Micro-electrode array (MEA) chips constitute electrochemical biosensors developed to investigate extracellular microenvironments and characterize cell processes based on voltage and/or current signals generated by physicochemical events occurred in the thin electrolyte layer formed between the cells and the surface of microelectrodes.

Table 3. Amperometric, voltammetric / potentiometric and capacitive methods used for monitoring SC differentiation.

Amperometric, voltammetric/potentiometric and capacitive methods			
Method	**Advantages**	**Shortcomings**	**Use in monitoring SC Differentiation**
Micro-electrode array	- simple fabrication - good biocompatibility	- measurements are affected by cell positioning with respect to electrodes	- detection of extracellular physicochemical events at the cell-microelectrode interface
Light-addressable potentiometric sensor	- selection of measuring area on the sensing surface by light addressability - spatial resolution	-low signal-to-noise ratio	- changes of cell shape and extracellular potentials of electrogenic cells
Field effect transistor	- real-time and non-invasive, long-time monitoring - high sensitivity	-dependence of sensor response on the characteristics of cell-FET interface	- changes in morphology and extracellular potentials of electrogenic cells

Depending on particular cases, the technical details regarding the design and functioning of MEA sensors may be very different, these aspects being already extensively presented in several review articles and books [92 - 95].

Electrical signals from a functionalized MEA system were used by Yuriko Furukawa to monitor the neural stem cells (NSC) differentiation in a culture obtained from rat embryo striatum [96]. The measured signals revealed the occurrence of neuron-like burst patterns, depending on the presence of nerve growth factor (NGF) or epidermal growth factor (EGF) in the culture and the measurement time (number of days *in vitro*). After four days in culture, the obtained electrical data revealed a clear difference between the NSCs cultured with NGF and those cultured with EGF, a burst pattern being observed in the first case (well differentiated neurons), while only a noise signal was generated in the presence of EGF (cells were not differentiated from NSCs). At longer times however (eleven days *in vitro*), cortical neuron-like burst patterns were recorded in both cases. These findings support the possibility to analyze the formation of neural networks by differentiated neurons, based on MEA electrical measurements.

Neural differentiation probed by bursting behavior was also studied using MEA technology by Stephens *et al.* [97]. These authors investigate neural activity and network formation during the differentiation of rat primary neurons and rat hippocampal neural progenitor cells.

In the case of neural progenitors, data indicate asynchronous differentiation among the induced neurons, which fire action potentials but do not show activity bursts characteristic to neural networks. Primary neurons however exhibit synchronized activity, generating network bursts with rates that increase with time. Starting with the fourth week after plating, data revealed the occurrence of long clusters of bursts separated by minutes of basal activity, called "superbursts". During the fifth week after plating, the activity reached a plateau [97]. The addition of progenitor cells to MEA induced the reappearance of superbursting in the already mature neural network.

The third distinct experimental case described by Stephens *et al.* [97] concerns progenitor-primary neural cells co-cultures. The obtained results reveal the enhancement of differentiation and maturation of progenitor cultures, induced by the active primary neural cultures.

The performed studies prove the efficiency of MEA sensors in detecting electrophysiological events used to monitor neural differentiation.

b) Light-Addressable Potentiometric Sensor (LAPS)

The light-addressable potentiometric sensor (LAPS) is a semiconductor device with a Si/SiO_2/sensitive layer (mainly Si_3N_4) structure which allows the detection of changes in ionic concentration or redox potentials induced by cells cultured on its surface (sensitive layer) [92] (Fig. **9**).

An external DC bias voltage can be applied on the LAPS chip to form a depletion layer at the interface between the silicon semiconductor and the SiO_2 insulator, an internal electric field being generated across the depleted region. The width of the depletion layer depends on the local surface potential of the LAPS. If an AC modulated light beam irradiates the bulk silicon, electron transitions occur and electron-hole carrier pairs are generated. The charge carriers are separated by the internal electric field and a photocurrent can be detected by means of an external circuit. The amplitude of this photocurrent depend on the strength of the internal electric field (and width of the depletion layer) and thus on the local surface potential of the LAPS. By recording the photocurrent of LAPS, its local surface potential can thus be determined [92].

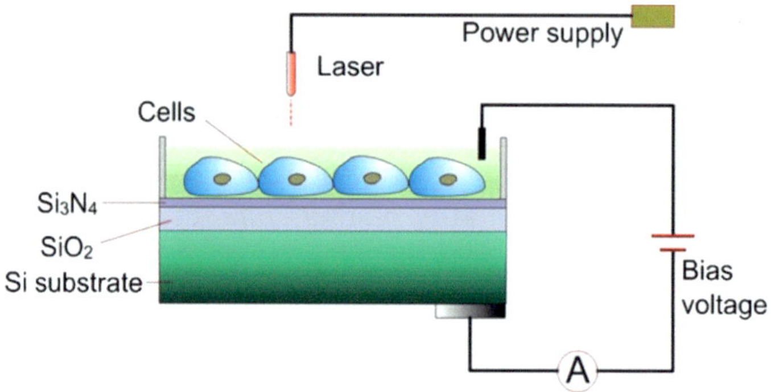

Fig. (9). Schematic representation of LAPS; Reused with permission (License No. 3505860744419) from [92].

When cells are cultured on the surface of LAPS they induce changes in its local surface potential due to the action of inorganic ions such as H^+, Na^+, K^+, Ca^{2+}, Cl^- or redox couples like CySS/CyS (cystine/cysteine) and GSSG/GSH (oxidized/reduced glutathione) [92]. The induced surface potential changes are accompanied by fluctuations in the measured photocurrent that can be related to changes in the ionic concentration or redox potential [92].

A LAPS system was used to monitor the beating of cardiomyocytes derived from mouse embryonic stem cells [98, 99]. The changes induced in cell shape and extracellular potentials by cardiomyocyte contractions led to the modulation of the recorded LAPS photocurrent (Fig. **10**). By means of light addressability, clusters of cells at different locations on the surface of the LAPS sensor can be monitored.

Besides assessing the differentiation of embryonic stem cells into cardiomyocytes, the described LAPS system was used to assess the cardiotoxic effects of different drugs by measuring the prolongation of ventricular action potentials of fully differentiated cardiomyocytes [99].

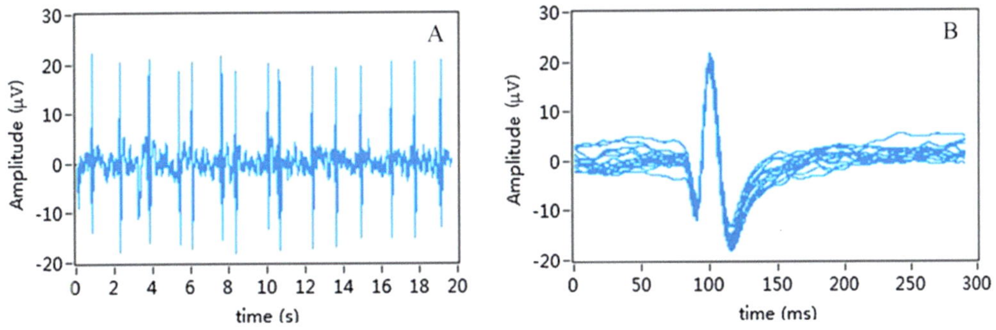

Fig. (10). Action potentials of cardiomyocytes recorded by LAPS: (**A**) original signals; (**B**) automatic clustering of the spikes; Reused with permission (License No. 3505860744419) from [99].

Based on their capability of monitoring a large variety of important cell physiological parameters, LAPS systems can be used to characterize stem cell differentiation as well as in drug effect analysis and toxin detection in a long-term and noninvasive manner

c) Field-Effect Transistor (FET)

The Field-effect transistor (FET) is a semiconductor device which can be used to monitor extracellular parameters such as redox potentials, concentration of ions, pH, CO_2 production as well as the extracellular potentials of electrogenic cells (neuronal and muscle cells) [100].

Constructive and functional details regarding different types of FET chips used for biological determinations can be found in [101, 102].

A schematic representation of a FET-type device used for biological determinations is shown in Fig. (**11**).

The FET allows the passing of a single carrier type current (either electrons or holes) in a drain channel between the *source* and *drain* terminals. The drain channel is covered by the *gate* insulator. Any change in the surface charge/potential of the gate insulator induced by the activity of cells cultured on its surface leads to a modulation of the drain channel conductance and consequently of the drain current of the FET device. The drain current changes are monitored by means of a peripheral circuit.

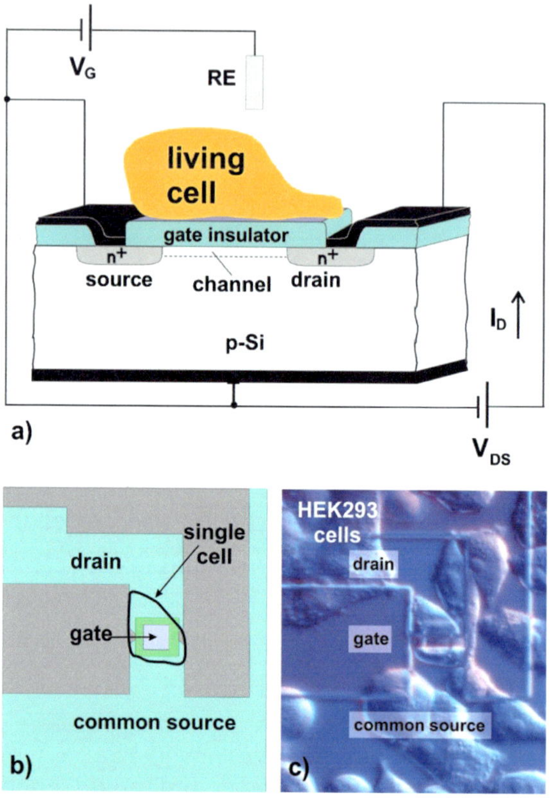

Fig. (11). FET-cell system; Reused with permission (License No: 3506120739954) from [101].

Although the use of FET-based devices in cell biology studies is not uncommon, only few published articles concern their use in monitoring SC differentiation [103, 104].

In a recent study conducted by Caserta *et al.* [103], embryonic stem cells (ESCs) were cultured on the surface of two distinct organic field-effect transistors (OFETs). On the tested organic films, the cells maintained morphologies similar to those on standard culture supports. The ESCs were differentiated into mature cardiomyocytes, their beats being recorded by the tested OFET systems. The obtained results support the use of FET-based biosensors to monitor the cardiac differentiation of embryonic stem cells.

Surface Plasmon Resonance (SPR)

Surface plasmon resonance (SPR) represents a highly sensitive and non-invasive method frequently employed to characterize the binding affinity of certain biomolecules to particular metal surfaces. Its underlying physical principles as well as its use in monitoring SC differentiation are outlined below. The most important positive and negative features of SPR are presented in Table **4**.

Table 4. Surface plasmon resonance method used to monitor SC differentiation.

Surface Plasmon Resonance		
Advantages	**Shortcomings**	**Use in Monitoring SC Differentiation**
- non-invasive - high sensitivity - label-free	- requires sensor chip functionalization - not efficient for high throughput assays - inaccurate kinetic analysis of small analytes - not suitable for concentration assays	- detection of phenotype specific molecules adsorbed on the sensor surface

Surface plasmons are excited when a p-polarized light beam strikes a nanometric metal film placed at the interface between two media with different refractive indices, light traveling from the higher to the lower refractive index medium, under total internal reflection (TIR) conditions. TIR occurs when the light incidence angle (with respect to the normal to the metal film surface) is larger than a critical angle, called resonance angle, for which a minimum intensity of the reflected light is observed. At the resonance angle, the energy of the incident photons is transferred to the outer shell and conduction electrons in the metal film and electron density waves traveling at the surface of the metal film, called surface plasmons, are excited. Excitation of surface plasmons leads to the drastic decrease in the intensity of the reflected light observed resonance conditions. For incidence angles larger than the critical angle, no light passes through the metal film. When mass is added by adsorption of biomolecules on the surface of the thin metal film, the refractive index at the interface is changed and a consequent shift in the critical angle is observed. The shift in the SPR angle provides valuable information regarding the kinetics of molecular adsorption on the sensor surface.

By using appropriate surface functionalization, SPR may be employed to record distinct spectra for specific cell types. This possibility was exploited by Kuo *et al.*

for the real-time monitoring of osteogenic differentiation in mesenchymal stem cells [105]. The authors used an antibody coated Au chip in order to detect variations in the expression level of OB-cadherin, increasingly expressed on the surface of cells during the osteogenic differentiation process. The obtained data indicated that the OB-cadherin expression level was proportional to the culture time, a larger resonance angle being observed as the cells were captured by the chip coating antibodies [105] (Fig. **12**).

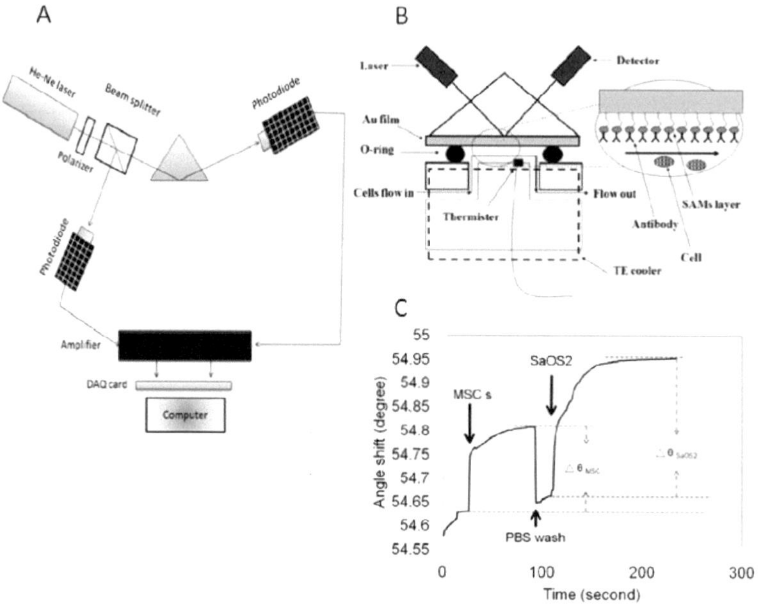

Fig. (12). (**A**) Schematic illustration of the SPR device; (**B**) The OB-cadherin expressing cells captured on the Au chip pre-coated with OB-cadherin antibodies, leading to the measured angle hift; (**C**) Graphical data from SPR measurements. Reused with permission (under the terms of Creative Commons Attribution License) from [105] (doi:10.1371/journal.pone.0022382.g001).

An inversely proportional linear relationship was found between the culture time and the difference in refractive angle shift, which approached to zero after fifteen days of differentiation suggesting a maximum level of OB-cadherin expression [105]. The outlined study confirms the potential of SPR to monitor SC differentiation in a noninvasive and label-free manner.

Piezoelectric Methods (Quartz Crystal Microbalance - QCM)

A Quartz Crystal Microbalance (QCM) measures the changes induced in the frequency of a quartz crystal resonator as a consequence of mass addition to its surface. The way in which QCM can be used to monitor the differentiation of stem cells is exemplified in the following. The main characteristics of QCM in comparison to SPR are briefly mentioned in Table **5**.

Table 5. Quartz crystal microbalance method used to monitor SC differentiation.

Quartz crystal microbalance		
Advantages	**Shortcomings**	**Use in Monitoring SC Differentiation**
non-invasive label-free	requires sensor chip functionalization less sensitive compared to SPR requires higher amounts of sample than SPR	characterization of cell-surface interactions changes in the mechanical and viscoelastic properties of cells adhered to the sensor surface

The working principle of QCM relies on the piezoelectric properties of quartz crystals. Due to piezoelectricity, when a properly cut quartz crystal is placed between two electrodes on which an alternating voltage is applied, the crystal can be made to oscillate at its resonant frequency. The oscillation frequency is modified however if a small mass is deposited on the surface of the crystal resonator. The determination of the added mass from the frequency change is performed based on appropriate theoretical models.

Although frequently used in cell biology studies [106], the QCM method was not widely applied in SC research.

A study regarding the use of a quartz crystal microbalance with dissipation (QCM-D) to monitor the spontaneously beating human pluripotent stem (hPS) cell-derived cardiomyocytes clusters (CMCs) was recently published by Tymchenko *et al.* [107]. The study confirms the potential of QCM-D to directly detect changes in the mechanical and viscoelastic properties of the cells adhered to the surface of the sensor. The measured beating rates were in the range 66-168 beats per minute, the beating profiles representing characteristic fingerprints of the fully differentiated hPS-CMCs. The obtained results point towards the possibility to develop real-time cardiotoxicity assays using QCM-D.

CONCLUSION

Involving micro- and nano-scale physical processes occurring at either, cell or culture level, the SC differentiation monitoring methods briefly outlined in the present chapter have the potential to offer significant accuracy and costs advantages over conventional biochemical methods. By studying such noninvasive and marker free methods for *in vitro* cell characterization, previously unknown fundamental physical aspects of SC differentiation may be uncovered.

CONFLICT OF INTEREST

The authors confirm that they have no conflict of interest to declare for this publication.

ACKNOWLEDGEMENTS

Author Lupu A.-R. acknowledges support under the PN-II-PT-PCCA-2013-4-1407, PN-II-PT-PCCA-2013-4-1386 and PN09_33.01.01 national grants.

Author Popescu T. acknowledges support from Romanian National Authority for Scientific Research, under the Core project PN09-450103.

REFERENCES

[1] Solanki A, Kim JD, Lee K-B. Nanotechnology for regenerative medicine: nanomaterials for stem cell imaging. Nanomedicine (Lond) 2008; 3(4): 567-78. [http://dx.doi.org/10.2217/17435889.3.4.567] [PMID: 18694318]

[2] Kitambi SS, Chandrasekar G. Stem cells: a model for screening, discovery and development of drugs. Stem Cells Cloning 2011; 4: 51-9. [http://dx.doi.org/10.2147/SCCAA.S16417] [PMID: 24198530]

[3] Arora N, Daley GQ. Pluripotent stem cells in research and treatment of hemoglobinopathies. Cold Spring Harb Perspect Med 2012; 2(4): a011841. [http://dx.doi.org/10.1101/cshperspect.a011841] [PMID: 22474618]

[4] Burande MA, Burande AR. Nanoparticle and Stem Cell Nanotechnology: Interdisciplinary Research Area Involving Pharmacology and Anatomy. International Journal of Recent Trends in Science and Technology 2012; 4(1): 2249-8109.

[5] Edmundson M, Thanh NT, Song B. Nanoparticles based stem cell tracking in regenerative medicine. Theranostics 2013; 3(8): 573-82. [http://dx.doi.org/10.7150/thno.5477] [PMID: 23946823]

[6] Kim J, Kim HN, Lim K-T, *et al.* Designing nanotopographical density of extracellular matrix for

controlled morphology and function of human mesenchymal stem cells. Sci Rep 2013; 3: 3552. [PMID: 24352057]

[7] Blagovic K, Kim LY, Voldman J. Microfluidic perfusion for regulating diffusible signaling in stem cells. PLoS One 2011; 6(8): e22892. [http://dx.doi.org/10.1371/journal.pone.0022892] [PMID: 21829665]

[8] Medema JP, Vermeulen L. Microenvironmental regulation of stem cells in intestinal homeostasis and cancer. Nature 2011; 474(7351): 318-26. [http://dx.doi.org/10.1038/nature10212] [PMID: 21677748]

[9] Ivanovic Z. Hypoxia or in situ normoxia: The stem cell paradigm. J Cell Physiol 2009; 219(2): 271-5. [http://dx.doi.org/10.1002/jcp.21690] [PMID: 19160417]

[10] Kusuda Furue M, Tateyama D, Kinehara M, Na J, Okamoto T, Sato JD. Advantages and difficulties in culturing human pluripotent stem cells in growth factor-defined serum-free medium. In Vitro Cell Dev Biol Anim 2010; 46(7): 573-6. [http://dx.doi.org/10.1007/s11626-010-9317-z] [PMID: 20428962]

[11] Frangioni JV, Hajjar RJ. *In vivo* tracking of stem cells for clinical trials in cardiovascular disease. Circulation 2004; 110(21): 3378-83. [http://dx.doi.org/10.1161/01.CIR.0000149840.46523.FC] [PMID: 15557385]

[12] Baumann K. Stem cells: Tracking regenerative behaviour. Nat Rev Mol Cell Biol 2012; 13: 541. [http://dx.doi.org/10.1038/nrm3428]

[13] Zhang Y, Ruel M, Beanlands RS, deKemp RA, Suuronen EJ, DaSilva JN. Tracking stem cell therapy in the myocardium: applications of positron emission tomography. Curr Pharm Des 2008; 14(36): 3835-53. [http://dx.doi.org/10.2174/138161208786898662] [PMID: 19128236]

[14] Nguyen PK, Riegler J, Wu JC. Stem cell imaging: from bench to bedside. Cell Stem Cell 2014; 14(4): 431-44. [http://dx.doi.org/10.1016/j.stem.2014.03.009] [PMID: 24702995]

[15] Gao Y, Cui Y, Chan JK, Xu C. Stem cell tracking with optically active nanoparticles. Am J Nucl Med Mol Imaging 2013; 3(3): 232-46. [PMID: 23638335]

[16] Turksen K, Ed. Imaging and Tracking Stem Cells Methods and Protocols, Methods in Molecular Biology, 1052. New York: Springer Science + Business Media 2013. ISBN: 978-1-62703-558-3 (Print) 978-1-62703-559-0 (Online). [http://dx.doi.org/10.1007/978-1-62703-559-0]

[17] Fang T-C, Poulsom R. Cell-based therapies for birth defects: A role for adult stem cell plasticity? Birth Defect Research Part C: Embryo Today: Reviews 69(3): 238-49. [http://dx.doi.org/10.1002/bdrc.10019]

[18] Ebert AD, Liang P, Wu JC. Induced pluripotent stem cells as a disease modeling and drug screening platform. J Cardiovasc Pharmacol 2012; 60(4): 408-16. [http://dx.doi.org/10.1097/FJC.0b013e318247f642] [PMID: 22240913]

[19] Merkle FT, Eggan K. Modeling human disease with pluripotent stem cells: from genome association

to function. Cell Stem Cell 2013; 12(6): 656-68.
[http://dx.doi.org/10.1016/j.stem.2013.05.016] [PMID: 23746975]

[20] Itzhaki I, Maizels L, Huber I, *et al.* Modelling the long QT syndrome with induced pluripotent stem cells. Nature 2011; 471(7337): 225-9.
[http://dx.doi.org/10.1038/nature09747] [PMID: 21240260]

[21] Guzman ML, Allan JN. Concise review: Leukemia stem cells in personalized medicine. Stem Cells 2014; 32(4): 844-51.
[http://dx.doi.org/10.1002/stem.1597] [PMID: 24214290]

[22] Sharpe ME, Morton D, Rossi A. Nonclinical safety strategies for stem cell therapies. Toxicol Appl Pharmacol 2012; 262(3): 223-31.
[http://dx.doi.org/10.1016/j.taap.2012.05.007] [PMID: 22617430]

[23] Zhu J, Zhou L, XingWu F. Tracking neural stem cells in patients with brain trauma. N Engl J Med 2006; 355(22): 2376-8.
[http://dx.doi.org/10.1056/NEJMc055304] [PMID: 17135597]

[24] Hsu JW, Vogelsang G, Jones RJ, Brodsky RA. Bone marrow transplantation in Shwachman-Diamond syndrome. Bone Marrow Transplant 2002; 30(4): 255-8.
[http://dx.doi.org/10.1038/sj.bmt.1703631] [PMID: 12203143]

[25] Alter BP. Diagnosis, genetics, and management of inherited bone marrow failure syndromes. Hematology (Am Soc Hematol Educ Program) 2007; 2007(1): 29-39.
[http://dx.doi.org/10.1182/asheducation-2007.1.29] [PMID: 18024606]

[26] Copelan EA. Hematopoietic stem-cell transplantation. N Engl J Med 2006; 354(17): 1813-26.
[http://dx.doi.org/10.1056/NEJMra052638] [PMID: 16641398]

[27] Dalle J-H. HSCT for Fanconi anemia in children: factors that influence early and late results. Bone Marrow Transplant 2008; 42 (Suppl. 2): S51-3.
[http://dx.doi.org/10.1038/bmt.2008.284] [PMID: 18978745]

[28] Goldberg JL, Laughlin MJ, Pompili VJ. Umbilical cord blood stem cells: Implications for cardiovascular regenerative medicine. Journal of Molecular and Cellular Cardiology 2007; 42(5): 912-20.

[29] van de Ven C, Collins D, Bradley MB, Morris E, Cairo MS. The potential of umbilical cord blood multipotent stem cells for nonhematopoietic tissue and cell regeneration. Exp Hematol 2007; 35(12): 1753-65.
[http://dx.doi.org/10.1016/j.exphem.2007.08.017] [PMID: 17949892]

[30] Goldstein G, Toren A, Nagler A. Transplantation and other uses of human umbilical cord blood and stem cells. Curr Pharm Des 2007; 13(13): 1363-73.
[http://dx.doi.org/10.2174/138161207780618759] [PMID: 17506721]

[31] Ali H, Bahbahani H. Umbilical cord blood stem cells - potential therapeutic tool for neural injuries and disorders. Acta Neurobiol Exp (Warsz) 2010; 70(3): 316-24.
[PMID: 20871652]

[32] Buzanska L, Machaj EK, Zab,ocka B, Pojda Z, Domaska-Janik K. Human cord blood-derived cells attain neuronal and glial features *in vitro*. J Cell Sci 2002; 115(Pt 10): 2131-8.

[PMID: 11973354]

[33] Daldrup-Link HE, Nejadnik H. MR Imaging of Stem Cell Transplants in Arthritic Joints. J Stem Cell Res Ther 2014; 4(2): 165.
[PMID: 25068075]

[34] Arien-Zakay H, Lecht S, Nagler A, Lazarovici P. Human umbilical cord blood stem cells: rational for use as a neuroprotectant in ischemic brain disease. Int J Mol Sci 2010; 11(9): 3513-28.
[http://dx.doi.org/10.3390/ijms11093513] [PMID: 20957109]

[35] Chachques JC. Development of bioartificial myocardium using stem cells nanobiotechnology templates. Article, ID: Car Res Prac 2011; p. 806795.

[36] Louvi A, Artavanis-Tsakonas S. Notch signalling in vertebrate neural development. Nat Rev Neurosci 2006; 7(2): 93-102.
[http://dx.doi.org/10.1038/nrn1847] [PMID: 16429119]

[37] Imayoshi I, Sakamoto M, Yamaguchi M, Mori K, Kageyama R. Essential roles of Notch signaling in maintenance of neural stem cells in developing and adult brains. J Neurosci 2010; 30(9): 3489-98.
[http://dx.doi.org/10.1523/JNEUROSCI.4987-09.2010] [PMID: 20203209]

[38] Shimojo H, Maeda Y, Ohtsuka T, Kageyama R. Dynamic Notch Signaling in Neural Progenitors Cells Cortical Development. Springer Japan 2013; pp. 1-17.

[39] Bani-Yaghoub M, Tremblay RG, Lei JX, *et al.* Role of Sox2 in the development of the mouse neocortex. Dev Biol 2006; 295(1): 52-66.
[http://dx.doi.org/10.1016/j.ydbio.2006.03.007] [PMID: 16631155]

[40] Bani-Yaghoub M, Kubu CJ, Cowling R, *et al.* A switch in numb isoforms is a critical step in cortical development. Dev Dyn 2007; 236(3): 696-705.
[http://dx.doi.org/10.1002/dvdy.21072] [PMID: 17253625]

[41] Bani-Yaghoub M, Bechberger JF, Naus CC. Reduction of connexin43 expression and dye-coupling during neuronal differentiation of human NTera2/clone D1 cells. J Neurosci Res 1997; 49(1): 19-31.
[http://dx.doi.org/10.1002/(SICI)1097-4547(19970701)49:1<19::AID-JNR3>3.0.CO;2-G] [PMID: 9211986]

[42] Bani-Yaghoub M, Felker JM, Sans C, Naus CC. The effects of bone morphogenetic protein 2 and 4 (BMP2 and BMP4) on gap junctions during neurodevelopment. Exp Neurol 2000; 162(1): 13-26.
[http://dx.doi.org/10.1006/exnr.2000.7294] [PMID: 10716885]

[43] Bani-Yaghoub M, Amundsen DE. Turing type instabilities in a mathematical model of Notch and Retinoic Acid pathways. WSEAS Transactions on Biology and Biomedicine 2006; 3(2): 89-96.

[44] Bani-Yaghoub M, Amundsen DE. Study and simulation of reaction-diffusion systems affected by interacting signaling pathways. Acta Biotheor 2008; 56(4): 315-28.
[http://dx.doi.org/10.1007/s10441-008-9062-9] [PMID: 18941903]

[45] Naus CC, Bani-Yaghoub M. Gap junctional communication in the developing central nervous system. Cell Biol Int 1998; 22(11-12): 751-63.
[http://dx.doi.org/10.1006/cbir.1998.0391] [PMID: 10873289]

[46] Jacobs S, Lie DC, DeCicco KL, *et al.* Retinoic acid is required early during adult neurogenesis in the dentate gyrus. Proc Natl Acad Sci USA 2006; 103(10): 3902-7.

[http://dx.doi.org/10.1073/pnas.0511294103] [PMID: 16505366]

[47] Tonge PD, Andrews PW. Retinoic acid directs neuronal differentiation of human pluripotent stem cell lines in a non-cell-autonomous manner. Differentiation 2010; 80(1): 20-30. [http://dx.doi.org/10.1016/j.diff.2010.04.001] [PMID: 20427117]

[48] Segklia A, Seuntjens E, Elkouris M, *et al.* Bmp7 Regulates the Survival, Proliferation, and Neurogenic Properties of Neural Progenitor Cells during Corticogenesis in the Mouse. PloS ONE 2012; 7(3): 9.

[49] Choe Y, Kozlova A, Graf D, Pleasure SJ. Bone morphogenic protein signaling is a major determinant of dentate development. J Neurosci 2013; 33(16): 6766-75. [Year published:DOI]. [http://dx.doi.org/10.1523/JNEUROSCI.0128-13.2013] [PMID: 23595735]

[50] Paschaki M, Cammas L, Muta Y, *et al.* Retinoic acid regulates olfactory progenitor cell fate and differentiation. Neural Dev 2013; 8: 13. [Year published:DOI]. [http://dx.doi.org/10.1186/1749-8104-8-13] [PMID: 23829703]

[51] McGivern JV, Ebert AD. Exploiting pluripotent stem cell technology for drug discovery, screening, safety, and toxicology assessments. Adv Drug Deliv Rev 2014; 69-70: 170-8. [http://dx.doi.org/10.1016/j.addr.2013.11.012] [PMID: 24309014]

[52] Kawada J, Kimura H, Akutsu H, Sakai Y, Fujii T. Spatiotemporally controlled delivery of soluble factors for stem cell differentiation. Lab Chip 2012; 12(21): 4508-15. [http://dx.doi.org/10.1039/c2lc40268h] [PMID: 22968416]

[53] Ertl P, Sticker D, Charwat V, Kasper C, Lepperdinger G. Lab-on-a-chip technologies for stem cell analysis. Trends Biotechnol 2014; 32(5): 245-53. [http://dx.doi.org/10.1016/j.tibtech.2014.03.004] [PMID: 24726257]

[54] Cohen DM, Chen CS. Mechanical control of stem cell differentiation, StemBook [Internet], http://www.ncbi.nlm.nih.gov/books/NBK27063. Cambridge (MA): Harvard Stem Cell Institute 2008. PMID:20614621

[55] Matsumoto Y, Iwasaki H, Suda T. Maintenance of Adult Stem Cells: Role of the Stem Cell Niche Adult Stem Cells: Biology and Regenerative Medicine. Springer Science + Business Media, LLC 2011. [http://dx.doi.org/10.1007/978-1-61779-002-7_2]

[56] Li D, Zhou J, Chowdhury F, Cheng J, Wang N, Wang F. Role of mechanical factors in fate decisions of stem cells. Regen Med 2011; 6(2): 229-40. [http://dx.doi.org/10.2217/rme.11.2] [PMID: 21391856]

[57] Aratyn-Schaus Y, Gardel ML. Transient frictional slip between integrin and the ECM in focal adhesions under myosin II tension. Curr Biol 2010; 20(13): 1145-53. [http://dx.doi.org/10.1016/j.cub.2010.05.049] [PMID: 20541412]

[58] Engler AJ, Sen S, Sweeney HL, Discher DE. Matrix elasticity directs stem cell lineage specification. Cell 2006; 126(4): 677-89. [http://dx.doi.org/10.1016/j.cell.2006.06.044] [PMID: 16923388]

[59] Teo R, Mohrlen F, Plickert G, MA1/4ller WA, Frank U. An evolutionary conserved role of Wnt signaling in stem cell fate decision. Dev Biol 2006; 289(1): 91-9. [http://dx.doi.org/10.1016/j.ydbio.2005.10.009] [PMID: 16309665]

[60] Liu G, Vijayakumar S, Grumolato L, *et al.* Canonical Wnts function as potent regulators of osteogenesis by human mesenchymal stem cells. J Cell Biol 2009; 185(1): 67-75.
[http://dx.doi.org/10.1083/jcb.200810137] [PMID: 19349579]

[61] Lee J, Abdeen AA, Kilian KA. Rewiring mesenchymal stem cell lineage specification by switching the biophysical microenvironment. Sci Rep 2014; 4: 5188.
[PMID: 24898422]

[62] Vazin T, Freed WJ. Human embryonic stem cells: derivation, culture, and differentiation: a review. Restor Neurol Neurosci 2010; 28(4): 589-603.
[PMID: 20714081]

[63] Ben-Yehudah A, White C, Navara CS, *et al.* Evaluating protocols for embryonic stem cell differentiation into insulin-secreting I-cells using insulin II-GFP as a specific and noninvasive reporter. Cloning Stem Cells 2009; 11(2): 245-57.
[http://dx.doi.org/10.1089/clo.2008.0074] [PMID: 19508115]

[64] Knapp DJ, Eaves CJ. Control of the hematopoietic stem cell state. Cell Res 2014; 24(1): 3-4.
[http://dx.doi.org/10.1038/cr.2013.139] [PMID: 24100350]

[65] Lucas D, Frenette PS. Stem cells: Reprogramming finds its niche. Nature 2014; 511(7509): 301-2.
[http://dx.doi.org/10.1038/nature13516] [PMID: 25030164]

[66] Bull E, Madani SY, Sheth R, Seifalian A, Green M, Seifalian AM. Stem cell tracking using iron oxide nanoparticles. Int J Nanomedicine 2014; 9: 1641-53.
[PMID: 24729700]

[67] Chang YK, Liu YP, Ho JH, Hsu SC, Lee OK. Amine-surface-modified superparamagnetic iron oxide nanoparticles interfere with differentiation of human mesenchymal stem cells. J Orthop Res 2012; 30(9): 1499-506.
[http://dx.doi.org/10.1002/jor.22088] [PMID: 22337660]

[68] Budde MD, Frank JA. Magnetic tagging of therapeutic cells for MRI. J Nucl Med 2009; 50(2): 171-4.
[http://dx.doi.org/10.2967/jnumed.108.053546] [PMID: 19164242]

[69] Li L, Jiang W, Luo K, *et al.* Superparamagnetic iron oxide nanoparticles as MRI contrast agents for non-invasive stem cell labeling and tracking. Theranostics 2013; 3(8): 595-615.
[http://dx.doi.org/10.7150/thno.5366] [PMID: 23946825]

[70] Emerit J, Beaumont C, Trivin F. Iron metabolism, free radicals, and oxidative injury. Biomed Pharmacother 2001; 55(6): 333 9.
[http://dx.doi.org/10.1016/S0753-3322(01)00068-3] [PMID: 11478586]

[71] Pethig R. Dielectric Properties of Biological Materials: Biophysical and Medical Applications. IEEE Trans Electr Insul 1984; EI-19(5): 453-74.
[http://dx.doi.org/10.1109/TEI.1984.298769]

[72] Tirado M, Grosse C, Schrader W, Kaatze U. Broad frequency range dielectric spectroscopy of aqueous suspensions of phospholipid vesicles. J Non-Cryst Solids 2002; 305: 373-8.
[http://dx.doi.org/10.1016/S0022-3093(02)01132-8]

[73] Angstmann M, Brinkmann I, Bieback K, Breitkreutz D, Maercker C. Monitoring human mesenchymal stromal cell differentiation by electrochemical impedance sensing. Cytotherapy 2011; 13(9): 1074-89.

[http://dx.doi.org/10.3109/14653249.2011.584863] [PMID: 21619493]

[74] Bagnaninchi PO, Drummond N. Real-time label-free monitoring of adipose-derived stem cell differentiation with electric cell-substrate impedance sensing. Proc Natl Acad Sci USA 2011; 108(16): 6462-7.
[http://dx.doi.org/10.1073/pnas.1018260108] [PMID: 21464296]

[75] Park HE, Kim D, Koh HS, Cho S, Sung JS, Kim JY. Real-time monitoring of neural differentiation of human mesenchymal stem cells by electric cell-substrate impedance sensing. J Biomed Biotechnol 2011; 2011: 485173.
[http://dx.doi.org/10.1155/2011/485173] [PMID: 21716652]

[76] Reitinger S, Wissenwasser J, Kapferer W, Heer R, Lepperdinger G. Electric impedance sensing in cell-substrates for rapid and selective multipotential differentiation capacity monitoring of human mesenchymal stem cells. Biosens Bioelectron 2012; 34(1): 63-9.
[http://dx.doi.org/10.1016/j.bios.2012.01.013] [PMID: 22366376]

[77] Zhang Y, Wang H, Wang L, Cheng J. Real-Time, Label-Free Monitoring of 3T3-L1 Preadipocyte Differentiation by a Bioelectrical Impedance Assay. Anal Lett 2012; 45: 1045-57.
[http://dx.doi.org/10.1080/00032719.2012.670791]

[78] Benson K, Cramer S, Galla H-J. Impedance-based cell monitoring: barrier properties and beyond. Fluids Barriers CNS 2013; 10(1): 5.
[http://dx.doi.org/10.1186/2045-8118-10-5] [PMID: 23305242]

[79] Skommer J, Darzynkiewicz Z, Wlodkowic D. Cell death goes LIVE: technological advances in real-time tracking of cell death. Cell Cycle 2010; 9(12): 2330-41.
[http://dx.doi.org/10.4161/cc.9.12.11911] [PMID: 20519963]

[80] Cheran LE, Cheung S, Al Chawaf A, *et al.* Label-free detection of neuron-drug interactions using acoustic and Kelvin vibrational fields. Analyst (Lond) 2007; 132(3): 242-55.
[http://dx.doi.org/10.1039/B615476J] [PMID: 17325758]

[81] Cheran LE, Thompson M. Neural Regenerative Medicine and Detection Strategies for Neurochemistry. J Nanoneurosci 2009; 1: 1-16.
[http://dx.doi.org/10.1166/jns.2009.1005]

[82] Cheran LE, Cheung S, Wang X, Thompson M. Probing the bioelectrochemistry of living cells. Electrochim Acta 2008; 53: 6690-7.
[http://dx.doi.org/10.1016/j.electacta.2008.01.053]

[83] Cheran LE, Benvenuto P, Thompson M. Coupling of neurons with biosensor devices for detection of the properties of neuronal populations. Chem Soc Rev 2008; 37(6): 1229-42.
[http://dx.doi.org/10.1039/b712830b] [PMID: 18497935]

[84] Yang M, Thompson M. Acoustic network analysis and equivalent circuit simulation of the thickness-shear mode acoustic wave sensor in the liquid phase. Anal Chim Acta 1993; 282: 505.
[http://dx.doi.org/10.1016/0003-2670(93)80114-Z]

[85] Ellis JS, Thompson M. Viscoelastic modeling with interfacial slip of a protein monolayer electrode-adsorbed on an acoustic wave biosensor. Langmuir 2010; 26(13): 11558-67.
[http://dx.doi.org/10.1021/la100798c] [PMID: 20394431]

[86] Aksoy C, Severcan F. Role of Vibrational Spectroscopy in Stem Cell Research. Spectroscopy. Int J 2012; 27(3): 167-84.

[87] Sathuluri RR, Yoshikawa H, Shimizu E, Saito M, Tamiya E. Gold nanoparticle-based surface-enhanced Raman scattering for noninvasive molecular probing of embryonic stem cell differentiation. PLoS One 2011; 6(8): e22802.
[http://dx.doi.org/10.1371/journal.pone.0022802] [PMID: 21829653]

[88] Huefner A, Kuan WL, Barker RA, Mahajan S. Intracellular SERS nanoprobes for distinction of different neuronal cell types. Nano Lett 2013; 13(6): 2463-70.
[http://dx.doi.org/10.1021/nl400448n] [PMID: 23638825]

[89] Tanthanuch W, Thumanu K, Lorthongpanich C, Parnpai R, Heraud P. Neural differentiation of mouse embryonic stem cells studied by FTIR spectroscopy. J Mol Struct 2010; 967(1?"3): 189-95.
[http://dx.doi.org/10.1016/j.molstruc.2010.01.007]

[90] Ami D, Neri T, Natalello A, *et al.* Embryonic stem cell differentiation studied by FT-IR spectroscopy. Biochimica and Biophysica Acta 1783; 1783: 98-106.

[91] Thumanu K, Tanthanuch W, Ye D, *et al.* Spectroscopic signature of mouse embryonic stem cell-derived hepatocytes using synchrotron Fourier transform infrared microspectroscopy. J Biomed Opt 2011; 16(5): 057005.
[http://dx.doi.org/10.1117/1.3580253] [PMID: 21639583]

[92] Wang J, Wu C, Hu N, Zhou J, Du L, Wang P. Microfabricated electrochemical cell-based biosensors for analysis of living cells *in vitro*. Biosensors (Basel) 2012; 2(2): 127-70.
[http://dx.doi.org/10.3390/bios2020127] [PMID: 25585708]

[93] Chen R, Li Y, Huo K, Chu PK. Microelectrode arrays based on carbon nanomaterials: emerging electrochemical sensors for biological and environmental applications, RCS Adv 2013; 3: 18698-715.

[94] Ghane-Motlagh B, Sawan M. Design and Implementation Challenges of Microelectrode Arrays: A Review. Materials Sciences and Appliccations 2013; 4: 483-95.
[http://dx.doi.org/10.4236/msa.2013.48059]

[95] Whitson J, Kubola D, Shimono K, Jia Y, Taketani M. Multi-Electrode Arrays: Enhancing Traditional Methods and Enabling Network Physiology. In: Baudry M, Taketani M, Eds. Advances in Network Electrophysiology Using Multi-Electrode Arrays. New York: Spring Press 2006; pp. 36-8.
[http://dx.doi.org/10.1007/0-387-25858-2_3]

[96] Furukawa Y, Shimada A, Kato K, Iwata H, Torimitsu K. Monitoring neural stem cell differentiation using PEDOT?"PSS based MEA. Biochimica et Biophysica Acta 2013; 1830: 4329-33.

[97] Stephens CL, Toda H, Palmer TD, DeMarse TB, Ormerod BK. Adult neural progenitor cells reactivate superbursting in mature neural networks. Exp Neurol 2012; 234(1): 20-30.
[http://dx.doi.org/10.1016/j.expneurol.2011.12.009] [PMID: 22198136]

[98] Liu Q, Cai H, Xiao L, Li R, Yang M, Wang P. Embryonic Stem Cells Biosensor and Its Application in Drug Analysis and Toxin Detection. Sensors Journal, IEEE 2007; 7(12): 1625-31.
[http://dx.doi.org/10.1109/JSEN.2007.908439]

[99] Liu Q, Yu H, Tan Z, *et al. In vitro* assessing the risk of drug-induced cardiotoxicity by embryonic stem cell-based biosensor. Sens Actuators B Chem 2011; 155: 214-9.

[http://dx.doi.org/10.1016/j.snb.2010.11.050]

[100] Cagnin S, Cimetta E, Guiducci C, Martini P, Lanfranchi G. Overview of micro- and nano-technology tools for stem cell applications: micropatterned and microelectronic devices. Sensors (Basel) 2012; 12(11): 15947-82.
[http://dx.doi.org/10.3390/s121115947] [PMID: 23202240]

[101] Poghossian A, Ingebrandt S, Offenhausser A, Schoning MJ. Field-effect devices for detecting cellular signals. Semin Cell Dev Biol 2009; 20(1): 41-8.
[http://dx.doi.org/10.1016/j.semcdb.2009.01.014] [PMID: 19429490]

[102] Schoning MJ, Poghossian A. Recent advances in biologically sensitive field-effect transistors (BioFETs). Analyst (Lond) 2002; 127(9): 1137-51.
[http://dx.doi.org/10.1039/B204444G] [PMID: 12375833]

[103] Caserta S, Barra M, Manganelli G, *et al.* Cardiomyocyte differentiation of embryonic stem cells on the surface of organic semiconductors. Int J Artif Organs 2013; 36(6): 426-33.
[PMID: 23653301]

[104] Cramer T, Chelli B, Murgia M, *et al.* Organic ultra-thin film transistors with a liquid gate for extracellular stimulation and recording of electric activity of stem cell-derived neuronal networks. Phys Chem Chem Phys 2013; 15(11): 3897-905.
[http://dx.doi.org/10.1039/c3cp44251a] [PMID: 23400105]

[105] Kuo YC, Ho JH, Yen TJ, Chen HF, Lee OK. Development of a surface plasmon resonance biosensor for real-time detection of osteogenic differentiation in live mesenchymal stem cells. PLoS One 2011; 6(7): e22382.
[http://dx.doi.org/10.1371/journal.pone.0022382] [PMID: 21818317]

[106] Xi J, Chen JY, Garcia MP, Penn LS. Quartz Crystal Microbalance in Cell Biology Studies, Biochip Tissue Chip 2013. S5: 001

[107] Tymchenko N, Kunze A, Dahlenborg K, Svedhem S, Steel D. Acoustical sensing of cardiomyocyte cluster beating. Biochem Biophys Res Commun 2013; 435(4): 520-5.
[http://dx.doi.org/10.1016/j.bbrc.2013.04.070] [PMID: 23643814]

SUBJECT INDEX

O

P

R

S

T

W

Made in the USA
Las Vegas, NV
04 November 2026

57425978R10202